HPE ATP - Storage Solutions V2
OFFICIAL CERTIFICATION STUDY GUIDE
(EXAM HPE0-J74)

First Edition

Radek Zima

HPE Press
660 4th Street, #802
San Francisco, CA 94107

HPE ATP - Storage Solutions V2
Official Certification Study Guide (Exam HPE0-J74)
Radek Zima

Published by:

Hewlett Packard Enterprise Press
660 4th Street, #802
San Francisco, CA 94107

ISBN: 978-1-942741-30-5

Printed in Mexico

WARNING AND DISCLAIMER
This book provides information about the topics covered in the HPE ATP - Storage Solutions V2 (HPE0-J74) certification exam. Every effort has been made to make this book as complete and as accurate as possible, but no warranty or fitness is implied.

The information is provided on an "as is" basis. The author, and Hewlett Packard Enterprise Press, shall have neither liability nor responsibility to any person or entity with respect to any loss or damages arising from the information contained in this book or from the use of the discs or programs that may accompany it.

The opinions expressed in this book belong to the author and are not necessarily those of Hewlett Packard Enterprise Press.

Materials developed prior to the Hewlett-Packard Company separation may contain branding, logos, web page links, and other elements/information that has not been updated for each HP Inc. and Hewlett Packard Enterprise. The general knowledge and skills are still considered of value to HP Inc. and Hewlett Packard Enterprise employees (partners/customers) respectively, so these legacy materials are being made available here. Plans are in place for updating the most highly-used content for each HP Inc. and Hewlett Packard Enterprise.

TRADEMARK ACKNOWLEDGEMENTS
All third-party trademarks contained herein are the property of their respective owner(s).

GOVERNMENT AND EDUCATION SALES
This publisher offers discounts on this book when ordered in quantity for bulk purchases, which may include electronic versions. For more information, please contact U.S. Government and Education Sales 1-855-447-2665 or email sales@hpepressbooks.com.

Feedback Information

At HPE Press, our goal is to create in-depth reference books of the best quality and value. Each book is crafted with care and precision, undergoing rigorous development that involves the expertise of members from the professional technical community.

Readers' feedback is a continuation of the process. If you have any comments regarding how we could improve the quality of this book, or otherwise alter it to better suit your needs, you can contact us through email at hpepress@epac.com. Please make sure to include the book title and ISBN in your message.

We appreciate your feedback.

Publisher: Hewlett Packard Enterprise Press

HPE Contributors: Ralph Luchs, Ian Selway, Chris Smit

HPE Press Program Manager: Michael Bishop

About the Author

Radek Zima is an independent consultant who specializes in IT infrastructure design, implementation and maintenance for HPE servers, storage, networking, management and cloud software. He develops and delivers trainings, workshops, demonstrations, and conference presentations for HPE channel partners, customers and employees at training centers and events around the world. Radek has a Bachelor's degree and a Master of Science degree from the Faculty of Informatics and Statistics, University of Economics in Prague.

Introduction

Based on the Foundations of HPE Storage Solutions Design course, this self-study guide helps you prepare for the HPE ATP - Storage Solutions V2 certification exam (HPE0-J74). The certification validates your ability to design and propose storage solutions for small to medium sized businesses (SMB). It addresses the fundamental and key technologies of storage and storage area networks along with principles of backup and data protection. The exam also verifies your understanding of the HPE Storage portfolio and how this relates to a converged infrastructure strategy and HPE's holistic vision for IT transformation.

Note

This guide can also be used to prepare for the Foundations of HPE Storage Solutions delta exam (HPE0-J75).

Interactive embedded Aurasma links and Videos

Hewlett Packard Enterprise Partner Ready Certification and Learning is driving innovation across our portfolio of technologies to enhance the way you learn. We have formed a partnership with Aurasma to provide embedded digital content as an additional benefit for customers who purchase this study guide.

Within this study guide you will see the My Learning HPE icon, which indicates that additional digital content is available to you. To view this content, you have to download the My Learning HPE app, powered by Aurasma, open the app, and then point the viewfinder at the page wherever you see the icon.

What do you need to do?

1. Locate the free My Learning HPE App in the app store for your iPhone or Android device (smartphone or tablet).
2. Download the app to your device.
3. Open the app and point your device on an image in this guide marked with the HPE My Learning icon until the video starts playing.
4. Once the video begins to play, you can continue watching even when you do not have your device close to the image.

QR Codes

To expand your print to digital reading experience, Quick Response (QR) codes are also used throughout this book. These two-dimensional bar codes can be read by a smart phone or tablet that has a QR reader installed. Here is an example:

Note

If you do not have a smart phone or tablet with a QR reader, you will see the URL underneath the code so that you can access the page from any device.

Scan this with a Smart Phone app that reads QR Codes.

Certification and Learning

Hewlett Packard Enterprise Partner Ready Certification and Learning provides the continuous learning programs and professional accreditations that can advance your career and increase your value to customers, colleagues and your company. Our technical and sales training enable you with the critical skills needed to identify new opportunities and to design, sell, manage and implement solutions that enable business advantage today. We provide continuous learning activities and guided learning plans in the most sought-after technology and IT transformation areas to help you succeed.

As a Partner Ready Certification and Learning certified member, your skills, knowledge, and real-world experience are recognized and valued in the marketplace. To continue your professional and career growth, you have access to our large HPE community of world-class IT professionals, trend-makers and decision-makers. Share ideas, best practices, business insights, and challenges as you gain professional connections globally.

To learn more about HPE Partner Ready Certification and Learning certifications and continuous learning programs, please visit http://certification-learning.hpe.com

Audience

This book is designed for presales or solution architects involved in supporting the sale of HPE storage solutions primarily to SMB customers.

Assumed Knowledge

To achieve the HPE ATP - Storage Solutions V2 certification, it is assumed that you have a minimum of six months of experience in storage technologies.

Relevant Certifications

After you pass the exam, your achievement may be applicable toward more than one certification. To determine which certifications can be credited with this achievement, log in to The Learning Center and view the certifications listed on the exam's More Details tab. You might be on your way to achieving additional certifications.

Preparing for Exam HPE0-J74

This self-study guide does not guarantee that you will have all the knowledge you need to pass the exam. It is expected that you will also draw on real-world experience and would benefit from completing the hands-on lab activities provided in the instructor-led training.

Recommended HPE Training

Recommended training to prepare for each exam is accessible from the exam's page in The Learning Center. See the exam attachment, "Supporting courses," to view and register for the courses.

Obtain Hands-on Experience

To pass the exam, Hewlett Packard Enterprise strongly recommends a combination of training, thorough review of additional study references, and sufficient on-the-job experience.

Exam Registration

To register for an exam, go to http://certification-learning.hpe.com/tr/certification/learn_more_about_exams.html.

CONTENTS

1 The Storage Market and HPE's Transformation Area Market Strategy

OBJECTIVES

In this chapter, you should learn to:

- ✓ Introduce and describe HPE Storage vision and strategy
- ✓ Describe the value propositions of HPE Converged Infrastructure solutions
- ✓ Describe the HPE Transformation Area market strategy
- ✓ Describe the role of software-defined storage (SDS) and the hyper-converged market

INTRODUCTION

This chapter provides an overview of HPE's approach to the storage market. It begins with the HPE Storage vision and strategy for small and medium-size businesses and HPE Converged Infrastructure solutions. Next, it introduces the HPE transformation areas, which are designed to generate revenue and profitable growth, increase agility and flexibility, deliver remarkable customer experience, amplify employee productivity, and reduce cost and risk. Lastly, it explores the rapidly growing category of storage-related products: software-defined storage and hyper-converged solutions.

HPE Storage vision and strategy for small and medium-size businesses

Getting more value from infrastructure

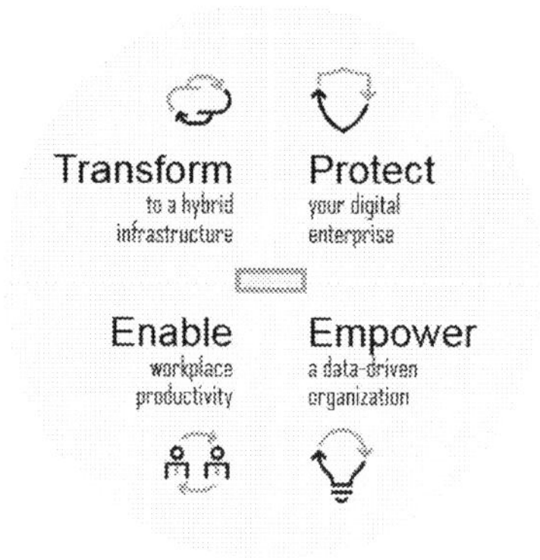

Figure 1-1 HPE Transformation Areas

HPE believes that storage must change to support on-demand computing and solve the major challenges facing customers in addressing cloud, big data, mobility, and security (Figure 1-1).

On-demand computing is centered around the delivery of virtualized IT-as-a-Service implemented on new models like Converged and Hyper-converged Infrastructure, Hybrid Cloud, and the Software-defined Data Center.

To support this transformation, storage must be:

- **Simple**—To support growth without complexity
- **Agile**—To handle unpredictable demand associated with on-demand computing gracefully
- **Fast**—To deliver responses to business users at the speed they need them
- **Efficient**—To enable IT organizations to offer high service levels, but at a cost structure that retains their competitiveness, with cloud alternatives that might be available to business units
- **Secure**—To reduce and manage the business risk associated with applications availability and the integrity of your data
- **Timeless**—To enable new storage to be integrated seamlessly into the data center today and to provide investment protection for the technology changes that will occur in the future

HPE's strategy is choice without compromise

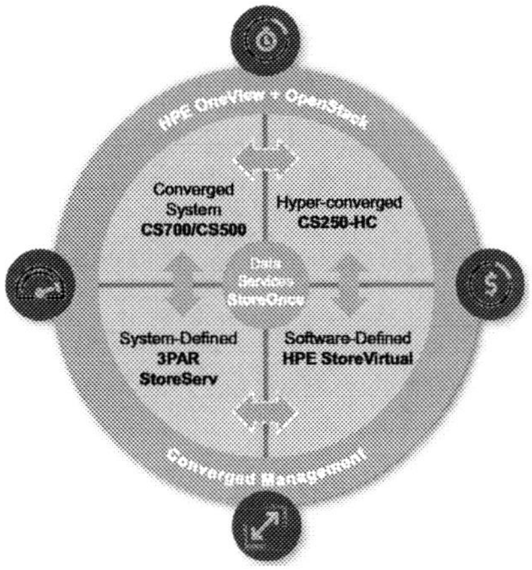

Figure 1-2 HPE's storage strategy

HPE's strategy is choice without compromise. It starts with the right storage architectures available in different delivery models, with a common set of federated data services and administered through common interfaces (Figure 1-2).

HPE Converged Storage can also be managed using other de facto standard management environments from VMware and Microsoft.

Considerations for next-generation (next-gen) storage

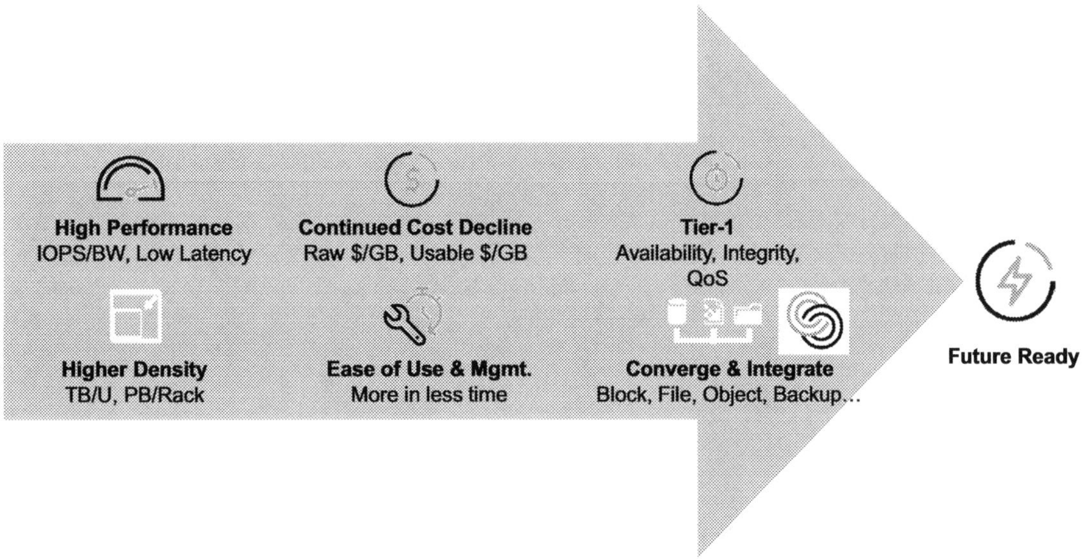

Figure 1-3 Next-gen storage

These are the requirements for next-generation storage (Figure 1-3):

- High performance
- Continued cost decline
- Tier-1 reliability
- Higher density
- Ease of use and management
- Converged and integrated

Gartner Magic Quadrants

Gartner named HPE a leader in General-Purpose and Solid State Disk Arrays in 2014 and 2015 (Figures 1-4 and 1-5).

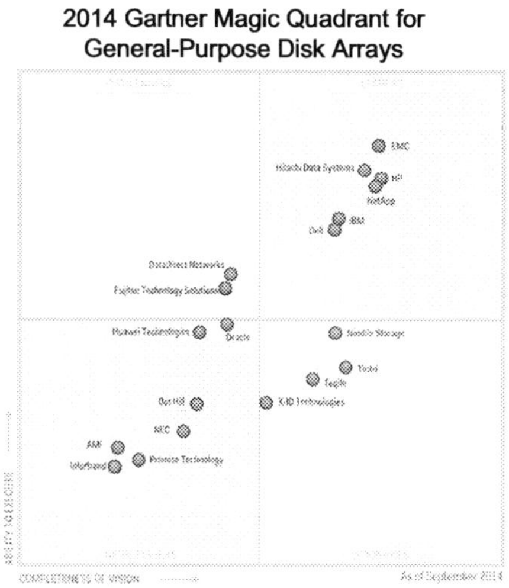

Figure 1-4 2014 Gartner Magic Quadrant for General-Purpose Disk Arrays

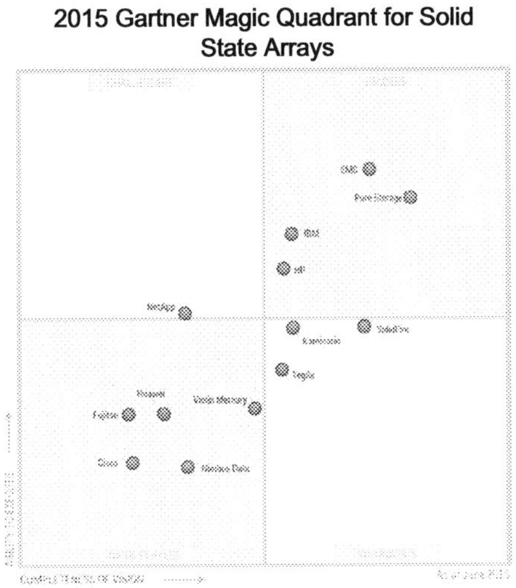

Figure 1-5 2015 Gartner Magic Quadrant for Solid State Arrays

---

Note

This graphic was published by Gartner, Inc. as part of a larger research document and should be evaluated in the context of the entire document. The Gartner document is available upon request from HPE.

Gartner does not endorse any vendor, product, or service depicted in its research publications and does not advise technology users to select only those vendors with the highest ratings or other designation. Gartner research publications consist of the opinions of Gartner's research organization and should not be construed as statements of fact. Gartner disclaims all warranties, expressed or implied, with respect to this research, including any warranties of merchantability or fitness for a particular purpose.

Small and medium-size business (SMB) key drivers

Figure 1-6 SMB key drivers

Small and medium-size business customers are driven by cost, cash flow, and time to market, and they typically have limited or no IT expertise (Figure 1-6).

Smaller businesses tend to make quick decisions to satisfy immediate business needs, and they expect quick results. However, they also expect many enterprise features in a "just good enough" solution. They see the purchase as a cost.

Midsized businesses are generally driven by the same business issues but tend to work on projects. They are more forward-looking than just focusing on solving today's issues.

Larger businesses tend to have enterprise behaviors and expectations. They consider making an investment to satisfy a business need rather than just to drive down a purchasing cost.

The three different business types all expect high availability, high performance, and no-nonsense solutions that are simple to use, at a reasonable cost.

In the small and medium-size business market, there are limited resources in terms of the numbers of IT staff, expertise, money, time, and opportunity. The market is changing more rapidly than before, leaving little time for lengthy decisions or implementations.

Small and medium-size businesses increasingly expect instant-on as a standard feature. Everyone wants IT to provide any workload, anywhere, anytime—agilely and efficiently. To be instant-on, a company must remain focused on innovation, not maintenance. The instant-on company must respond rapidly to changing business and client needs. It must be efficient—achieving better return on investment (ROI), lower costs, and better operational efficiency than its competitors. It must effectively manage risks posed by security threats, regulations, and the unknown.

Additionally, workloads have changed and are continually changing at an ever-increasing pace. Previously, applications were predictable and could fit in a traditional storage system set up for block storage. Planners could measure and predict capacity, I/O performance and bandwidth, file size, and change rate.

Most storage today was designed 17 to 20 years ago for predictable workloads on physical computers. Rapidly, the market is moving toward unpredictable storage, with applications running in virtual machines (VMs) with multitenancy. As the volume of content continues to increase rapidly, the high volume of data is also driving unpredictable workloads.

The dramatic increase in the storage of files of different types, sizes, and formats, and the addition of duplication, makes predicting the capacity, the performance, and of course, the associated cost a major challenge. Analyzing, cataloging, and indexing this data for business use becomes a major task.

Today's businesses and workforce expect to have secured access to this data anywhere, anytime, at minimal cost. Today's technology is making this possible.

Figure 1-7 Today's megatrends require IT to be always on

Not all trends apply to small and medium-size businesses. The mission-critical computing world is facing many challenges in an always-on world. These challenges are caused by several factors (Figure 1-7).

For example, the advancing technology of the megatrends (cloud, big data, mobility, and security) is challenging enterprises everywhere, but these trends also present unique opportunities.

User demands on the enterprise continue to escalate at an amazing rate. Every 60 seconds, the volume of credit card transactions, the number of new mobile subscribers, and the volume of stock trades and tweets grow at astounding rates.

Note

Following are the sources for these statistics:

- 225,000 credit card transactions worldwide: https://www.worldpaymentsreport.com/, November 2015 (estimates)
- 7.89 million stock trades at NYSE: Ask.com, 2014
- 250 new mobile subscribers worldwide: International Telecommunication Union, November 2015
- 360,000 tweets worldwide: http://www.internetlivestats.com/, November 2015

The infrastructure providers that are available to mission-critical customers have shifted and expanded dramatically from only the traditional IT vendors to include cloud service providers and the emerging "roll your own" infrastructure with Original Design Manufacturers (ODMs, or "white box" vendors) plus Open Source "free" software.

Despite all of this, IT must continue to deliver on the demands of the business and the users for the highest availability, efficiency, and proven stability for the most critical, high-value applications.

Small and medium-size companies face big challenges

Figure 1-8 Challenges of small and medium-size companies

Small and medium-size businesses face these common challenges (Figure 1-8):

- **Not enough time**—IT generalists are dealing with ever-increasing workloads and complexity.
- **Not enough money**—Flat to decreasing budgets are at odds with more demanding end-user requirements.
- **Too much risk**—Disasters, equipment failures, and human errors can destroy a business.

Small and medium-size business: Just Right IT

Figure 1-9 Components of Just Right IT

The right IT meets the top four business priorities (Figure 1-9):

- **Increase productivity**—Automated business processes help to control costs, simplify management, and boost performance.
- **Grow your business**—The best prospects for growth live in customer data.
- **Reduce costs**—Affordable IT can lower costs today and keep the business competitive for tomorrow.
- **Stay up and running**—IT is the heart of business operations. Minimize downtime with a reliable data disaster and recovery plan.

More information can be found here: http://www8.hp.com/us/en/business-solutions/smb/index.html.

HPE Storage is focused on three key customer benefits for small and medium-size businesses

Figure 1-10 Three key customer benefits for small and medium-size business market

For this market, HPE Storage focuses on three key customer benefits (Figure 1-10):

- **Simple to manage**—Deliver immediate value. Solve problems quickly and use the skills administrators already have.
- **Affordable to own**—Provide cost-effective solutions. Stretch budgets further and share resources across more projects.
- **Reliable to operate**—Keep the business running smoothly and securely. Confidently work with the market leader to protect data and improve availability.

HPE Simply StoreIT solutions

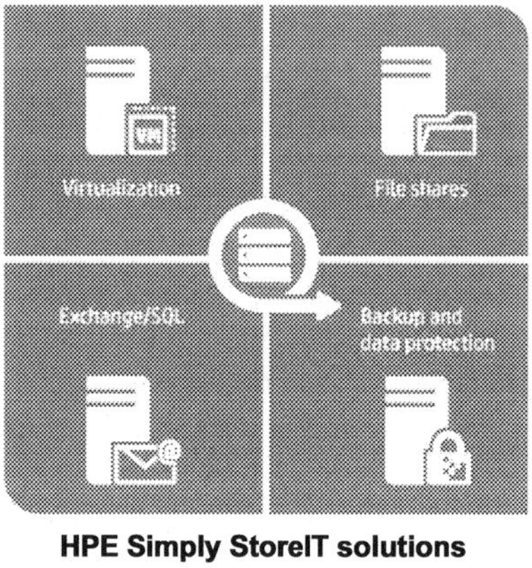

Figure 1-11 HPE Simply StoreIT solutions

HPE Simply StoreIT provides a framework to cut through the confusion of storage for small and medium-size businesses and helps them define a storage strategy that satisfies their needs today and meets their growth plans (Figure 1-11).

HPE Simply StoreIT is an HPE Storage program designed to take the stress out of storage with solutions that are easy to manage, affordable to own, and reliable to operate. These solutions help new businesses as well as growing businesses.

A business that is just **starting** might have core IT functions in place; however, it is likely that the business will have little or no dedicated IT staff. Decisions are made quickly, and IT works with many functions that make the company run. The company is adding employees and might be faced with growing facilities challenges. The budget for IT is limited at this point.

In the **building momentum** phase, the business is growing and adding employees, sales teams, and more customers. The IT capability is evolving, including IT personnel and IT infrastructure. The business is adding mission- and business-critical applications, and it needs technologies and processes that will increase efficiency. Demands are growing for productivity and business analysis tools.

In addition, facilities are expanding and might include multiple sites. Facility costs are becoming an issue for the company. There is a need to build a more robust IT infrastructure that includes data protection and security.

In the **business expansion** phase, the IT capabilities continue to evolve to meet the needs of the business. These capabilities include adding more applications such as business intelligence, analytics, collaboration, and productivity tools. The IT staff might be working on greater system integration or be building more rigorous Disaster Recovery and Business Continuity capabilities. The security of information is becoming an even greater priority. Business expansion is creating challenges for IT as more sites, products, customers, and applications are added. The IT staff would like to provide more agility to quickly meet the growing business needs.

Also at this stage, the business sees IT as a tool for competitive advantage. The IT function needs to implement technologies and processes that will keep up with today's demand and anticipate future requirements. IT becomes an enabler for the business, not a roadblock.

HPE has done the research and has created a framework to help determine the best solution to meet the business needs.

The Simply StoreIT framework includes four solutions with choices that map back to where customers are in the growth model:

- Shared storage for virtualization
- Storage for Microsoft Exchange and SQL
- Storage for file sharing
- Backup and data protection solutions

The way to market

HPE has a large channel and uses it to market products and solutions. HPE servers have a leading market share in most, if not all, regions.

The convergence of storage, networks, servers, power and cooling, management, and services enables HPE to position solutions that are integrated, proven combinations of HPE products and that can be designed and implemented by the channel partners.

It is the role of the HPE presales storage architect to assist in the discovery of opportunities and to position solutions to solve customers' business challenges. HPE provides marketing, technical training, and certification to support this effort.

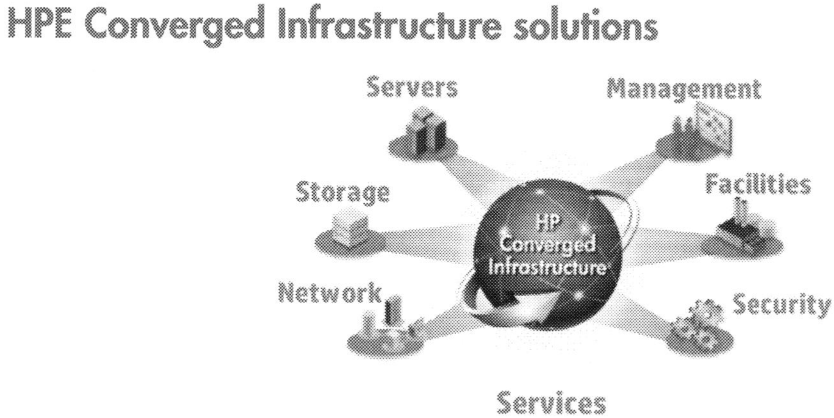

Figure 1-12 HPE Converged Infrastructure portfolio for SMB

Only HPE enables businesses to start small and grow to enable new business at a pace that meets business demands. HPE Converged Infrastructure technology is designed to accelerate the provisioning of IT services through shared pools of interoperable resources (Figure 1-12).

HPE Converged Storage is a storage architecture designed to eliminate the physical, logical, and management boundaries that have traditionally separated storage from the rest of IT.

The HPE Converged Storage product portfolio was designed around technologies such as deduplication, compression, metadata search, and object APIs for cloud applications.

Smaller SMBs tend to buy a solution when they need it, rather than working on initiatives that might show a greater benefit in the longer term. HPE presales and sales consultants should work with SMBs to help them understand the benefits of working on initiatives.

By developing a longer-term overall plan, SMBs would be better prepared when they need to make a decision. Long-term plans would prevent them from making point-in-time decisions that could lead to disparate stand-alone silos and reduce the efficiency of IT.

Such long-term initiatives include the following:

- Consolidation
- Virtualization
- Automation
- Managed services
- Hosting
- Converged Infrastructure
- Cloud

Consolidation drives down the complexity and amount of hardware, reducing capital expenditures (CAPEX) and operational expenditures (OPEX).

Virtualization is happening at the server and operating system levels, and in storage, networking, and on the desktop, increasing the efficiency of an infrastructure, also reducing CAPEX and OPEX.

Converged infrastructure accelerates the provisioning of IT services and applications through shared pools of interoperable resources.

HPE Converged Storage strategy

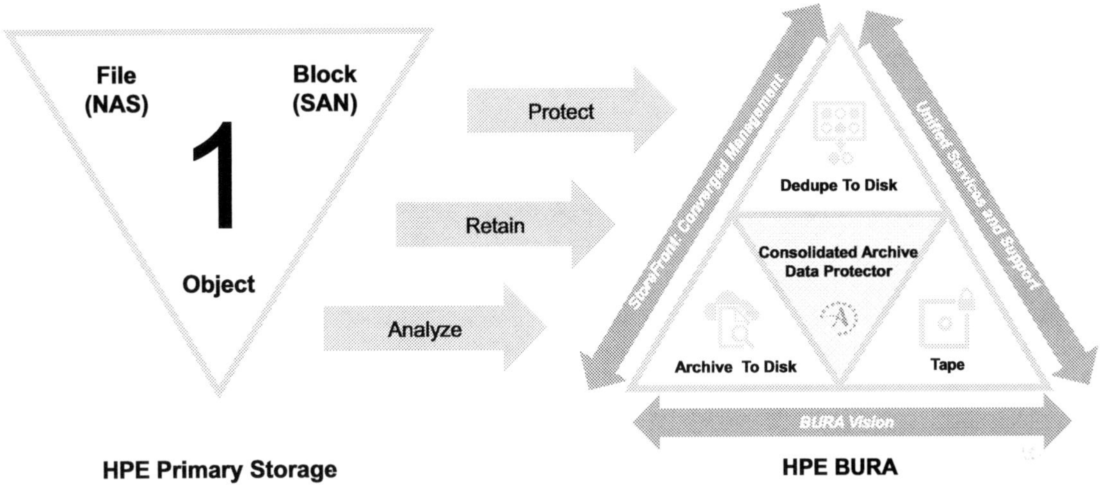

Figure 1-13 Two core pillars of the HPE Converged Storage strategy

HPE has two core pillars in its Converged Storage strategy (Figure 1-13):

- **HPE Primary Storage**—Direct-attached storage (DAS), Network-attached storage (NAS), and Storage area network (SAN) products
- **HPE BURA**—Backup, recovery, and archiving products

HPE BURA vision

BURA stands for backup, recovery, and archive. It is also called data protection, but people often associate "data protection" with just backup.

The HPE BURA vision is

A converged, federated data protection and retention portfolio, helping customers protect, retain, and analyze information while reducing risk and cost.

HPE data protection products share these main features:

- Federated deduplication and high availability
- Intelligent archiving and rapid data extraction of data
- LTO-6 and LTO-7 support
- Integrated protection through Data Protector and Consolidated Archive
- Extensive ISV partnership ecosystem

HPE Transformation Area market strategy

The idea economy is here

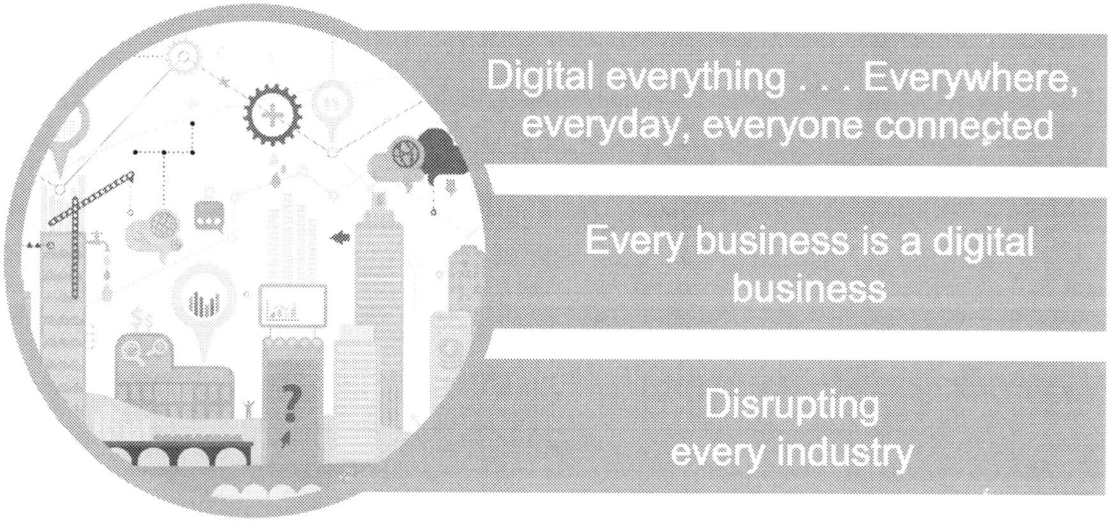

Figure 1-14 The idea economy is here

Ideas have always fueled business success. Ideas have built companies, markets, and industries. However, there is a difference today (Figure 1-14).

Businesses operate in the idea economy, which is also called the digital, application, or mobile economy. Doing business in the idea economy means turning an idea into a new product, capability, business, or industry, and this has never been easier or more accessible—for you and for your competitors.

Today, an entrepreneur with a good idea has access to the infrastructure and resources that a traditional Fortune 1000 company would have. That entrepreneur can rent compute capacity on demand, implement a Software-as-a-Service enterprise resource planning system, use PayPal or Square for transactions, market products and services using Facebook or Google, and have FedEx or UPS run the supply chain.

Companies such as Vimeo, One Kings Lane, Dock to Dish, Uber, Pandora, Salesforce, and Airbnb used their ideas to change the world with little start-up capital. For example, Uber had a dramatic impact after launching its application, which connects riders and drivers, in 2009. Without owning a single car, Uber now serves 352 cities in 65 countries (as of 15 November 2015). This company has completely disrupted the taxi industry. The San Francisco Municipal Transportation Agency reported that cab use in San Francisco has dropped 65% in two years.

In a technology-driven world, it takes more than just ideas to be successful. Success is defined by how quickly ideas can be turned into value.

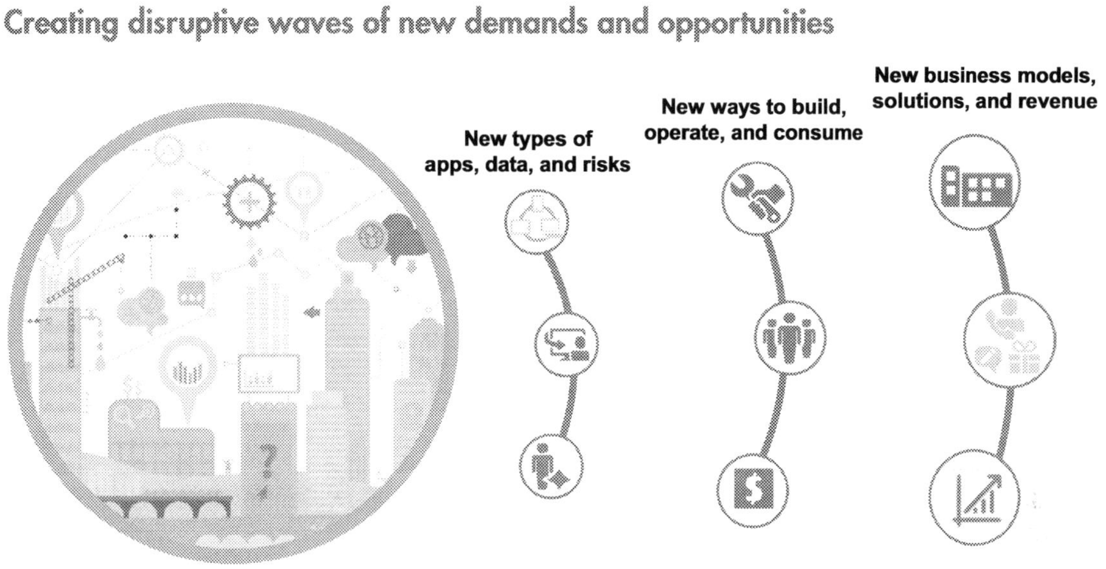

Figure 1-15 Opportunities and challenges

The idea economy presents an opportunity and a challenge for most enterprises. On the one hand, cloud, mobility, big data, and analytics give businesses the tools to accelerate time to value. This increased speed enables organizations to combine applications and data to create dramatically new experiences, even new markets (Figure 1-15).

On the other hand, most organizations were built with rigid IT infrastructures that are costly to maintain. This rigidity makes it difficult, if not impossible, to implement new ideas quickly.

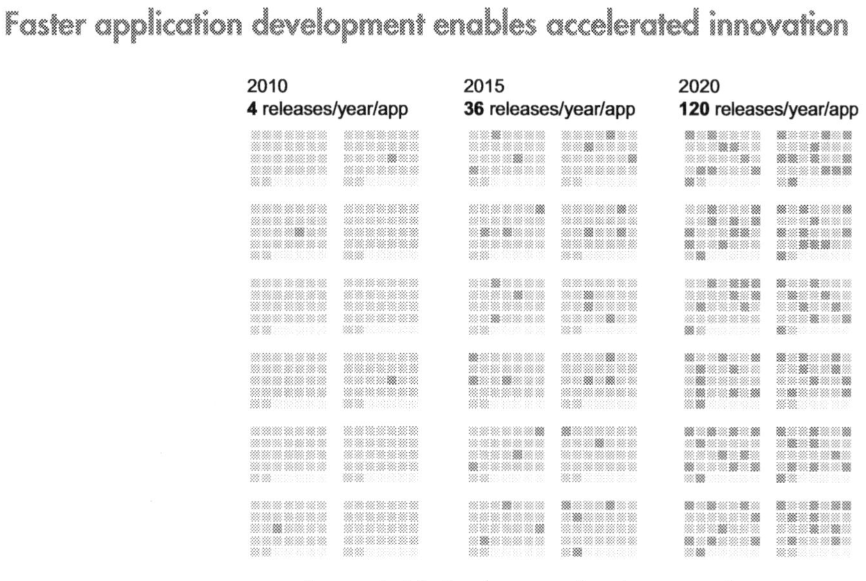

Figure 1-16 Application development changes

From 2010 to 2015, much has changed from an application development perspective. In 2010, the average application release cycle was four releases per year, per application. In 2015, this number rose to 36 releases per year, per application. It is projected that by 2020, there will be 120 releases per year, per application (30 times more releases than in 2010; Figure 1-16).

Considerations from the Forrester Thought Leader Paper commissioned by HPE, *Better Outcomes, Faster Results: Continuous Delivery and the Race for Better Business Performance*, help summarize this trend:

- Agility is paramount.
- "Even when delivering at cadences of less than a week, 20% of organizations want to go even faster."
- Developers need flexibility.
- Companies want open, flexible architectures for application portability and lock-in prevention.

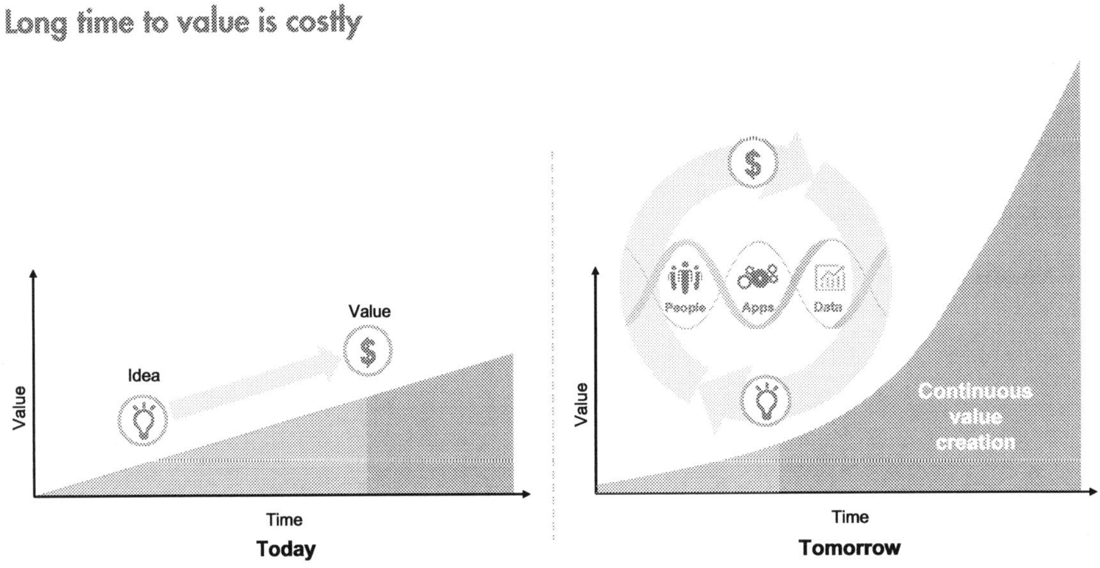

Figure 1-17 Time to value changes

Success today is defined by how quickly an enterprise can turn ideas into value—how quickly a business can experiment, learn, test, tune, and make things better. Speed is a key differentiator in all industries (Figure 1-17).

Uber did not invent a new technology. Instead, the company took advantage of the explosion of smartphones and mobile applications to design a compelling customer experience, ultimately creating a new way of doing business.

This example is not only about Uber executing a good idea. It is also about the taxicab industry's inability to act quickly to transform its business models to compete. Examples such as Uber serve as a warning. Every Fortune 1000 company is at risk of missing a market opportunity, failing to secure its enterprise, and being disrupted by a new idea or business model.

Timelines for IT projects and new applications used to be planned over years and months. Today, these projects take weeks or days. Increasingly, it is shrinking to hours. Now is the time for company leaders to ask questions like the following:

- How quickly can we capitalize on a new idea?
- How rapidly can we seize a new opportunity?
- How fast can we respond to a new competitor that threatens our business?

The good news for established companies is that the same technologies that are making it easy for new companies to get started are also enabling existing companies to adapt quickly to changing business models and achieve faster time to value.

Thriving in the idea economy requires companies to adopt a fresh approach that:

- **Is experience and outcome driven**—Rapidly compose new services from any source to meet the evolving needs of customers and citizens.
- **Proactively manages risks**—Remain safe and compliant in a world with a rapidly changing threat to landscape.
- **Is contextually aware and predictive**—Harness 100% of data to generate real-time instant insights for continuous improvement, innovation, and learning.
- **Is hyperconnected to customers, employees, and the ecosystem**—Deliver experiences that enable employees and engage customers in a persistent, personalized way.

In the idea economy, applications and information are the products.

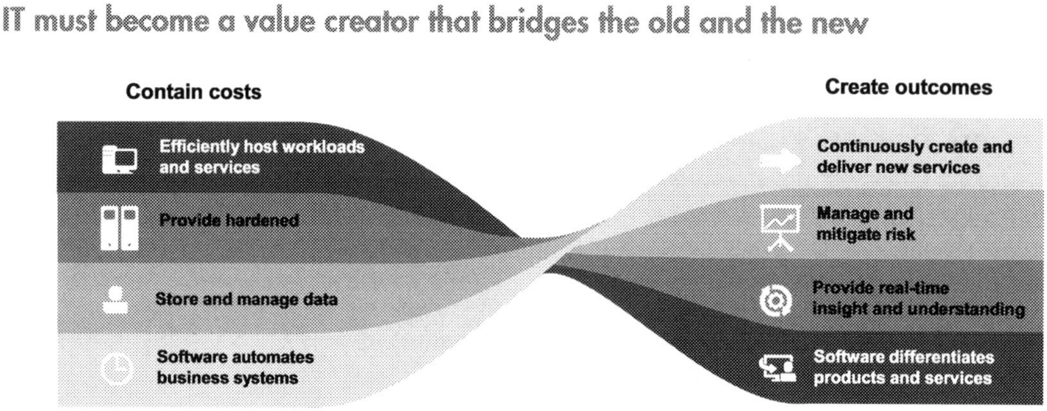

Figure 1-18 IT must shift its focus

To respond to the disruptions created by the idea economy, IT must transform from a cost center to a value creator (Figure 1-18). To evolve, IT must shift its focus in several ways:

- From efficiently hosting workloads and services to continuously creating and delivering new services
- From simply providing hardened systems and networks to proactively managing and mitigating risks
- From just storing and managing data to providing real-time insight and understanding
- From using software to automate business systems to differentiating products and services

Companies need to make their IT environments more efficient, productive, and secure as they transition to a value creator. These organizations must be able to act rapidly on ideas by creating, consuming, and reconfiguring new solutions, experiences, and business models.

One of the first steps in achieving this kind of agility is to break down the old infrastructure silos that make enterprises resistant to new ideas internally and vulnerable to new ideas externally. Designing compelling new experiences and services does not work if the infrastructure cannot support them.

The right compute platform can make a significant impact on business outcomes and performance. Examples include storage that "thinks" as much as it stores, networking that moves information faster and more securely than ever before, and orchestration and management software that provides predictive capabilities.

Each company is on a unique journey to the cloud, custom-made for the way it consumes and allocates resources, transforms to the changing landscape, implements its financial models, and achieves its desired outcomes.

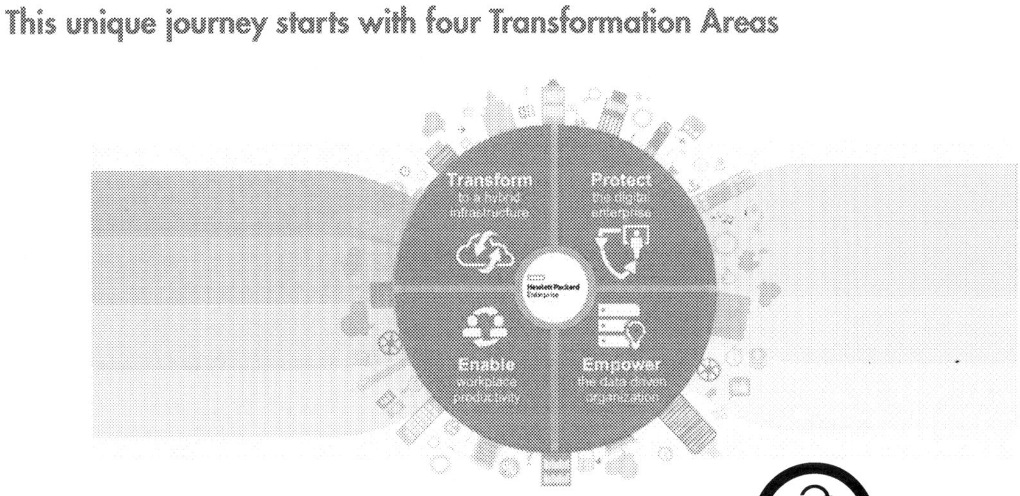

Figure 1-19 Transformation Areas

Note

To view a 2-minute video introducing HPE Transformation Areas, scan the above graphic into the HPE My Learning app on a mobile device.

The HPE Transformation Areas are designed to (Figure 1-19):

- Generate revenue and profitable growth
- Increase agility and flexibility
- Deliver remarkable customer experiences
- Amplify employee productivity
- Reduce cost and risk

These transformation areas reflect what customers consider most important:

- **Transforming to a hybrid infrastructure**—A hybrid infrastructure enables customers to get better value from the existing infrastructure and to deliver new value quickly and continuously from all applications. This infrastructure should be agile, workload optimized, simple, and intuitive.

- **Protecting the digital enterprise**—Customers consider it a matter of when, not if, their digital walls will be breached. The threat landscape is wider and more diverse than ever before. A complete risk management strategy involves security threats, backup and recovery, high availability, and disaster recovery.
- **Empowering the data-driven organization**—Customers are overwhelmed with data; the solution is to derive value from the information that exists. Data-driven organizations generate real-time, actionable insights.
- **Enabling workplace productivity**—Many customers are increasingly focused on enabling workplace productivity. Delivering a great digital workplace experience to employees and customers is a critical step.

Protecting the digital enterprise

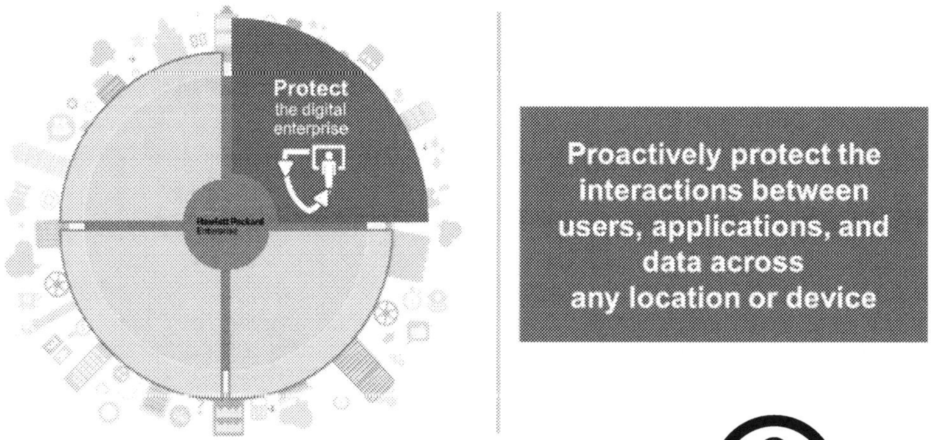

Figure 1-20 Protecting the digital enterprise

Note

To view a 5-minute video explaining the "Protect the digital enterprise" transformation area, scan the "Protecting the digital enterprise " graphic into the HPE My Learning app on a mobile device (Figure 1-20).

Which problems can be solved?

Today's challenges	Transformation objectives
Growing vulnerabilities: Lack of integrated protection mechanisms, inconsistent patch management and recoverability testing	**Built-in resilience:** Automated maintenance and robust management processes, high-availability components across the stack
Reactive strategy: Uncoordinated spending, compliance issues, and underinvestment in broader capabilities to handle emerging threats and data protection gaps	**Planned ecosystem:** Strategic planning that enables investment in the new school of cyber architecture and the latest protection topologies as needed
Rigid operations: Manual and siloed allocation of backup jobs to target devices, managed separately from business applications	**Adaptive and federated:** Elastic pools of protection capacity and analytics-based optimization to balance performance and cost
Over-reliance on silver bullets: Limited impact of tools due to inadequate training of staff and suboptimal security processes	**Integrated solutions:** People and processes as secure as your technology with regularly assessed capabilities

Figure 1-21 IT challenges and transformation objectives

All businesses must manage the emerging risks created by the proliferation of apps, new consumption models, and the shift to mobile and cloud capabilities. With the right strategy, organizations can access all the benefits of an app-centric, hybrid world and proactively protect their network (Figure 1-21).

Many customers struggle with these issues:

- **Growing threats and vulnerabilities**—Lack of integrated protection mechanisms and inadequate technology maintenance and testing
- **Reactive strategies**—Uncoordinated spending, compliance issues, and underinvestment to handle emerging threats and data protection gaps
- **Rigid operations**—Manual and siloed allocation of backup jobs to target devices, managed separately from business applications with fragmented security controls
- **Over-reliance on silver bullets**—Limited impact of tools because of insufficient integration, inadequately trained staff, and suboptimal security processes

To go beyond these challenges, customers must focus on these concepts:

- **Built-in resilience**—Automated and integrated data protection and security controls, robust security governance, and high-availability infrastructure

- **Planned ecosystem**—Strategic planning and investment in sophisticated enterprise security, the latest protection topologies, and tools for compliance
- **Adaptive and federated systems**—Integrated tools, elastic pools of protection capacity, and analytics-based optimization to balance performance
- **Integrated solutions**—Regular assessments of capabilities to ensure people, technology, and processes are aligned to deliver better business outcomes

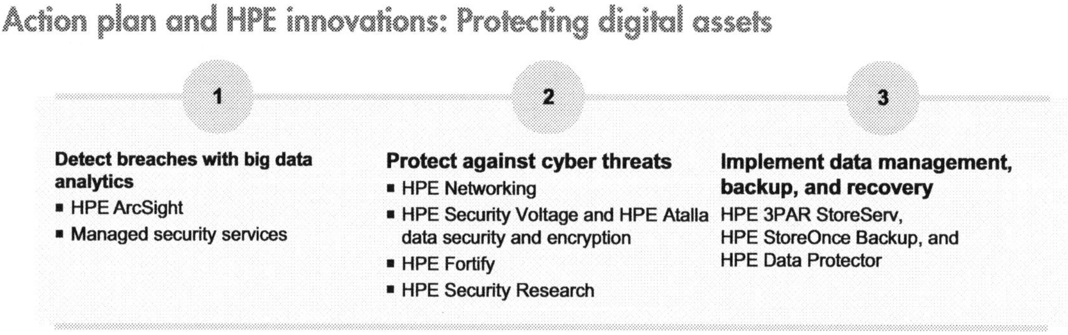

Figure 1-22 HPE innovations: Protecting the digital enterprise

Security and risk protection should be integrated when the infrastructure is set up. Enhancing security after the infrastructure is in place can cost 10 times more than initial prevention. Businesses need a single solution that balances regulatory requirements, cyber threats, asset protection, and business change. The key is to protect the most important part of the business and understand how people access it and then to create policies and tools for those users (Figure 1-22).

These HPE products, solutions, and services align with each step of protecting enterprise digital assets:

1. **Detect breaches with big data analytics**—HPE ArcSight and managed security services identify potential and successful security and compliance breaches.
2. **Protect against cyber threats**—HPE Networking, HPE Security Voltage, HPE Atalla data security and encryption, HPE Fortify, and HPE Security Research deploy next-generation vulnerability analysis, encryption, and intrusion protection using the latest threat intelligence.
3. **Implement data management, backup, and recovery**—HPE 3PAR StoreServ, HPE StoreOnce Backup, and HPE Data Protector ensure business continuity during a crisis and simplify regulatory compliance.

Empowering the data-driven organization: Learner activity

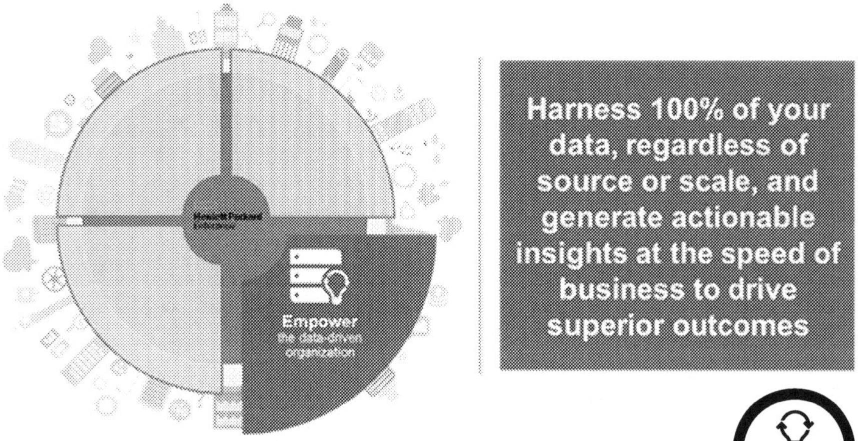

Figure 1-23 Empowering the data-driven organization

Note

To watch a 5-minute video explaining the "Empower the data-driven organization" transformation area, scan the above graphic into the HPE My Learning app on a mobile device (Figure 1-23).

Which problems can be solved?

Today's challenges	Transformation objectives
Lack of flexibility: Inadequate IT investment planning and expensive proprietary systems constrain ability to scale out or extend to new data types	**Agility and scalability:** An investment road map enables the rapid deployment of powerful open hardware and software at a lower cost with more flexibility to scale
Narrow analytics: Insights are backward-looking and generated in silos, with limited relevance to future business decisions	**Actionable analytics:** Predictive insights should be constantly refined and highly relevant to multiple facets of the business
Limited relevance: Analytics output is not always useful because search queries are too slow and draw from only a fraction of available data	**Data-driven decisions:** Powerful analytics apps (traditional or cloud based) connect to virtually any data source quickly and easily
Information as a liability: Inadequate tracking and indexing of information creates compliance and business risks	**Information as an asset:** Information is governed in a secure end-to-end life cycle balancing value, cost, and risk

Figure 1-24 IT challenges and transformation objectives

Rapid evolution in technology has created a distributed digital world—data is everywhere. It presents new opportunities to capture value, as well as new sources of risk. To compete, businesses must generate actionable insights that can drive better business outcomes (Figure 1-24).

Many customers struggle with these issues:

- **Lack of flexibility**—Inadequate IT investment planning and expensive proprietary systems constrain ability to scale out or extend to new data types.
- **Narrow analytics**—Insights are backward-looking and generated in silos, with limited relevance to future business decisions.
- **Limited relevance**—Analytics output is not always useful because search queries are too slow and draw from only a fraction of available data.
- **Information as a liability**—Inadequate tracking and indexing of information creates compliance and business risks.

To meet these challenges, customers must focus on these concepts:

- **Agility and scalability**—An investment road map enables the rapid deployment of powerful open hardware and software at a lower cost with more flexibility to scale.
- **Actionable analytics**—Predictive insights should be constantly refined and highly relevant to multiple facets of the business.
- **Data-driven decisions**—Powerful analytics solutions (traditional or cloud-based) connect to virtually any data source quickly and easily.
- **Information as an asset**—Information is governed in a secure end-to-end life cycle, balancing value, cost, and risk.

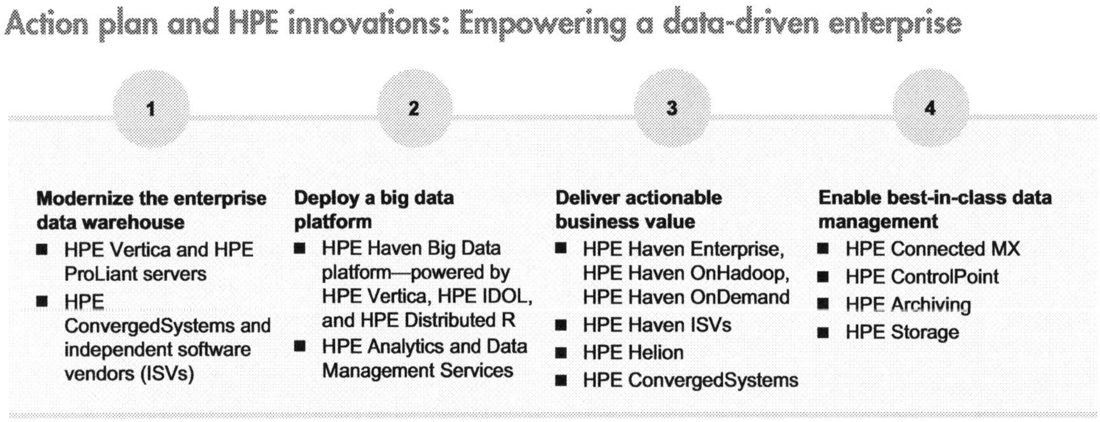

Figure 1-25 HPE innovations: Empowering a data-driven enterprise

According to a survey of HPE customers, companies realize only 10%–15% of the expected value on their big data investments (Figure 1-25). Three main lessons have been learned from past HPE customer engagements:

- Customers must optimize for their existing data. Optimizing the core infrastructure and hardware allows for evolving data sources such as media and text.
- Customers need to drive continuous analytics into business processes. Insights must happen in real time and be embedded into the decision flow, not created and processed as separate events.
- Customers must monetize business intelligence (BI) sources. Modern BI sources are not properly monetized for two reasons:
    - They do not integrate all new data sources
    - They do not properly combine data from existing warehouses

These HPE products, solutions, and services align with each step of empowering a data-driven enterprise:

1. **Modernize the enterprise data warehouse**—HPE Vertica, HPE ProLiant servers, HPE Converged Systems, and ISVs improve scalable performance and responsiveness by adopting a more effective cost model.
2. **Deploy a big data platform**—HPE Haven Big Data platform powered by HPE Vertica, HPE IDOL, HPE Distributed R, and HPE Analytics and Data Management Services help develop analytics apps and services on-premises and in the cloud.
3. **Deliver actionable business value**—HPE Haven Enterprise, HPE Haven OnHadoop, HPE Haven OnDemand, HPE Helion, and HPE ConvergedSystems deliver simple insights that are responsive to business needs.
4. **Enable best-in-class data management**—HPE Connected MX, HPE ControlPoint, HPE Archiving, and HPE Storage integrate data management and collaboration tools to maximize efficiency and effectiveness.

Enabling workplace productivity: Learner activity

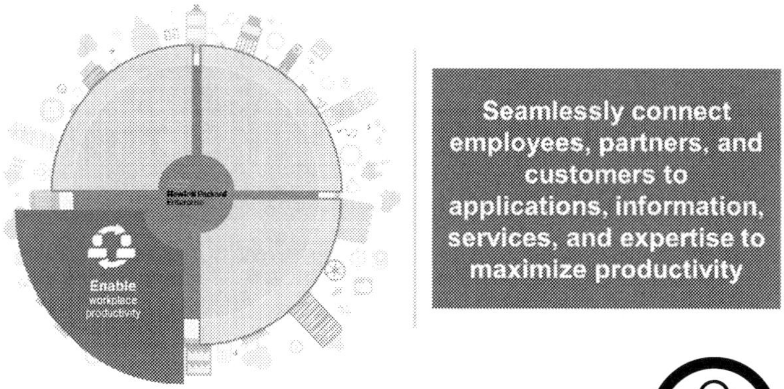

Figure 1-26 Enabling workplace productivity

Note

To watch a 5-minute video explaining the "Enable workplace productivity" transformation area, scan the above graphic into the HPE My Learning app on a mobile device (Figure 1-26).

Which problems can be solved?

Today's challenges	Transformation objectives
Increasing costs: Meeting user expectations is more and more costly and time-consuming due to aging, rigid infrastructure	**Greater efficiency:** More efficient software-defined infrastructure and user-based management reduce costs and improve user experience
Deskbound workers: Wired networks and desktop devices are necessary, limiting opportunities for creative collaboration	**Universal accessibility:** High-performance wireless available for any device, at any time, with seamless communication and collaboration
Constrained mobility: Inadequate support for mobile devices stifles productivity and prompts employee workarounds that create risk	**Anywhere workforce:** Flexibility to work anywhere, accessing resources on any device at any time, with secure, tested, and monitored apps
Legacy investment limitations: Options for technology refresh are constrained by past purchasing decisions	**Adaptable investment strategy:** Flexibility to evolve capabilities, devices, and applications in line with business needs

Figure 1-27 IT challenges and transformation objectives

Delivering a superior user experience to customers, employees, and partners is a major driver of productivity. To be competitive, the modern enterprise needs to support ubiquitous access, seamless communication, and high-performing applications—without jeopardizing data security and corporate assets (Figure 1-27).

Many customers struggle with these issues:

- **Increasing costs**—Meeting user expectations is more costly and time-consuming because of an aging, rigid infrastructure.
- **Deskbound workers**—Wired networks with separate voice and data capabilities make desktop devices a necessity and limit opportunities for creative collaboration.
- **Constrained mobility**—Inadequate support for mobile devices stifles productivity and prompts employee workarounds that create risk.
- **Legacy investment limitations**—Options for technology refreshes are constrained by past purchasing decisions.

To meet these challenges, customers must focus on these concepts:

- **Greater efficiency**—Software-defined infrastructure and user-based management reduce costs and improve the user experience.
- **Universal accessibility**—High-performance wireless devices, new working practices, and cross-device collaboration improve communication.
- **An anywhere workforce**—The flexibility to work anywhere means accessing resources on any device with secure, tested, and monitored apps.
- **An adaptable investment strategy**—Capabilities, devices, and applications must evolve in line with business needs.

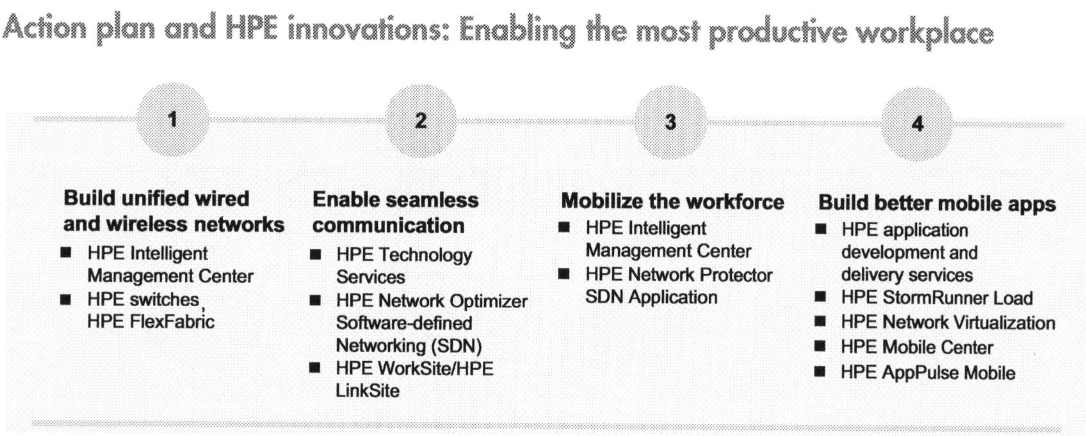

Figure 1-28 HPE innovations: Enabling a more connected community

From an infrastructure perspective, plan the full technology stack from end to end. From the initial touch point to the core data, the infrastructure must be performance-ready for more devices, more features, and smarter environments. It is not just a matter of adding more switches and Wi-Fi nodes. It is about providing real-time access to information (Figure 1-28).

Additionally, think about how to constantly improve the user experience. That means optimizing across mobile apps and infrastructure (designing, testing, and securing every aspect) and exploiting the full power of analytics into the feedback loop.

These HPE products, solutions, and services align with each step of enabling workplace productivity:

1. **Build unified wired and wireless networks**—HPE Intelligent Management Center, HPE switches, and HPE FlexFabric reduce costs and improve the user experience.
2. **Enable seamless communication**—HPE Technology Services, HPE Network Optimizer Software-Defined Networking (SDN), and HPE WorkSite/HPE LinkSite deploy the latest productivity applications.
3. **Mobilize the workforce**—HPE Intelligent Management Center and the HPE Network Protector SDN Application enable bring-your-own-device (BYOD) capability without compromising security.
4. **Build better mobile apps**—HPE application development and delivery services, HPE StormRunner Load, HPE Network Virtualization, HPE Mobile Center, and HPE AppPulse Mobile can be used to build, test, and monitor mobile apps for optimal user experiences.

Transforming to a hybrid infrastructure: Learner activity

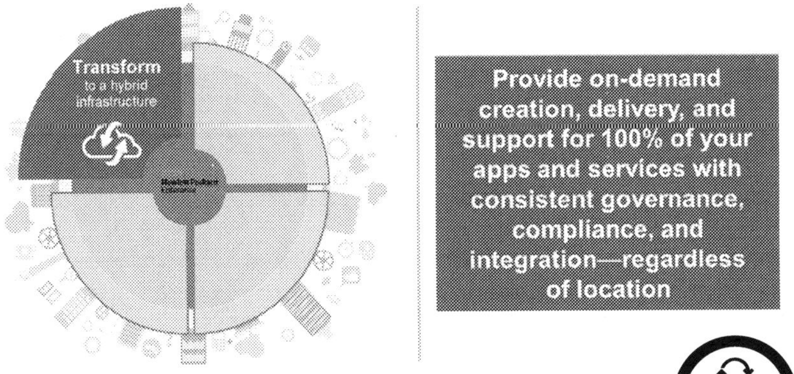

Figure 1-29 Transforming to a hybrid infrastructure

Note

To watch a 5-minute video explaining the "Transform to a hybrid infrastructure" transformation area, scan the above graphic into the HPE My Learning app on a mobile device (Figure 1-29).

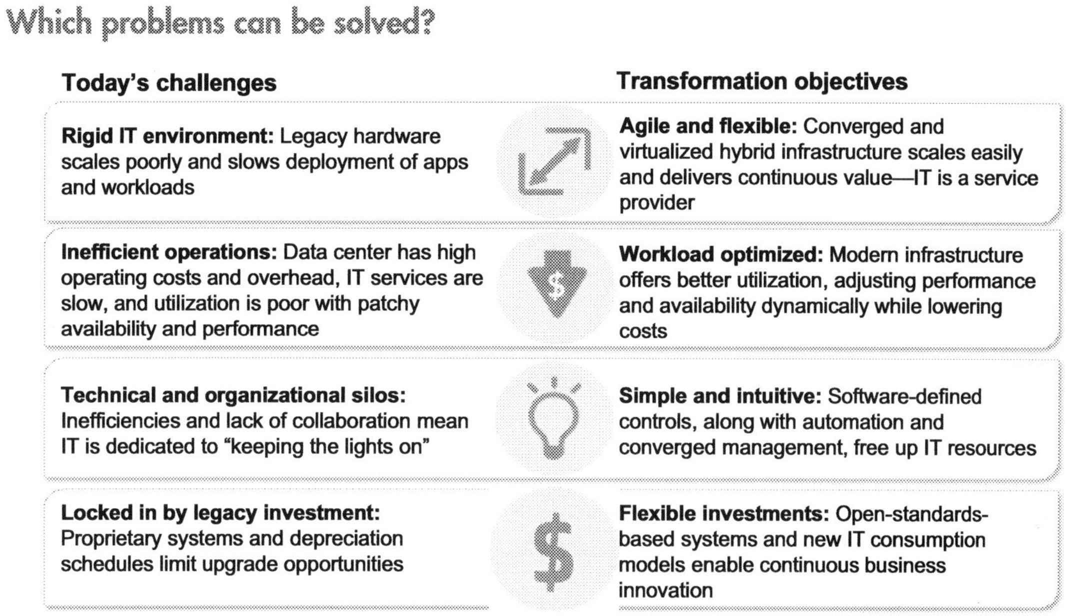

Figure 1-30 IT challenges and transformation objectives

Hybrid is a reality in which extracting optimum performance and efficiency from applications is essential. The best environment for applications, whether traditional, mobile, or cloud native, is unique to each business (Figure 1-30).

Many customers struggle with these issues:

- **Rigid IT environment**—Legacy hardware scales poorly and slows the deployment of apps and workloads.
- **Inefficient operations**—The data center has high operating costs and overhead, slow IT services, and poor utilization with patchy availability and performance.
- **Technical and organizational silos**—Inefficiencies and a lack of collaboration mean IT is dedicated to "keeping the lights on."

- **Being locked in by legacy investment**—Proprietary systems and depreciation schedules limit upgrade opportunities.

To go beyond these challenges, customers must focus on these concepts:

- **Agility and flexibility**—A converged and virtualized hybrid infrastructure scales easily and delivers continuous value to make IT a service provider.
- **Workload optimization**—Modern infrastructure offers better utilization, adjusting performance and availability dynamically.
- **Simplicity and intuitiveness**—Software-defined controls, along with automation and converged management, free up IT resources.
- **Flexible investments**—Open-standards-based systems and new IT consumption models enable continuous business innovation.

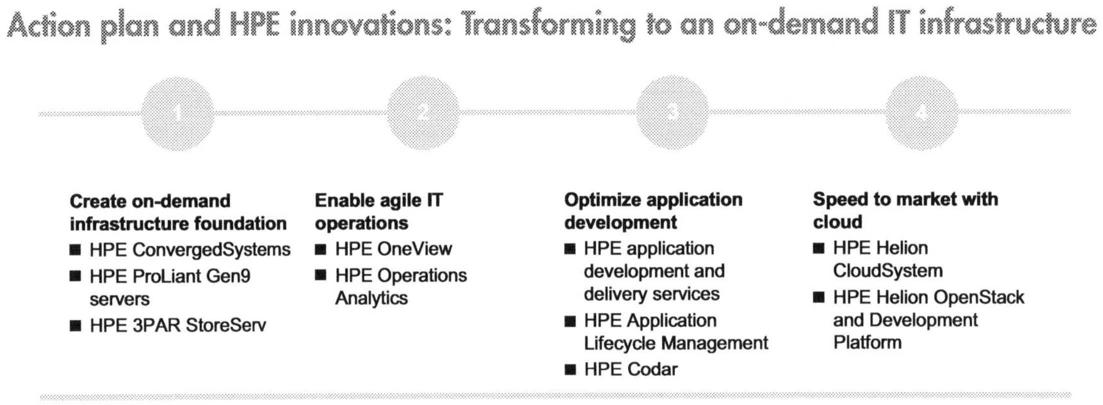

Figure 1-31 HPE innovations: Transforming to a hybrid infrastructure

The journey to hybrid is difficult, nonlinear, and different for every customer. More than 90% of enterprises say their implementation of hybrid is behind company goals and lagging behind their competitors (based on HPE Enterprise Services interviews with customers; Figure 1-31).

The most successful journeys often require bold moves such as shifting to new platforms and software-as-a-service. Sometimes, it is best to move straight to cloud, even from old, nonvirtualized environments.

These HPE products, solutions, and services align with each step of transforming to a hybrid infrastructure:

1. **Create an on-demand infrastructure foundation**—HPE ConvergedSystems, HPE ProLiant Gen9 servers, and HPE 3PAR StoreServ improve efficiency and create agility for the next generation of applications and services.
2. **Enable agile IT operations**—HPE OneView and HPE Operations Analytics transform the management of infrastructure and clouds with analytics and automation.
3. **Optimize application development**—HPE application development and delivery services, HPE Application Lifecycle Management, and HPE Codar deliver high-quality applications across legacy, cloud, and mobile environments.
4. **Speed to market with cloud**—HPE Helion CloudSystem and HPE Helion OpenStack and Development Platform access unlimited scale and speed in a secure way.

Software-defined storage and hyper-converged market

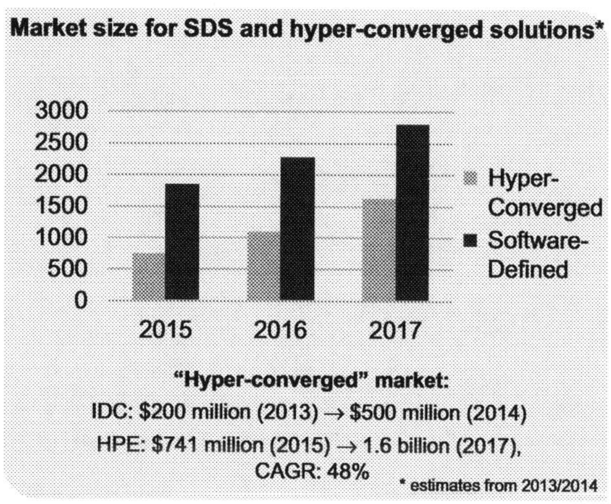

Figure 1-32 Software-defined storage and hyper-converged market growth

The hyper-converged category is growing rapidly. Hyper-converged technology began in virtual desktop infrastructure (VDI), streaming media, and dev/test environments. Now VDI has less than 50% of the market. There still is growth in generalized virtualization and in SQL and database consolidation (Figure 1-32).

What is software-defined storage?

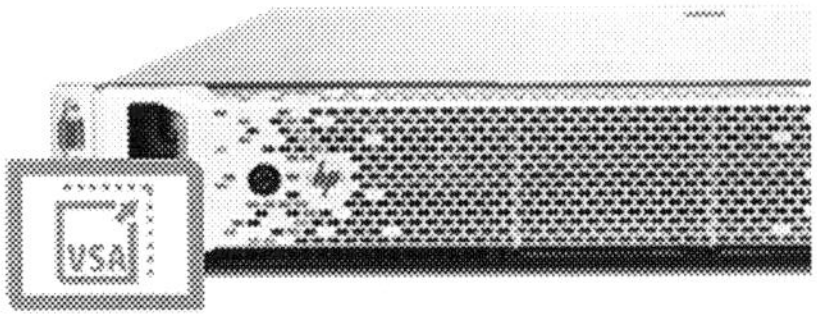

Figure 1-33 HPE VSA software-defined storage

Software-defined storage (SDS) has three main characteristics (Figure 1-33):

- Hardware agnostic:
    - Any x86 server or storage platform
    - An open pool of shared capacity
- Hypervisor independent:
    - VMware, Microsoft Hyper-V, and Linux KVM
    - Spans multiple hypervisors
- Federated and autonomic:
    - Scale-out and move data easily
    - Enable nondisruptive response to demands

The focus of SDS is to provide great value by focusing on cost optimization and maximum flexibility within the data center.

SDS solutions are hardware agnostic. By taking advantage of hypervisor virtualization technologies, SDS can run on virtually any x86 based hardware and any form of storage, allowing for the reuse of older equipment or the efficient optimization of newly acquired technologies.

SDS solutions are hypervisor independent. True SDS technologies provide long-term investment protection by not locking you into any one proprietary hypervisor technology. HPE StoreVirtual VSA supports VMware vSphere, Microsoft Hyper-V, and KVM environments. HPE delivers common management and data services across these hypervisors.

Lastly, SDS solutions seamlessly scale out as your needs grow over time, providing federation and autonomic management from day 1. SDS solutions do not restrict you to isolated towers of storage that do not interact easily with the rest of your environment; instead they provide nondisruptive response to constantly changing demands.

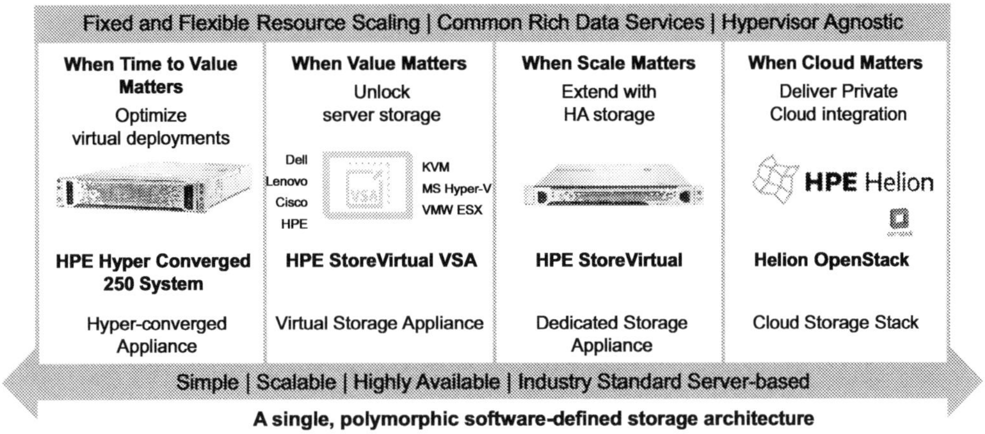

Figure 1-34 HPE SDS solutions

SDS provides flexible software deployments in a single, polymorphic architecture. SDS products include (Figure 1-34) the following:

- **HPE Hyper Converged 250 System**—Hyper-converged appliance
- **HPE StoreVirtual VSA**—Virtual storage appliance
- **HPE StoreVirtual**—Dedicated storage appliance
- **HPE Helion OpenStack**—Cloud storage stack
- **HPE Scality RING**—Software-defined storage platform

HPE ConvergedSystem

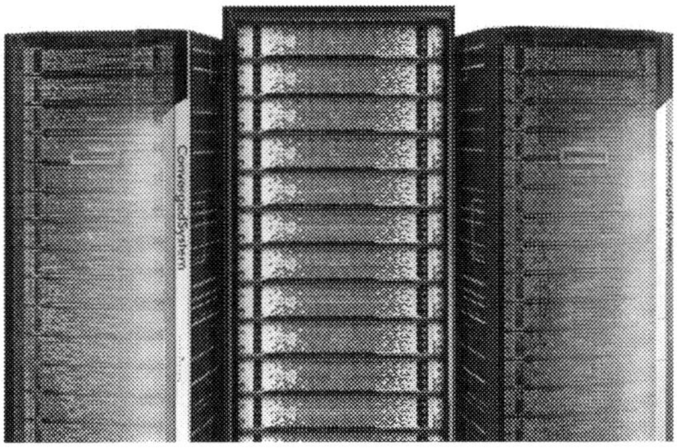

Figure 1-35 HPE ConvergedSystem

HPE ConvergedSystem solutions combine compute, storage, networking, and software in a box. These solutions are integrated and optimized for today's key workloads. The HPE ConvergedSystem portfolio simplifies the infrastructure life cycle, enabling you to quickly build a flexible, integrated, optimized system that includes hardware, infrastructure management, support, and service options (Figure 1-35).

- **CloudSystem**—Complete OpenStack-based integrated solutions
- **Big Data System**—Converged power for data analytics and optimized for SAP HANA or Microsoft Analytics platforms
- **Client Virtualization System**—Optimized for VDI
- **General-purpose system**—Designed to deliver on-demand IT and Infrastructure-as-a-Service to the business
- **Hyper-converged system**—Streamlined infrastructure for small businesses and branch offices

---

More information can be found here: https://www.hpe.com/us/en/integrated-systems/converged.html.

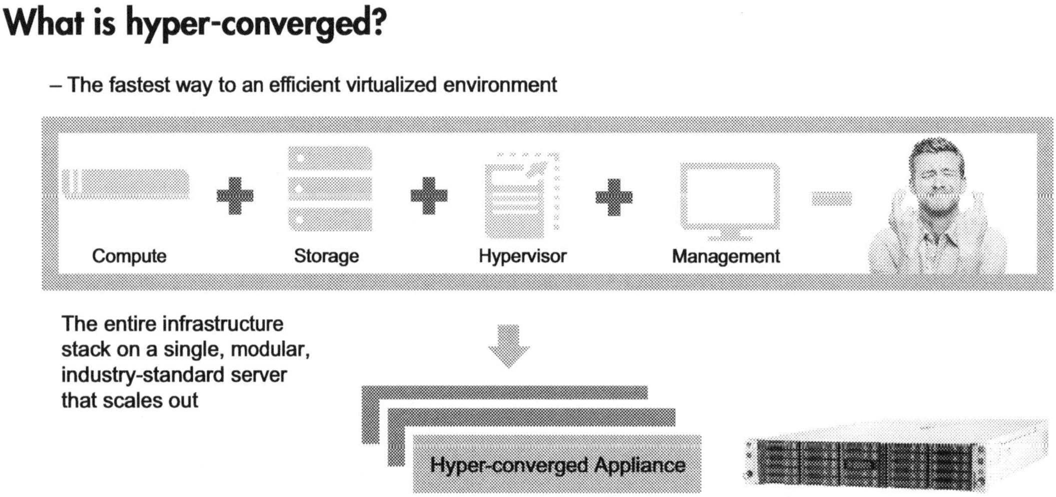

Figure 1-36 Hyper-converged system diagram

The HPE family of hyper-converged systems is preconfigured with servers, storage, networking, and VMware vSphere to enable complete deployment of a virtualized environment in less than 15 minutes and at nearly half the cost of other hyper-converged approaches (Figure 1-36).

You can manage everything simply and easily from HPE OneView for VMware vCenter, including multiple clusters across multiple sites. Proven 99.999% high availability, transparent failover in the event of failure, and inherent disaster recovery capabilities provide worry-free business protection.

The benefits of HPE hyper-convergence

- **Superfast deployment**—Deploy and expand in minutes. HPE OneView InstantOn handles all IP addressing, server and storage clustering, system startup, and updates.
- **Simplified management**—Reduce administrative overhead with provisioning and monitoring directly from the HPE OneView for vCenter management console.
- **Optimized for virtualization**—Boost performance and application resiliency with advanced all-inclusive HPE StoreVirtual data services deeply integrated with VMware and managed through VM policies.

Hyper-converged systems from HPE

HPE offers three proven hyper-converged choices for achieving better business outcomes:

- **HPE Hyper Converged 250 System**—Better flexibility and more value
    - Combines powerful compute and highly available storage and management capabilities into a single, scale-out appliance.
- **HPE Helion CloudSystem on HPE Hyper Converged 250**—An easy path to hybrid cloud
    - Accelerate your provisioning with ready-to-use, workload-optimized infrastructure along with the most complete, integrated, and open cloud in the industry.
- **HPE Hyper Converged 250 for Microsoft**—Virtualization for your Microsoft environment
    - HPE built-in resiliency pairs with add-on cloud services from Microsoft in a package that is easy to acquire, quick to set up, and simple to operate.

Learning check

The following questions will help you measure your understanding of the material presented in this chapter. Read all the choices carefully, because there might be more than one correct answer. (Select or write the correct answer for each question).

1. Describe today's idea economy.

   ....

   ....

   ....

   ....

2. What should organizations focus on to create better business outcomes through IT transformation? (Select three.)

   a. Continuously creating and delivering new services
   b. Hosting workloads and services
   c. Software differentiating products and services
   d. Providing real-time insight and understanding
   e. Storing and managing data

3. Which solution protects against cyber threats?

   a. HPE OneView
   b. HPE Fortify
   c. HPE ArcSight
   d. HPE Data Protector

4. Complete the action plan to empower a data-driven enterprise.

   a. Modernize the ________ ________ warehouse
   b. Deploy a _______ _______ platform
   c. Deliver actionable _______ _______
   d. Enable best-in-class data ___________

5. What matters when transforming to a hybrid infrastructure? (Select three.)
   a. Open-standards-based solutions
   b. IT dedicated to "keeping the lights on"
   c. Rigid IT environment
   d. Easy control of infrastructure and apps
   e. The right capabilities across people, processes, and governance
   f. Technical and organizational silos

For answers, see Chapter 1 Answers on page 437.

2 SAN Basics—Technical Background

OBJECTIVES

In this chapter, you should learn to:

- ✓ Explain the basics of storage technologies
- ✓ Describe Fibre Channel (FC) technology and differentiate the topologies
- ✓ Describe the Internet Small Computer System Interface (iSCSI) protocol
- ✓ Explain the HPE Virtual Connect (VC) technology and capabilities

INTRODUCTION

This chapter provides an overview of technical background required for understanding storage area networks (SANs). It begins with basics of storage technologies and Fibre Channel (FC) topologies. Then it introduces the Internet Small Computer System Interface (iSCSI) protocol that is used in SANs. Lastly, it explores the VC technology and its relationship to storage products.

Basic storage technologies

Parameters of storage systems and drives

Figure 2-1 3PAR StoreServ 8450

When a storage system or a drive is evaluated, the following parameters can be used to describe the product (Figure 2-1):

- Protocol used
  - Serial Attached SCSI (SAS)
  - Serial ATA (SATA)
  - Fibre Channel
  - Protocol generation such as SAS-1,SAS-2, SATA revision 3.2, and so on
- Type
  - Flash or rotating drives
  - Performance
  - Entry, Midline, or Enterprise
- Capacity and number of drives
- Rotational speed
  - 7.2K, 10K, and 15K
- Form
  - Large form factor (LFF)
  - Small form factor (SFF)

Note
This list contains only the most common characteristics.

Serial Attached SCSI protocol

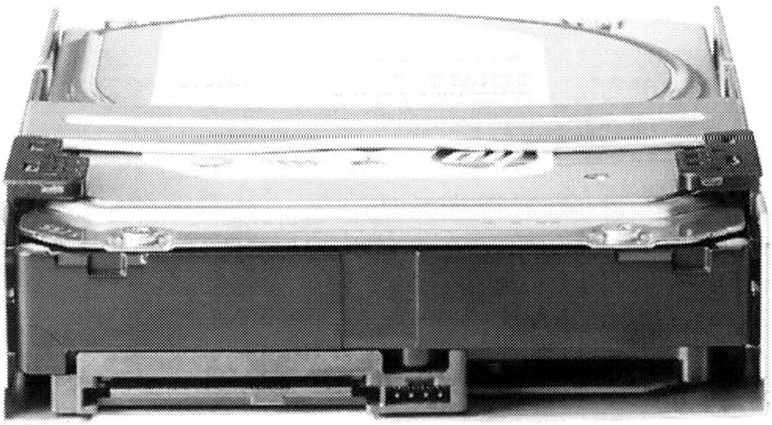

Figure 2-2 SAS interface

SAS has become the de facto hard disk drive (HDD) standard for mission-critical applications (Figure 2-2).

SAS uses a point-to-point architecture that transfers data to and from SCSI storage devices by using serial communication. SAS devices use differential signaling to achieve reliable, high-speed serial communication. SAS inherits its command set from parallel SCSI and its frame formats and full-duplex communication from FC. SAS also supports SATA targets.

SCSI/SAS communication

The three types of SAS devices are initiators, targets, and expanders.

- Initiator devices include host bus adaptors (HBAs) and controllers. The initiator attaches to one or more targets, forming a SAS domain. The initiator is also responsible for providing commands and data.
- Target devices include SAS HDDs or solid-state drives (SSDs), SATA HDDs or SSDs, and SAS tape drives. The target waits for commands, manages the commands provided by the initiator, and performs the operations.
- Expanders are low-cost, high-speed switches. Using expanders, you can increase the number of targets attached to an initiator to create a larger SAS domain.

Commands

SCSI commands are designated as mandatory, optional, or vendor-specific (vendor-unique). Mandatory commands are required of all devices.

Examples of SCSI commands:

- Read
- Write
- Erase
- Report LUNs

Note

For more information about SCSI commands, go to: https://en.wikipedia.org/wiki/SCSI_command.

SAS generations

SAS is entering into its fourth generation with increasing speeds:

- **SAS-1**—3.0 Gb/s, introduced in 2005
- **SAS-2**—6.0 Gb/s, available since February 2009
- **SAS-3**—12.0 Gb/s, available since March 2013
- **SAS-4**—22.5 Gb/s, under development by International Committee for Information Technology Standards (INCITS) and expected in 2017

SAS-2 includes zoning capabilities to improve resource deployment flexibility, security, and data traffic management. SAS-2 maintains backward compatibility with SAS-1.

SAS-2 devices (initiators, targets, or expanders) can support more than one communication speed. If any two linked devices support multiple speeds, the devices use the highest mutually supportable speed. The linked devices determine that speed during a speed negotiation process at startup. A sequential series of speed negotiation windows (SNW) characterizes this process. In SNW-1 and SNW-2, linked devices test established combinations of SAS speeds, transmission amplitude, slew rate, de-emphasis, and spread spectrum clocking (SSC). In SNW-3, the linked devices negotiate link speed and SSC settings.

Unlike SAS-1, SAS-2 allows for training of the transceiver mechanism (PHY) and for exchanging parameters. After SNW-3 has negotiated the speed and settings, a training-speed negotiation window (Train-SNW) tests the fastest mutually supported speed.

The SAS-2.1 standard defines active cables, storage power management, and additional connectors. Also, SAS-2.1 splits the protocol layer into a separate standard, SAS Protocol Layer (SPL).

SAS-3 doubled the rate to 12.0 Gb/s in 2013; SAS-4 is planned for 2017, with more efficient signal encoding that allows 24 Gb/s.

Active cables

The SAS-2.1 standard supports active cables, which are thin cables with active circuitry to reduce cable weight, improve cable management, and improve airflow.

Active circuitry includes built-in drivers, repeaters, and an equalizing filter. The equalizer removes intersymbol interference (ISI), a form of signal distortion. The drivers and repeaters increase the signal-to-noise ratio by boosting the received signal and reducing near-end crosstalk (NEXT). NEXT occurs when two wires are close enough for the signal traveling in one wire to interfere with the signal traveling in the other.

Active cables include a low-power equalizing filter to compensate for dielectric and conductor losses, and the SAS-3 specification increased the rate to 12 Gb/s. SAS-4 will introduce 22.5 Gb/s signaling with a more efficient 128b/150b encoding scheme to realize a usable data rate of 2400 MB/s while retaining compatibility with 6 Gb/s and 12 Gb/s.

Storage power management

SAS-2.1 devices can turn off SAS physical links when they are idle. Any initiator can target and use power management functions. Each SAS transceiver consumes about 200 mW. Turning off all SAS physical links saves a little less than 1 W for a dual-ported drive with two transceivers and a controller (or an attached SAS expander) with two transceivers.

SAS devices

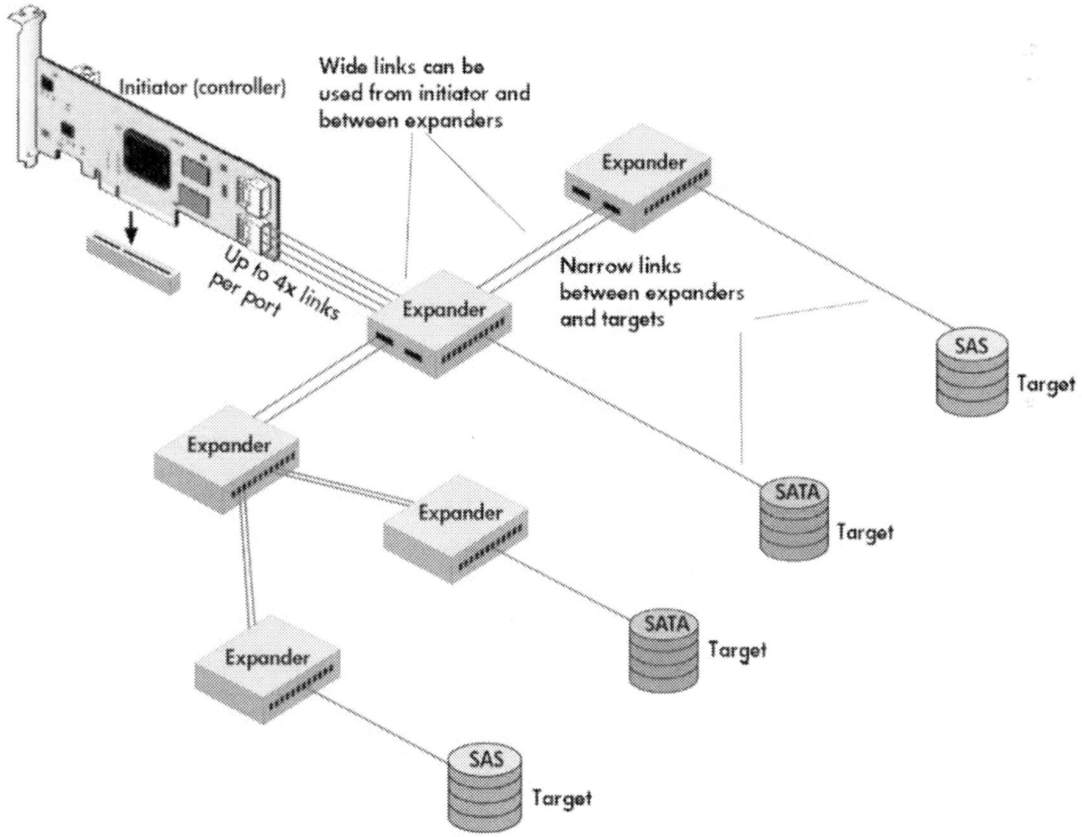

Figure 2-3 Typical SAS domain

The typical SAS domain contains the three types of communication devices: initiators, targets, and expanders. Expanders are optional and not used in every case (Figure 2-3).

Initiators

SAS initiators have multiple ports for connecting to internal and external targets. Each initiator port can have a single physical link (a narrow port) or two, four, or eight physical links (a wide port). You can connect SAS initiator ports to separate domains for failover redundancy. A typical initiator is an HBA or array controller.

Targets

SAS drives (both enterprise-class and midline) have two narrow ports. SAS drives use the same electrical and physical connection interface as SATA drives; however, SATA drives have a single narrow port. You can have SAS and SATA devices in a single domain. The size of the routing tables of the expanders determines how many initiators and targets you can have in a domain.

Expanders

Expanders connect initiators, targets, and other expanders. They receive commands and data in one port and route them to another port based on the SAS address of the target. Expanders use three routing methods: direct, table, and subtractive.

- Direct routing forwards the commands and data to targets directly attached to the expander.
- Table routing forwards the commands and data to another expander.
- When an expander receives a SAS address that it does not recognize, it uses subtractive routing to forward the commands and data to another expander that recognizes the address.

Each routing method uses routing tables that are maintained in each expander. The expander creates the routing table during the discovery process known as self-configuration.

Drive array basics

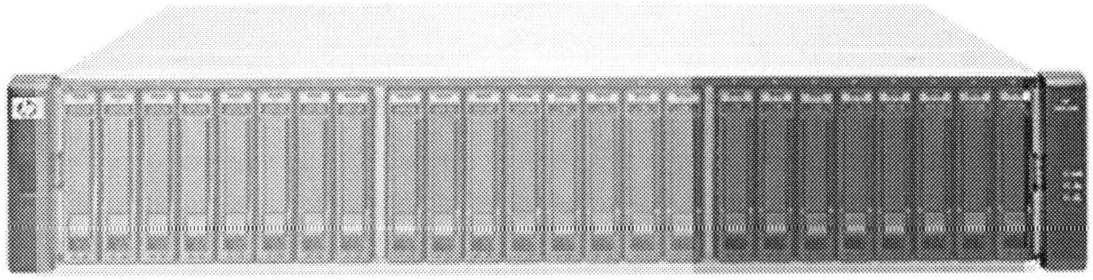

Figure 2-4 Drive array example

An array is a set of physical disk drives that can be combined into a single logical drive or subdivided into multiple logical drives that are distributed across all disks in the set (Figure 2-4).

Having several physical hard drives enables the controller to divide the data across multiple drives. A file is divided into a selected number of sectors, and then the file data is written concurrently across a series of drives in an array.

This approach of combining drives brings several advantages:

- **Performance**—Because multiple drives are accessed simultaneously, the process of writing (or reading) a file across multiple drives is much faster than writing to or reading from a single drive.
- **Redundancy**—If configured properly, one or more drives can fail without affecting the data accessibility on the array.
- **Capacity**—Combining multiple drives together enables you to create logical unit numbers (LUNs) that are bigger than individual drives.
- **Management**—Drive arrays enables central management for all hard drives and operational tasks.

RAID levels

Storage solutions usually support a subset from the following RAID levels.

RAID 0—Disk striping

In RAID 0, a file is divided into stripes and then written across multiple disks. Data is striped across all drives. This greatly decreases disk latency (the amount of time a disk head has to wait for the target sector to move under the head).

RAID 0 offers several advantages over other RAID types:

- All of the disk space is available for data.
- RAID 0 is the least costly.
- Overall disk performance is improved, especially the speeding up of operations that retrieve data from disk storage.
- Read and write performance is excellent.

RAID 0 also has some disadvantages compared to other RAID types:

- RAID 0 is not fault tolerant and provides no redundancy (and therefore no hot-plug capability).
- All data is lost if one of the drives fails.

By definition, RAID 0 requires two or more drives for a true stripe set. However, with some array controllers, a RAID 0 logical volume can be created with a single drive.

Important
Data striping is faster than conventional file writing to a single disk; however, there is no fault tolerance if any single drive fails. If one disk should fail, all data on the array would be lost.

RAID 1—Disk mirroring

With disk mirroring, data is written twice to two mirrored drives. If one drive fails, the mirrored drive is the backup. A RAID 1 implementation requires an even number of disks. Drives must be added in pairs to achieve a RAID 1 expansion.

A RAID 1 implementation mirrors the entire data structure on different drives and allows split seeks. The drives with the requested data nearest to the read/write heads are used for the read, which slightly improves read performance.

This viable, fault-tolerant RAID solution is considered expensive because it requires twice as much drive storage. Only 50% of the total disk space is available for data storage.

RAID 1+0—Mirroring of stripe sets

RAID 1+0 is mirroring with more than two drives. A stripe set (RAID 0) is created across each half of the mirrored drives (RAID 1), thereby both mirroring and striping the data.

Multiple disks can fail without data loss if the disks are in different mirror pairs. Disks 0 and 1 could fail, and all data would be intact on Disks 2 and 3. However, if two disks in the same mirrored pair failed, the data would be lost. RAID 1+0 cannot guarantee protection against this type of two-disk failure.

In a RAID 1+0 configuration, the array controllers can

- Sustain an entire bus failure if the drives are equally distributed across the buses
- Service I/O requests to all operational drives in a degraded condition
- Survive *n*/2 drive failures, where *n* is the number of drives in the array, as long as one member of each mirrored pair survives

RAID 1+0 requires an even number of drives.

This solution is fault tolerant but is considered expensive. It requires double the disk space because only 50% of the total disk space is available for data storage.

RAID 1+0 has good performance and redundancy, but it also has write penalties (two physical write requests for one logical write request).

Note

Some manufacturers refer to RAID 1+0 as RAID 10.

RAID 5—Distributed Data Guarding (data striping and error correction)

Concurrent access and distributed parity are properties of RAID 5. Data is striped across multiple drives and then its parity sum is calculated, which is also striped across multiple drives. Performance increases because the parity is spread across all drives, and there is no need to access a single parity drive after every write command.

RAID 5 is best suited for I/O-intensive applications and transaction processing, thereby making it an ideal solution for high-performance, fault-tolerant servers. RAID 5 requires four physical I/Os from controller to disk per single logical I/O from operating system to controller:

- Read old data
- Read old parity
- Write new data
- Write new parity

Any single drive can fail and the information from the lost drive can be recovered from the parity data stored on other drives.

A minimum of three drives is required, and $n + 1$ drives are needed, where n is the number of drives used for data.

The biggest limitation of RAID 5 is the increased read time in a failure. In RAID 5, regardless of which disk fails, data must be recalculated on each read from the remaining disks.

RAID 6—Advanced Data Guarding

RAID 6, also known as Advanced Data Guarding (ADG), provides high fault tolerance. It distributes two sets of parity data to protect against two drive failures (XOR parity and Reed–Solomon code). Parity (P) is written twice for each piece of data (D). These two sets of data are different, and each set occupies a capacity equivalent to that of one of the constituent drives.

RAID 6 provides high read performance and high data availability. Any two drives can fail without the loss of critical data.

RAID 6 provides

- Higher fault tolerance than RAID 5
- Lower implementation costs than RAID 1+0
- Greater usable capacity per U than RAID 1

RAID 6 requires a minimum of four hard drives, and $n + 2$ drives, where n is the number of drives used for data. Because there are two sets of parity data, RAID 6 provides a relatively low write performance when compared to RAID 5 and its one set of parity data.

RAID 50 (RAID 5+0)

RAID 50 (RAID 5+0) is a nested RAID method that uses RAID 0 block-level striping across RAID 5 arrays with distributed parity. RAID 50 tolerates one drive failure in each spanned array without data loss. RAID 50 configurations require a minimum of six drives and require less rebuild time than single RAID 5 arrays.

RAID 60 (RAID 6+0)

RAID 60 (RAID 6+0) is a nested RAID method that uses RAID 0 block-level striping across multiple RAID 6 arrays with dual distributed parity. With the inclusion of dual parity, RAID 60 tolerates the failure of two disks in each spanned array without loss of data. RAID 60 configurations require a minimum of eight drives.

Storage Networking Industry Association

The Storage Networking Industry Association (SNIA) is made up of 400 member companies spanning the global storage market. The SNIA connects the IT industry with end-to-end storage and information management solutions.

As a not-for-profit association, the SNIA enables members to develop robust solutions for storing and managing the massive volumes of information generated by today's businesses. For more than a decade, the SNIA has worked to bring recognition of storage issues to the IT world, making storage less complicated for the end user. As a result, the SNIA has adopted the role of industry catalyst for the development of storage solution specifications and technologies, global standards, and storage education.

From vendors, to channel partners, to end users, SNIA members are dedicated to providing the industry with a high level of knowledge exchange and thought leadership. SNIA members also share a common goal: To promote acceptance, deployment, and confidence in storage-related architectures, systems, services, and technologies, across the IT and business communities.

The SNIA works toward this goal by forming and sponsoring Technical Work Groups (TWGs), building and maintaining a vendor-neutral Technology Center in Colorado Springs, Colorado, and promoting activities that expand the breadth and quality of the storage and information management market. The SNIA attributes its ability to accomplish these goals to the dedication and hard work of hundreds of volunteers from its member companies.

The SNIA has its headquarters at the Technology Center in Colorado Springs, and it has seven regional affiliates spanning the globe.

HPE is a "large vendor" member of the SNIA with voting rights, and it has a member on the board of directors.

SMI-S support

SNIA validates that a member company's products (software or hardware) conform to a particular version of the Standard Management Initiative Specification (SMI-S) specification for storage management. By the implementation of the SMI-S Conformance Testing Program (SMI-S CTP), SNIA can provide an impartial validation of a storage management software product, or it can affirm that an item of storage networking infrastructure conforms to a version of SMI-S.

SMI-S defines a method for the interoperable management of a heterogeneous SAN and describes the information available to a Web-Based Enterprise Management (WBEM) client from an SMI-S compliant CIM server and an object-oriented, XML-based, messaging-based interface that is designed to support the specific requirements of managing devices in and through SANs. Developer support for SMI-S is available through the SNIA SMI-S Google group.

Note

For more information about the SMI-S, go to: http://www.snia.org/tech_activities/standards/curr_standards/smi.

SNIA mission statement

The SNIA aims to "Lead the storage industry worldwide in developing and promoting standards, technologies, and educational services to empower organizations in the management of information."

Note

For more information about the SNIA mission and vision, go to: http://www.snia.org/education/mission-vision.

Approaches to implementing storage

There are three principal and complementary ways to implement storage solutions.

Direct-attached storage (DAS) storage solutions—These are the most straightforward storage solutions, and they can be either internal in a single server (usually using the SAS protocol) or directly attached external storage. However, DAS storage cannot be shared among multiple servers, and the one-to-one connection between device and server can add management complexity.

Network-attached storage (NAS) devices—These devices are primarily used for unstructured data and file sharing. NAS storage adds storage management functionality, improved data protection, performance enhancements, and the ability to share data using industry-standard protocols across an Ethernet network. NAS devices can usually be installed easily and quickly, and they can be accessed by servers and nonserver devices.

SAN storage—SAN is a solution rather than a product, using shared network components. A SAN provides a dedicated network of storage systems that are accessible by multiple servers. In contrast to how NAS delivers data using file systems, a SAN allows block-level access and delivers even better levels of performance, management, and resilience. SANs offer business continuity with redundant components, automated failover, and centralized management. Common protocols used in SANs are FC, iSCSI, FC over Ethernet (FCoE), and InfiniBand.

Direct-attached storage

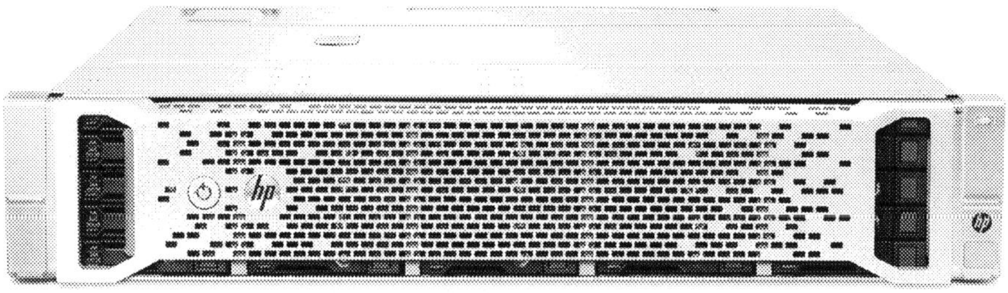

Figure 2-5 HPE D3700 Enclosure

DAS consists of an open-system server running any application with dedicated internal or external storage subsystems using a SAS protocol. DAS provides dedicated storage for multiple clients with a one-to-one server-to-storage ratio (Figure 2-5). DAS offers the easiest way to deploy incremental amounts of storage as needed without extensive planning. As RAID inside the server has become less expensive, DAS has grown in popularity. A high percentage of deployed storage is now DAS.

The traditional approach involves DAS solutions that attach RAID arrays or hard drives directly to a server. DAS is familiar, works well, and is less costly than initial SAN investments.

Using a DAS, many DAS customers have separate storage systems and storage management software products connected to individual servers.

Some advantages of DAS include the following:

- Ease of deployment
- Scalability
- Cost—relatively inexpensive to acquire, maintain, and expand
- High performance and reliability
- Fast server-to-storage data transfer

DAS disadvantages include the following:

- **Inefficient resources**—Storage space exists in isolated pools.
- **Unplanned redundancy**—Duplicate copies of the same file might reside on different servers.
- **Increased management**—Server-based management means that islands of data are difficult to bridge and require a significant amount of labor to manage.
- **Decreased access**—If the server becomes unavailable, access to the stored data is disrupted for clients.

Network-attached storage

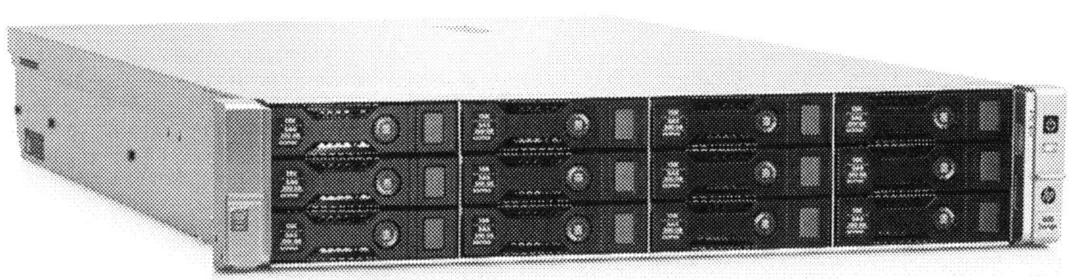

Figure 2-6 HPE StoreEasy 1650 Storage

NAS solutions consist of a specialized server-storage device that connects directly to the network. A file system is located and managed on the NAS device (Figure 2-6). Data is transferred to servers and other devices through the LAN and to clients, using industry-standard file-sharing protocols such as Common Internet File System/Server Message Block (CIFS/SMB) and Network File System (NFS). The intelligent NAS device enables data sharing among heterogeneous network clients.

NAS storage devices require storage cabinets that provide specialized file access, security, and network connectivity. A NIC is required on the server to access the storage. NAS provides file-to-disk block mapping and client access at the file level using network protocols.

NAS technology simplifies manageability and improves data access to clients and applications. A NAS solution generally works with a mix of clients and servers running different operating systems.

The dedicated NAS appliance can provide shared storage between heterogeneous clients. Disk arrays and other storage devices connect to the network through a traditional LAN interface such as Ethernet. Storage devices attach to network hubs similarly to the way servers and other network devices do. All network users have equal access to the stored data and do not have to go through the server. NAS makes storage resources more readily available and helps to alleviate the server bottlenecks commonly associated with accessing storage devices.

Some advantages of NAS include the following:

- Installs easily
- Is affordable
- Incorporates mature technologies
- Offers a scalable solution
- Increases network performance
- Supports remote management

Storage area network

Figure 2-7 Typical SAN components: FC switch, server HBA, and array

The SAN is a separate network that provides a storage repository that is attached to multiple host servers. SANs enable external storage to be shared by the servers without impacting system performance or the primary network (Figure 2-7).

A SAN solution supplies open-system servers running applications on an open operating system. The solution includes shared external storage resources, network infrastructure components (such as FC switches), and value-added software for enhanced storage and data management. This technology provides consolidated and virtualized storage with massive scalability and fault tolerance.

Storage interfaces include the following:

- iSCSI
- FC

The basic purpose of a SAN is to transfer data between computer systems and storage elements, and among storage elements.

SAN benefits include the following:

- Centralized storage management
- Data replication
- Easy expansion and storage reallocation
- Disaster recovery
- Scalability
- Facilitation of backup and disaster recovery

SAN types

This section differentiates the four common types of SANs.

Homogeneous operating system SAN

A homogeneous SAN is a network where the same operating system runs on all servers in the SAN. This configuration is suitable for small SANs that use a single operating system vendor, but it severely restricts the growth and consolidation of existing storage systems.

A homogeneous SAN storage topology consists of the following components:

- Shared, pooled storage is attached to multiple single-vendor servers running a homogeneous operating system
- Any-to-any connectivity between servers and I/O (disk and tape) using a common switched fabric and network switches
- A static allocation of server and I/O resources

Heterogeneous operating system SAN

In a heterogeneous SAN, multiple operating systems run on multiple platforms within the SAN. This type of SAN interconnects multiple shared storage environments into a single network. All storage can be managed from a single location or from multiple locations. All of the consolidated storage becomes available to any host server, regardless of physical location.

FC is the fundamental technology, or backbone, used with heterogeneous SANs. It is the transport mechanism that gives increased distance, higher speed, and enhanced management functionality to SANs.

An enhanced heterogeneous SAN storage topology consists of

- Shared, pooled storage attached to multiple servers running various operating systems
- Any-to-any connectivity between servers and I/O (disk and tape)
- A static allocation of server and I/O resources with improved management function

Homogeneous fabric SAN

This type of SAN has a fabric consisting of products from a single vendor such as HPE. Proprietary features are usually available.

Heterogeneous fabric SAN

This type of fabric contains open-fabric-compliant products from various vendors (e.g., HPE and third parties). The components must comply with industry standards.

General SAN topologies

This section differentiates the common SAN technologies and describes the most common standard topologies. Some topologies can only be designed for FC, not for iSCSI.

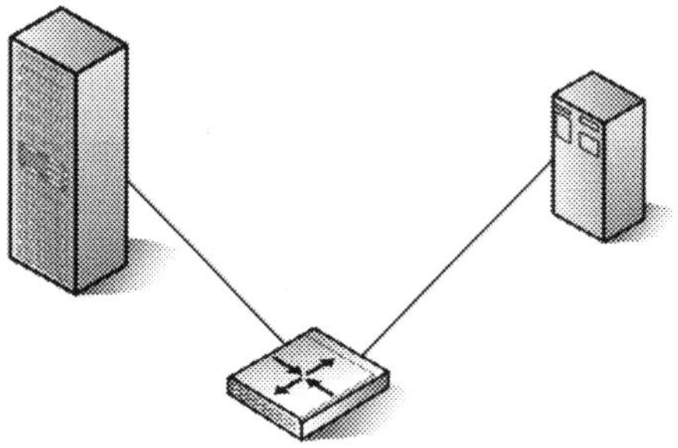

Figure 2-8 Single-switch fabric example

Single switch fabric

A single-switch fabric consists of a FC switch, a server, and a storage system (Figure 2-8). This topology forms the basis for all HPE standard topologies. For example, two single-switch fabrics could be connected to create a cascaded fabric or three or more single-switch fabrics could be connected to create a ring fabric or a core-edge fabric.

The benefits of a single-switch fabric include the following:

- Easy installation and configuration of servers and storage
- Maximum fabric performance because all communicating devices connect to the same switch
- Support for local, centralized, and distributed data access needs

Cascaded fabric

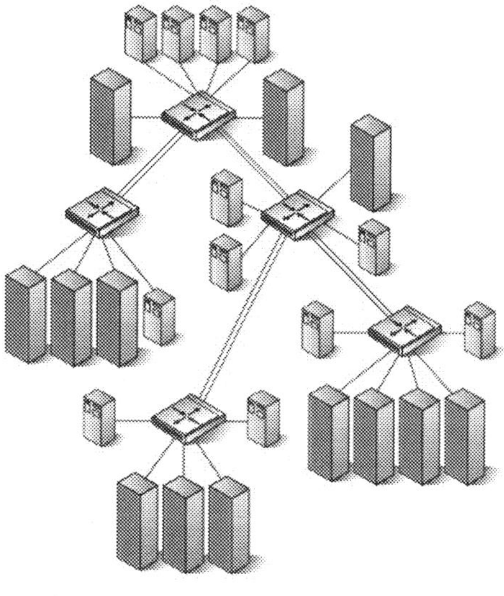

Figure 2-9 Cascaded fabric example

A cascaded fabric is a set of interconnected switches, arranged in a tree format, that have one or more interswitch links (ISLs; Figure 2-9). The administrator can connect one switch to one or more switches using a single ISL to each, or the administrator could connect a pair of ISLs between two switches. HPE recommends a minimum of two ISL connections on each switch to provide fabric path redundancy. A cascaded fabric topology is recommended when multiple groups of devices with localized intraswitch access are needed.

Cascading enables you to

- Achieve optimum I/O activity by connecting servers and storage to the same switch in the cascaded fabric

- Easily scale the fabric over time by adding cascaded switches

A cascaded fabric offers several benefits:

- The ability to connect SANs in diverse geographic locations
- Ease of scalability for increased server and storage connectivity
- Shared backup and management support
- Optimum local performance when communicating devices are connected to the same switch in the cascaded fabric
- Cost efficiency because of the large number of switch ports available

Meshed fabric

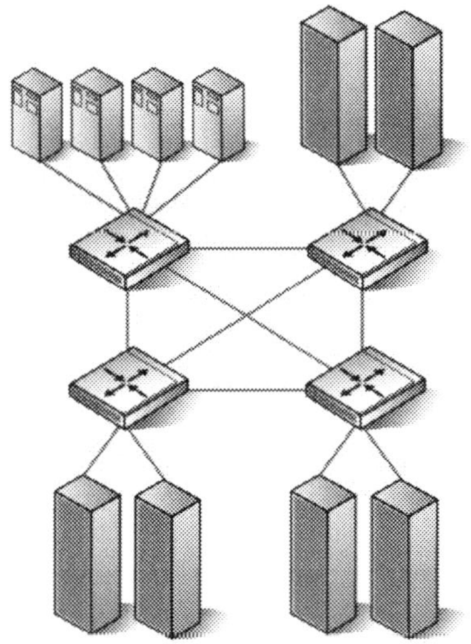

Figure 2-10 Meshed fabric example

A meshed fabric is a group of interconnected switches using multiple ISLs for fabric resiliency (Figure 2-10). If one ISL fails, the switch automatically reroutes data through an alternate path in the fabric. If the alternate path includes other switches, the data must pass through those switches to reach its destination.

A meshed fabric offers several benefits:

- The ability to meet multiple data access needs
- Multiple paths for internal fabric resiliency

- Ease of scalability
- Shared backup and management support
- Support for a mix of local and distributed data access
- Less impact on performance because of intraswitch traffic

Ring fabric

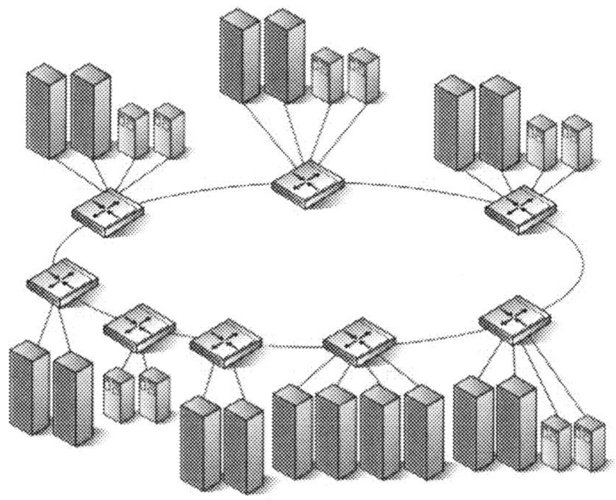

Figure 2-11 Ring fabric example

A ring fabric is a ring of interconnected switches. It provides a level of fabric resiliency similar to the meshed fabric, and it ensures full fabric connectivity with a minimum of two paths for each switch (Figure 2-11).

The ring fabric enables you to

- Scale the fabric in a modular fashion
- Achieve optimum I/O performance by connecting a group of servers and storage to one switch

---

Note

HPE does not recommend the ring fabric for applications requiring many-to-many connectivity.

A ring fabric offers several benefits:

- A modular design with ease of scalability by adding a switch and other devices
- Multiple paths for internal fabric resiliency
- Support for a mix of local data access and occasional centralized data access

Core-edge fabric

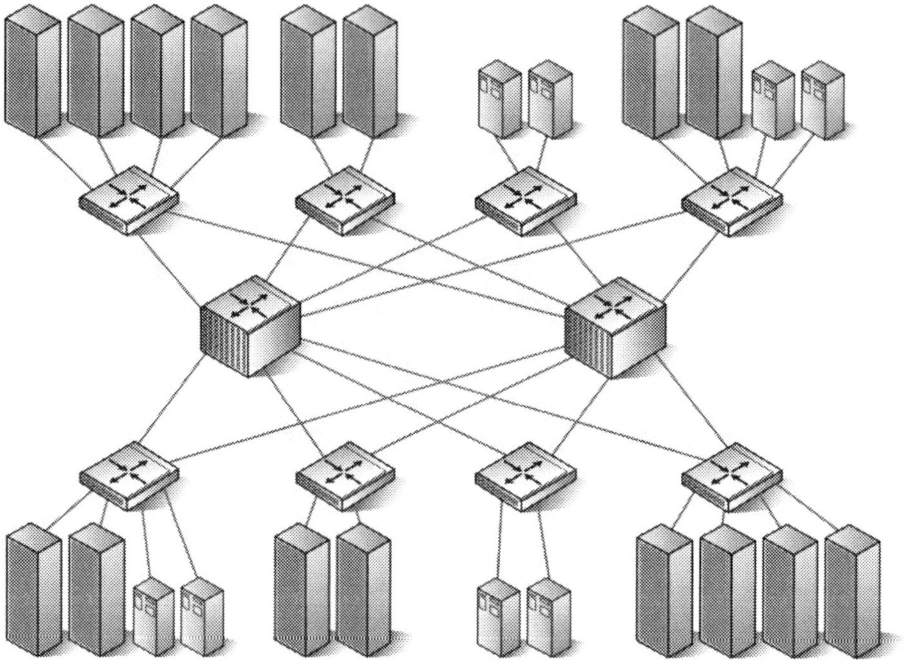

Figure 2-12 Core-edge fabric example

HPE recommends using a core-edge fabric wherever possible. A core-edge fabric has one or more FC switches (called core switches) that connect to edge switches in the fabric. The core switches provide high bandwidth and redundant connectivity to the edge switches (Figure 2-12). The edge switches provide user ports for servers and storage.

You can also connect centralized storage (disk or tape) to the core switches if centralized access is required.

The core-edge fabric is optimal for these conditions:

- Many-to-many connectivity environments that require high performance
- Unknown or changing I/O traffic patterns
- SAN-wide storage pooling

Core-edge fabric types

The number of ISLs between edge and core switches—typically expressed as a fan-in ratio such as 7:1—characterizes the core-edge fabric types. The first number (7) indicates the number of edge ports. The second number (1) indicates the number of ISLs used by the edge ports to connect to a core switch in the fabric.

There are two core-edge fabric topology types: fat tree and skinny tree.

- A fat tree topology has at least 50% of its edge ports dedicated as ISLs, resulting in an ISL ratio of 1:1.
- A skinny tree topology has less than 50% of its edge ports dedicated as ISLs, resulting in an ISL ratio of *x*:1, where *x* is 2 or more.

Fibre Channel

FC is an integrated set of standards developed by the American National Standards Institute (ANSI). FC technology is an industry-standard interconnect serial data transfer architecture that delivers a high level of reliability, throughput, and distance flexibility for the server industry.

FC protocol

FC uses fiber optics to provide an interconnection scheme among devices. It enables the rapid delivery of large amounts of data by supporting high-performance protocols such as Fiber Distributed Data Interface (FDDI) and Small Computer System Interface (SCSI).

A complete FC solution includes hardware platforms, storage systems, operating systems, and applications.

A FC solution is hardware intensive and operates among only a few devices with predefined addresses. Data communication on a FC system occurs over a direct or a switched point-to-point connection between the communicating devices. As a result, FC systems transport data at a high speed with low overhead.

In comparison, computer networks handle a more extensive range of tasks than FC systems because the operating system translates and converts data across several layers of the network. Because a network is software intensive, it has relatively high overhead.

FC speeds

- 1 Gb/s (200 MB/s)
- 2 Gb/s (400 MB/s)
- 4 Gb/s (800 MB/s)
- 8 Gb/s (1600 MB/s)
- 16 Gb/s (3200 MB/s)
- 32 Gb/s and 128 Gb/s on the road map for 2016

Why use FC?

FC is currently the most appropriate protocol for the operation of SANs. FC has many benefits:

- Overcomes the physical limitations of SCSI and SAS
- Offers improved distances

- Has superior throughput and average bandwidth
- Provides enhanced address space
- Is a network topology rather than a bus-based solution

FC technology has these key features:

- It allows for up to 16 million devices, which can include host computers, disk drives, drive arrays, and tape storage, to be combined in a single network that is connected to interconnect devices such as switches.
- FC is scalable.
- Distances of 10 km between devices can be achieved using single-mode, fiber-optic cables. With the use of Extended Long-Wavelength laser (ELWL) distances, far greater distances can be achieved.
- Data transfer rates in the original implementation were 1.0625 Gb/sec, but these rates are now up to 16 Gb/sec.
- Data reliability is assured with a bit error rate of 10—12.
- FC supports multiple topologies, providing for more flexibility in achieving higher availability and performance.

FC components

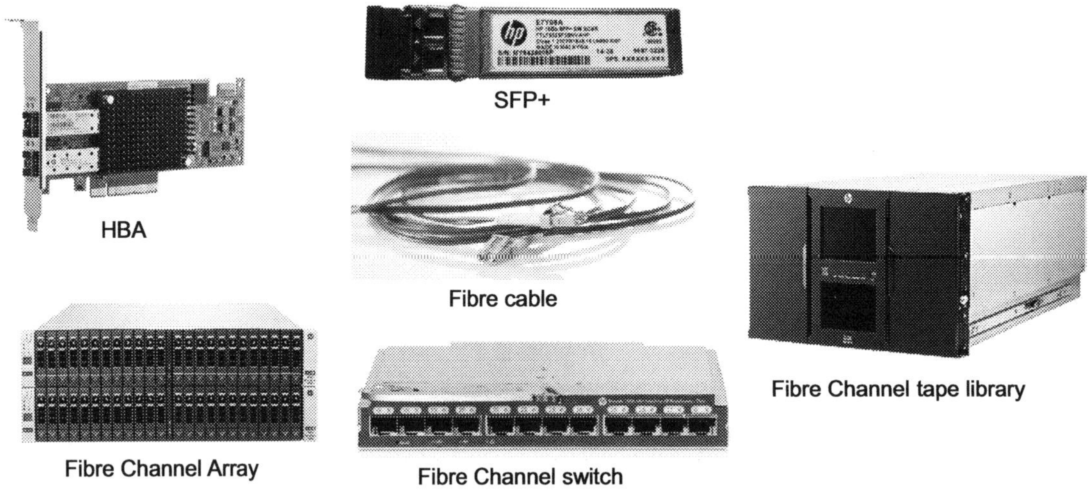

Figure 2-13 16 Gb HBA, 16 Gb SFP+, Brocade 16 Gb SAN Switch for HPE BladeSystem, 3PAR StoreServ 7450, and HPE StoreEver MSL6480 Tape Library

A FC implementation can include these hardware components (Figure 2-13):

- HBAs
- Small form-factor pluggable (SFP) transceivers

- Storage and FC drive arrays
- FC array controllers
- FC switches
- Tape libraries
- FC cables

FC topologies

FC supports the following three topologies:

- Point to point
- FC Arbitrated Loop (FC-AL)
- FC Switched Fabric (FC-SW)

Each node must have at least one port to provide access to other devices if it is connected to a FC topology. The port on the node is known as an N_port. In the port-naming convention, N designates node, L designates loop, and F designates fabric. The FC port names, types, topologies, and descriptions are listed in the Table 2-1.

Table 2-1 FC port names, types, topologies, and descriptions

Port	Type	Associated topology
N_port	Node	Point-to-point or fabric
NL_port	Node	Node connected to an arbitrated loop
F_port	Fabric	Fabric port
FL_port	Fabric	Fabric connected to an arbitrated loop
L-port	Loop	Hub port on an arbitrated loop
L_C_F	Loop	Link Control Facility (L_C_F) is a hardware facility that attaches to each end of the link and manages transmission and reception of data. It is located within each port, contains a transmitter and a receiver, and provides the logical interface to the node.
Rx		The receiver (Rx) is the portion of the link control facility dedicated to receiving an encoded bit stream from the media, converting this bit stream into transmission characters, and decoding these characters using the rules specified by physical layers of Fibre Channel (FC-PH).
Tx		The transmitter (Tx) is the portion of the link control facility dedicated to converting valid data bytes and special codes into transmission characters using transmission code rules. The Tx then converts these characters into a bit stream and transmits the bit stream to the transmission media, which is either optical or electrical.
G_port		This is a generic switch port that operates in either E_port mode or F_port mode.
E_port		This interswitch expansion port is used to connect to an E_port of another switch to build a larger fabric.

Point-to-point topology

Figure 2-14 Point-to-point topology example

The simplest FC topology is point to point in which a single link connects only two ports (Figure 2-14). This topology is inexpensive because no switch is required.

To create larger point-to-point configurations, you can provide multiple ports on each node.

Each point-to-point connection provides the full bandwidth supported by the ports on the devices. Depending on the type of link (multimode or single-mode fiber), the two nodes can be separated by up to 500 m (multimode fiber) or 10 km (single-mode fiber).

FC-AL topology

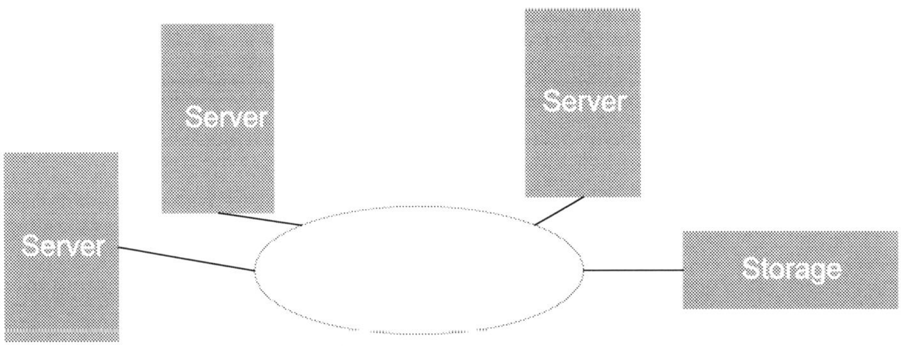

Figure 2-15 FC-AL topology example

By connecting nodes to a FC hub, an FC-AL topology is created (Figure 2-15).

FC-AL adds capacity (support for up to 126 nodes on a single loop) but not performance. The bandwidth is shared among all active nodes on the loop.

This topology allows a single connection between one pair of ports at any point in time. After a session (two devices communicating) is started, the other devices connected to the loop must wait until the connection ends.

Devices participating in the arbitrated loop share access, but the active link has the full bandwidth. Only one pair of nodes can communicate on the loop at one time.

Switched fabric topology

A network of switches in a FC environment is referred to as a fabric. Nodes connect to this fabric to access other nodes. A wide-open architecture uses intelligent switches to connect many ports.

The FC fabric was designed as a generic interface between a node and the physical layer. By adhering to this interface, FC nodes can communicate over the fabric with other nodes without knowing about that node.

A fabric is often referred to as a switch topology. Frames are routed through various switches by having the fabric elements interpret the destination address identifier in a frame when it arrives at each fabric element.

Ports on one node can communicate with ports on other nodes connected to the same fabric. With the fabric topology, many connections can be active at the same time.

The any-to-any connection service and peer-to-peer communication service provided by a fabric is fundamental to the FC architecture. FC can support both channel and network protocols simultaneously.

The class of service used influences the frame routing. The class of service request is generated at the start of the frame delimiter of each frame.

An FC-SW topology is usually more expensive than an FC-AL topology.

A switched fabric topology offers these benefits:

- Port failures are isolated.
- Multiple pairs of ports can communicate simultaneously.
- Up to 16 million nodes can be supported.
- Switches act like routers.
- It provides high-aggregate bandwidth.
- It enables electrical and logical isolation.

Internet Small Computer System Interface

iSCSI is an Internet Engineering Task Force (IETF) SCSI transport protocol for mapping block-oriented storage data over TCP/IP networks.

Overview of the iSCSI protocol

The iSCSI protocol enables universal access to storage devices and SANs over standard Ethernet-based TCP/IP networks.

These networks can be dedicated networks, or they can be shared with traditional Ethernet applications.

iSCSI maps SCSI to a network

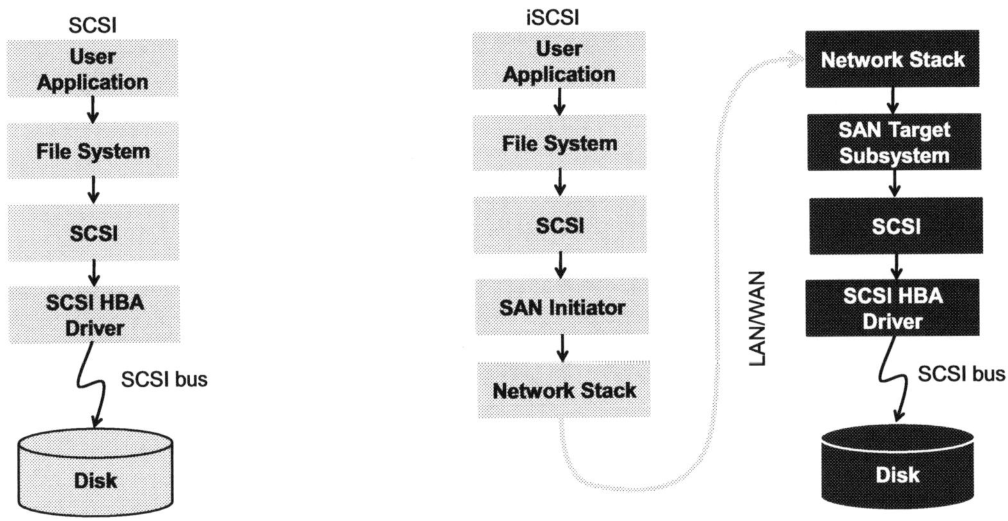

Figure 2-16 iSCSI maps SCSI to a network

SCSI is a long-established protocol for connecting disks to computers (Figure 2-16). All common operating systems contain drivers and logic for SCSI devices. By replacing the SCSI hardware driver with a SAN initiator, iSCSI creates a connection to the target SAN system. The user system sees the iSCSI connection as if it were a normal SCSI disk, so no further modifications are required to the accessing system or applications.

The SCSI protocol has been mapped over various transports, including Parallel SCSI, IPI, IEEE-1394 (firewire), and FC. These transports are I/O specific and have limited distance capabilities.

The iSCSI protocol is a means of transporting SCSI packets over TCP/IP and providing for an interoperable solution that can take advantage of existing Internet infrastructure, Internet management facilities, and address distance limitations.

The iSCSI protocol uses TCP/IP for its data transfer. Unlike other network storage protocols such as FC (which is the foundation of most SANs), iSCSI requires only the simple and ubiquitous Ethernet interface (or any other TCP/IP-capable network) to operate. This enables low-cost centralization of storage without all of the usual expense and incompatibility normally associated with FC SANs.

Critics of iSCSI expect worse performance than FC because the TCP/IP protocol adds overhead to the communication between the client and the storage. However, new techniques like the TCP Offload Engine (TOE) help in reducing this overhead. Tests have shown excellent performance of iSCSI SANs, whether TOEs or plain Gigabit Ethernet NICs were used. In fact, in modern high-performance servers, a plain NIC with an efficient network driver code can outperform a TOE card because fewer interrupts and direct memory access (DMA) memory transfers are required.

Initial iSCSI solutions were based on a software stack. The iSCSI market has been growing steadily and should improve in performance and usability as more organizations deploy Gigabit and 10 Gigabit networks and as manufacturers integrate iSCSI support into their operating systems, SAN products, and storage subsystems. iSCSI became even more interesting when Ethernet started to support higher speeds than FC.

iSCSI stack

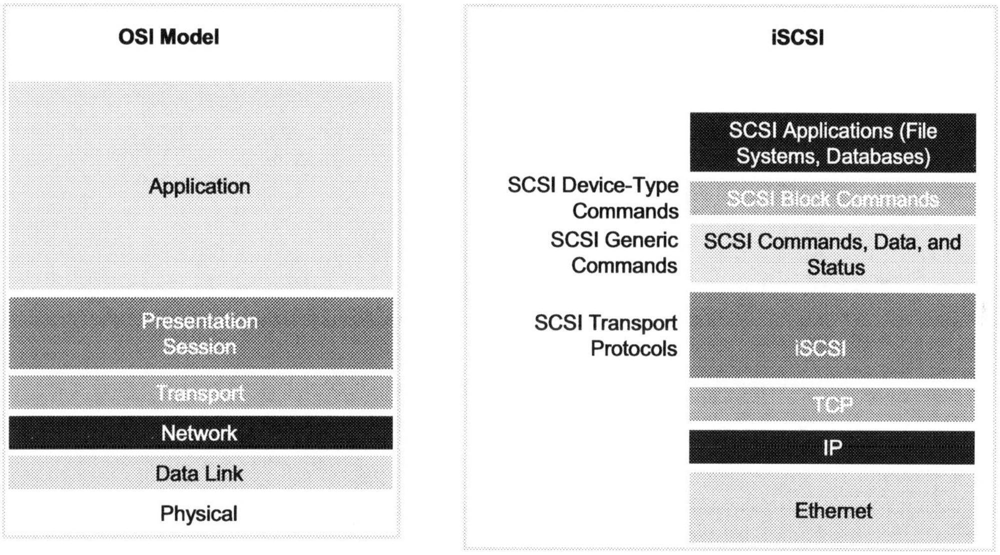

Figure 2-17 iSCSI stack

iSCSI uses TCP/IP for reliable data transmission over potentially unreliable networks (Figure 2-17). The iSCSI layer communicates with the operating system standard SCSI set and includes encapsulated SCSI commands, data, and status reporting capability.

When the operating system or the application requires a data write operation, the SCSI Command Descriptor Block (CDB) must be encapsulated for transport over a serial gigabit link and delivered to the target. The iSCSI protocol monitors the block data transfer and validates completion of the I/O operation. This occurs over one or more TCP connections between the initiator and the target.

In practical applications, an initiator can have multiple target resources over an IP network and, consequently, multiple concurrent TCP connections are active.

The iSCSI protocol maps the SCSI Remote Procedure Call model to the TCP/IP protocol and provides a conceptual layer completely independent of the SCSI CDB information. SCSI commands are transported by iSCSI request, and the SCSI response and status are handled by iSCSI responses. iSCSI protocol tasks are then carried by this same iSCSI request and response mechanism.

Following the pattern of the SCSI protocol, iSCSI uses the concepts of initiator, target, and communication messages called protocol data units (PDUs). Likewise, the iSCSI transfer direction is defined, respectively, to the initiator. As a means to improve performance, iSCSI allows a phase collapse that provides a command or a response and its associated data to be sent in a single iSCSI PDU.

iSCSI encapsulation

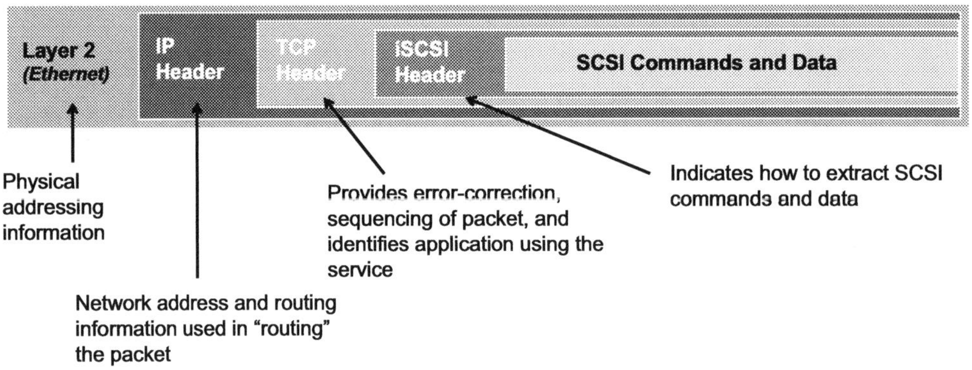

Figure 2-18 iSCSI encapsulation into a network packet

The iSCSI standard stipulates that the protocol must not require modification to the current IP and Ethernet infrastructure to support storage traffic (Figure 2-18). The iSCSI protocol standard must allow implementations to equal or to improve on the current state of the art for SCSI interconnects.

The iSCSI protocol has these requirements:

- It must provide low delay communications.
- It must provide high bandwidth and bandwidth aggregation.
- It must have low host CPU utilizations, equal to or better than the current technology.
- It must allow you to build I/O adapters to handle the entire SCSI task.
- It must permit zero-copy memory architectures.
- It must not impose complex operations on host software.
- It must be cost competitive with alternative storage networking technologies.

iSCSI initiators

The iSCSI initiator (IP host) is a system that starts the exchange of information with an iSCSI target. IP hosts access the iSCSI target storage systems as if they were directly attached. The two types of iSCSI initiators are software and hardware.

An IP host can access an iSCSI environment using one of the following initiators:

- **Software iSCSI initiator**—The iSCSI code runs on the host and allows an Ethernet NIC to handle iSCSI traffic. Software iSCSI offers low cost but with a performance penalty and CPU overhead. Software iSCSI initiators are available from many vendors.
- **TCP Offload Engine Network Interface Card (TOE NIC)**—This shifts the processing of the communications protocol stack (TCP/IP) from the server processor to the NIC, lowering CPU overhead and use.
- **Hardware iSCSI initiator (iSCSI HBA)**—A high-performance HBA integrates both TCP/IP and iSCSI functions. Although this integration adds cost to the HBA, it also provides high-speed iSCSI transport and minimal CPU overhead. The HBA transfers SCSI commands and data encapsulated by iSCSI directly to the host.

Virtual Connect

To implement a VC architecture, hardware components must be used inside the HPE BladeSystem c7000 or c3000 Enclosures. The key hardware components of a VC solution include these modules:

VC Ethernet modules:

- Can be the VC Flex-10/10D or the VC Flex-10
- Connect selected server Ethernet ports to specific data center networks
- Support aggregation and tagging of uplinks to the data center for connection to any data center switch

VC FC modules:

- Selectively aggregate multiple server FC HBA ports on an FC uplink using N_Port ID Virtualization (NPIV)
- Display as a set of HBA ports to external FC switches

VC FlexFabric modules:

- Manage and connect to both Ethernet and SAN storage

Note

Not all VC modules are working in HPE BladeSystem c3000 Enclosures, and HPE OneView cannot be used to manage VC infrastructure in these enclosures. Always verify support in the QuickSpecs.

HPE VC family

Figure 2-19 HPE BladeSystem c7000 Enclosure and VC modules

VC is a set of interconnect modules and embedded software for HPE BladeSystem c-Class enclosures that simplifies the setup and administration of server connections (Figure 2-19). The VC modules support the HPE BladeSystem c7000 Enclosure, the HPE BladeSystem c3000 Enclosure, and all the server blades and networks contained within the enclosure.

The VC Ethernet and FlexFabric modules can also be connected to other devices such as printers, laptops, rack servers, and storage devices. To connect to devices other than switches, create a VC network for that device and only connect uplinks for that network to that device. If you connect uplinks from that network to other devices, one of the uplinks becomes standby because of a loop avoidance algorithm.

The Virtual Connect Fibre Channel (VCFC) and FlexFabric modules enable connection to Brocade, Cisco, McDATA, or QLogic data center FC switches; the modules display as node ports to the FC fabric.

HPE VC currently includes the following components:

- FlexFabric 20/40 F8 (20/40 Ethernet and 8 Gb FC with support for multihop FCoE)
- FlexFabric 10 Gb/24-Port Module for BladeSystem c-Class
- HPE Flex-10/10D 10 Gb/30 Port Module for BladeSystem c-Class
- 16 Gb 24-Port FC Module for BladeSystem c-Class
- 8 Gb 20-Port FC Module for BladeSystem c-Class

- 8 Gb 24-Port FC Module for BladeSystem c-Class
- HPE VC Manager (VCM)
- HPE Virtual Connect Enterprise Manager (VCEM)
- HPE OneView

Note

FlexFabric modules are supported in Ethernet-only mode in the BladeSystem c3000 enclosures.

VC management

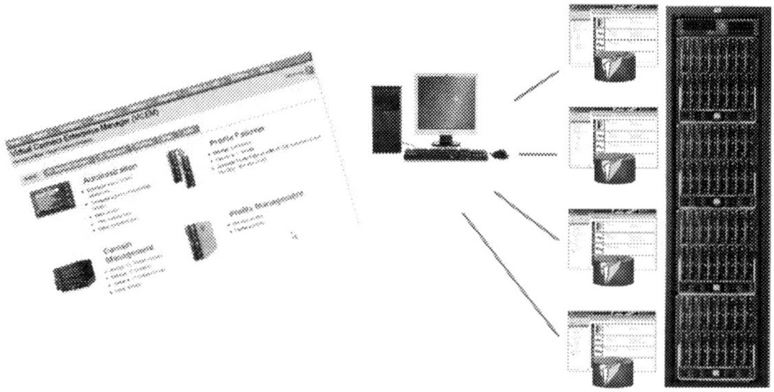

Figure 2-20 VC management

Virtual Connect Manager (VCM) is embedded in the VC Ethernet or FlexFabric module and is designed to configure and manage a single VC environment (Figure 2-20). This could be a single enclosure or a multienclosure domain containing up to four physically linked enclosures in the same rack. Users can access VCM through a web-based GUI or CLI. The Onboard Administrator provides a web link to the GUI. You can also use a Secure Shell (SSH) session to establish a console connection to the CLI. VCM runs on the primary module and supports 64 ranges of Media Access Control (MAC) addresses and World Wide Names (WWN).

VCEM is the primary application for managing VC across the data center. It provides centralized management for up to 250 VC environments, administers a total of 256,000 LAN and SAN addresses from a central pool, and is ideal for multiple-rack or distributed BladeSystem environments with more than one rack of enclosures.

VCEM also enables you to create domain groups that use a master profile to manage multiple VC domains connected to the same LANs and SANs. Within a domain group environment, administrators can quickly move profiles and associated server workloads between any servers that belong to the same group, which could be in the same rack, across the data center, or even at different physical locations. Because it is built on VC technology, you can do that without impacting the configuration and availability of production networks.

A domain group can also be used to push VC domain configuration changes such as network assignments or parameter modifications to multiple domains simultaneously. Essentially, any changes to the master profile are applied to all members of a domain group.

VCEM is designed to scale as the infrastructure grows, and it simplifies the addition of new enclosures. For growing businesses, HPE recommends using VCEM from the beginning to get the maximum benefits for bare-metal enclosure deployment and simplified infrastructure build-out.

VCEM is a plug-in for HPE Systems Insight Manager (HPE SIM) and is being replaced by HPE OneView. HPE OneView 2.00 can manage up to 64 enclosures. The number of MAC addresses and WWN pools are virtually unlimited.

Note

The HPE SIM and HPE Insight Control offerings will continue to be available for sale for the near future, including plans to support the next HPE ProLiant server generation, which is expected to start shipping in 2017. The HPE VCEM lifespan will also be extended.

VC domains

A VC domain is simply the term used to describe an HPE BladeSystem enclosure configured with VC Ethernet and FC modules. The VC domain can be further defined as a single enclosure domain that consists of one BladeSystem enclosure and up to 16 servers or as a multienclosure domain that can contain up to four physically linked enclosures and 64 servers managed as a single logical group. The domain simplifies the management of the enclosures in that domain and enables the configuration to be maintained consistently.

The VC domain contains the configuration for three main components:

- Networks (connecting to uplink ports)
- Profiles (connecting to downlink ports)
- Device bays

The domain is formed by giving VCM the administrator credentials of the target enclosure Onboard Administrator. This enables VCM to establish a trust relationship with the enclosure. This relationship gives VCM access to the servers through the HPE integrated Lights-Out (iLO) processor.

VC domain functions can be accessed through a GUI or by command line.

VC fundamentals

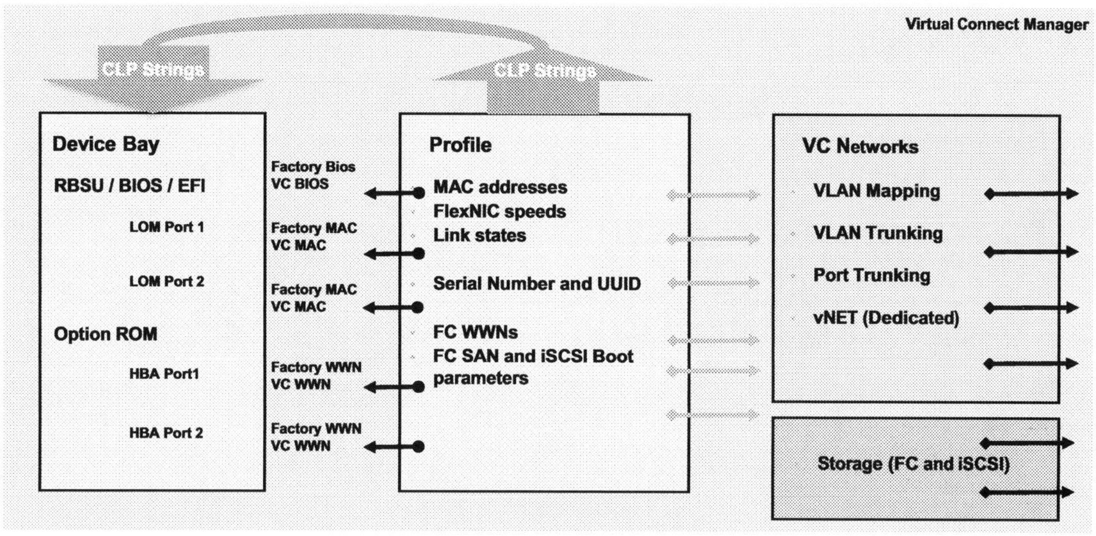

Figure 2-21 VC profile settings

The VC domain contains the networks and connections to the data center (Figure 2-21). Establishing the domain is the first step in VC configuration.

The profiles are assigned to a server, but they are associated with a device bay. This means that when every server installed within that bay is powered on, it will receive the assigned profile. The profile can contain any of the following attributes:

- MAC addresses
- Ethernet settings
- Ethernet connection port speed
- Link state
- PXE settings
- Serial number and universally unique identifier (UUID)
- FC WWNs
- Server boot configuration, either local or boot from SAN (iSCSI or FC)
- FC SAN connections

VC begins at the hardware level. Using the profile configuration, VCM builds an instruction set and, using the Command Line Processing (CLP) strings, it passes this set to the server ROM. The server BIOS implements the instruction set during the power-on self-test (POST).

Every I/O port has a second memory space to which VCM can assign an alternate MAC or WWN. VCM can modify the HBA BIOS settings.

HPE VCFC family

HPE VC includes the following components for connecting HPE BladeSystem solutions to the corporate SAN using VC (Figure 2-22):

Figure 2-22 HPE VCFC family

- HPE VC 16 Gb 24-Port FC Module for BladeSystem c-Class
- HPE VC 8 Gb 24-Port FC Module for BladeSystem c-Class
- HPE VC 8 Gb 20-Port FC Module for BladeSystem c-Class

VC Worldwide Names

A FC WWN is a 64-bit value used during login to uniquely identify a FC HBA port and to get a port ID. The naming convention is as follows:

- The first four bits identify the naming authority.
- The first two bytes are either hex 10:00 or xx:xx (where *x* is vendor specified).
- When the first nibble is either 5 or 6, it is then followed by a three-byte vendor identifier (IEEE OUI) and 4.5 bytes for a vendor-specified serial number.

HPE has reserved 64 dedicated ranges of FC WWNs.

- For example, 50:06:0B:00:00:C2:62:00 to 50:06:0B:00:00:C3:61:FF is one range.
- The company ID is the second through seventh hex digit in the WWN. So in the example of 50:06:0B:00:00:C2:6E:00, the HPE ID is 0060B0.

Note

The current registered list can be found here: http://standards.ieee.org/develop/regauth/oui/oui.txt.

Although each BladeSystem FC HBA ships with factory-default port and node WWNs for each port, the administrator can set each VC domain to either a VC-defined WWN or a factory-default WWN. VC-assigned WWNs will mask the factory-default WWNs, while the server remains in that VC domain.

When assigning WWNs, VC assigns both a port WWN and a node WWN. Because the port WWN is typically used for configuring fabric zoning, it is the WWN displayed throughout the VC user interface. The assigned node WWN is always the same as the port WWN incremented by one.

Configuring VC to assign WWNs in server blades maintains a consistent storage identity, even when the underlying server hardware is changed. This method enables you to replace server blades without affecting the external FC SAN administration.

VC fabric login sequence

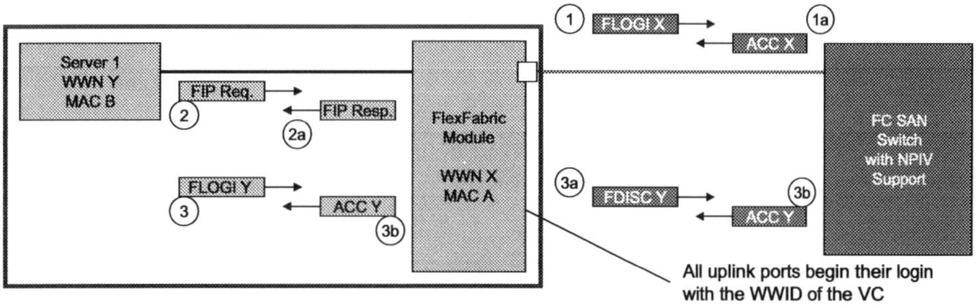

Figure 2-23 VC FLOGI sequence diagram

Fabric login (FLOGI) using the HBA aggregator WWN (represented by WWN X in Figure 2-23, the WWN of FlexFabric or VC FC module), establishes the buffer credits for the overall link and receives an overall Port ID. The server HBA logs in normally using its WWNs, and the server HBA fabric logins are translated to Fabric Discovery (FDISC). FDISC enables the exchange service parameters with the fabric without affecting the operating parameters between the N_Port and the fabric.

Traffic for all N_Port IDs is carried on the same link.

The VC FC uplink ports must log in to the fabric initially. The WWN is only exposed to the FC Name Server with the initial FLOGI request (by FC VC to acquire a 24-bit FC address). When done, those WWNs are not used in a zone scheme.

Note

FLOGI is one of three different types of logins for FC. The FLOGI process enables a node to log in to the fabric and receive an address from a switch, and it effectively sets up a session between the N_Port and the switch.

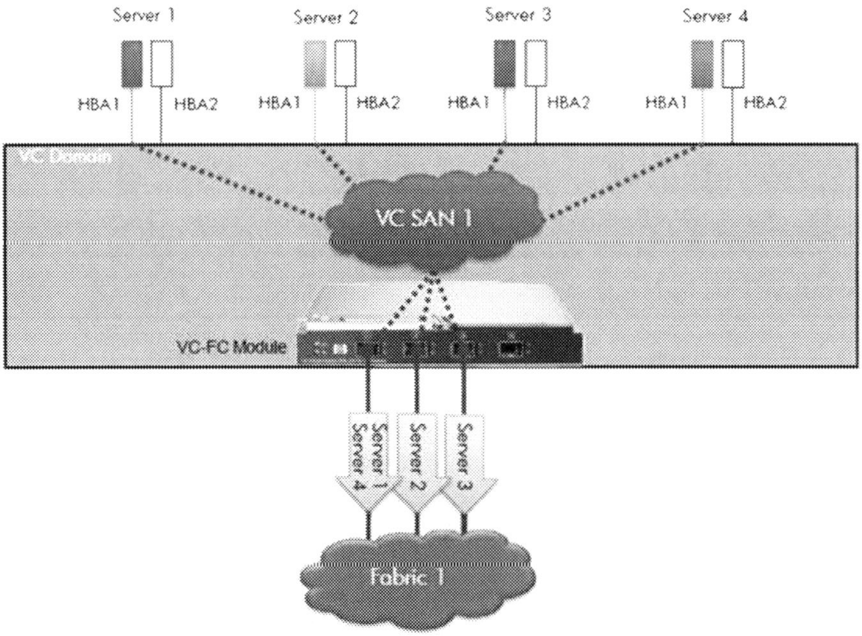

Figure 2-24 VC login distribution diagram

VC login distribution

VC FC does not do dynamic load balancing (Figure 2-24). It is an aggregator (physical port concentrator), which is why it needs NPIV to pass through the WWNs from the physical blade HBAs.

Before firmware 1.31, the login was static only. Beginning with VC 1.31, the login distribution can be set to Static or Dynamic. Dynamic means that as logins to a fabric occur, VCM distributes them equally in a round-robin assignment. Dynamic login distribution brings load balancing and increased redundancy. When VC Fabric uplinks are grouped into a single fabric, the module uses dynamic login distribution to load balance the server connections across all available uplink ports.

The VC module uses the port with the least number of logins across the VC SAN fabric or, when the number of logins is equal, VC makes a round-robin decision.VC modules with firmware version 3.0 and later does not offer Static Uplink Login Distribution.

There are two login redistribution modes for the Fabric-Attach fabric:

- **Manual Login Redistribution**—When configured, a user is expected to initiate a Login Redistribution request using the VC GUI or CLI interfaces.
- **Automatic Login Redistribution**—When configured, the VC FlexFabric module initiates Login Redistribution automatically when the specified time interval expires.

Learning check

The following questions will help you to measure your understanding of the material presented in this chapter. Read all the choices carefully because there might be more than one correct answer. Select or write the correct answers for each question.

1. Name the three main approaches for storing data.

   ....

   ....

   ....

   ....

2. Name at least three components of the FC solution.

   ....

   ....

   ....

   ....

   ....

3. The simplest FC topology is point to point.

   ☐ True

   ☐ False

4. The iSCSI stack uses UDP.

   ☐ True

   ☐ False

5. What is the name of the embedded VC management tool that is accessible by both GUI and CLI?

   ....

   ....

   ....

For answers, see Chapter 2 Answers on page 438.

3 Direct Attached Storage

OBJECTIVES

In this chapter, you should learn to:

- ✓ Identify the devices in the HPE direct attached storage (DAS) portfolio
- ✓ Explain unique features, target customers and describe specifications and performance limits for D2000 product family
- ✓ Explain unique features, target customers and describe specifications and performance limits for D3000 product family
- ✓ Explain unique features, target customers and describe specifications and performance limits for D6000 product family
- ✓ Locate sizing and reference information

INTRODUCTION

This chapter provides an overview of DAS. It begins with basics of HPE products in this category and then describes the unique features, capacity limits, performance figures, and target customers for the HPE D2000, D3000, and D6000 disk enclosures. It then describes reference documents and tools you can use for sizing and configuration.

Direct Attached Storage

DAS provides the most straightforward HPE Storage solutions for small- to-medium-size businesses. However, DAS storage cannot be shared among multiple servers, and the one-to-one connection between the device and the server can add management complexity.

DAS solutions typically are connected directly to the server using the Serial Attached SCSI (SAS) protocol. The storage devices can be either an internal part of the server or implemented as external enclosures.

Following are some typical components of a DAS solution (Figures 3-1 to 3-5):

- Server

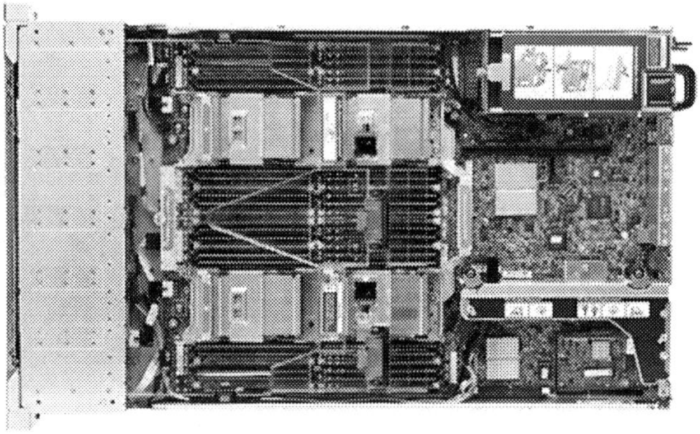

Figure 3-1 HPE ProLiant DL560 Gen8

- Storage or array controller

Figure 3-2 HPE Smart Array P841 Controller

- Connectivity and cables

Figure 3-3 HPE Mini-SAS HD 4x Cable

- Enclosure

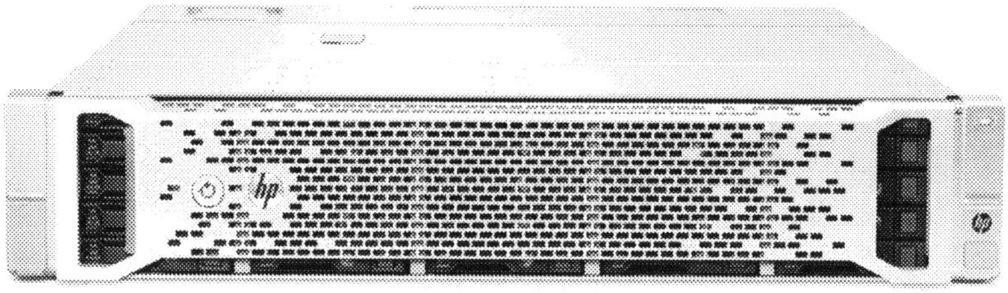

Figure 3-4 HPE D3700 Enclosure

HPE Storage product portfolio

HPE has a range of DAS, network-attached storage (NAS), and storage area network (SAN) products to satisfy small and medium-size business customers. HPE Storage solutions offer maximum scalability, industry-leading performance, a fully integrated suite of centralized management tools, and unmatched data protection and disaster tolerance features.

HPE Storage products can be divided into the following categories:

- Primary storage
- Backup, recovery, and archive
- Software-defined storage (SDS)
- Storage software
- Storage networking

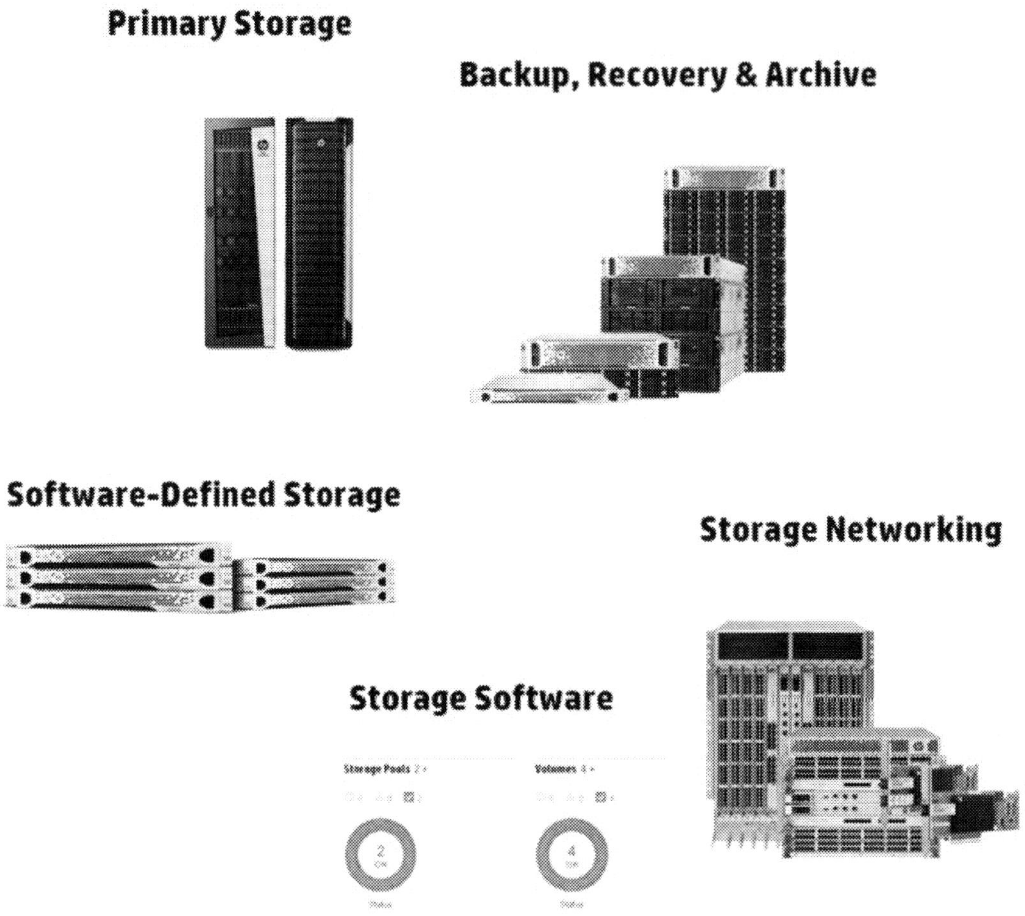

Figure 3-5 HPE Storage product portfolio

Primary storage

Tier-1 storage is for mission-critical applications, virtualization, and cloud. High-end and mid-range storage arrays are designed for IT-as-a-Service and leverage a single flash-optimized architecture with support for block, file, and object access.

HPE primary storage products include these devices:

- HPE 3PAR StoreServ
- HPE StoreVirtual
- HPE MSA
- HPE StoreEasy

- HPE XP
- Disk enclosures

Disk enclosures are typical DAS products belonging to the Primary Storage products category, and the following section looks at the HPE disk enclosures.

Disk enclosures

Disk enclosures can be used when the internal drive slots of an HPE ProLiant server or HPE BladeSystem enclosure are filled (Figure 3-6). Disk enclosures are flexible, modular solutions that simplify capacity expansion to address growing business needs.

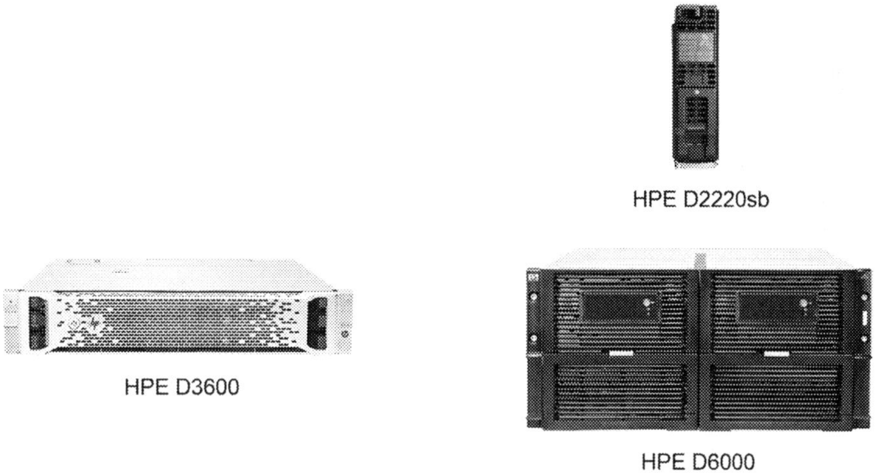

Figure 3-6 Disk enclosures

Why choose HPE disk enclosures?

- **Affordable**—Cost-effectively extend and grow your existing storage capacity.
- **Scalable**—Leverage HPE Smart Array technology with the latest SAS and Serial ATA (SATA) hard drives and Solid-State Drives (SSDs).
- **Simple**—Enhance productivity with intuitive setup and day-to-day interaction through ProLiant management tools.

The disk enclosure portfolio includes the following product families:

- HPE D2000 disk enclosures
- HPE D3000 enclosures
- HPE D6000 Disk Enclosure

Factors influencing disk enclosure performance

Processor and memory performance has grown in step with Moore's law, getting faster and smaller. However, storage performance has lagged far behind, creating a significant bottleneck to system performance. Today, more IT dollars are spent on storage-centric applications such as database, data warehousing, and virtualized workloads than on compute-centric applications.

Choosing the best server storage options for a computing environment requires an understanding of storage devices and storage needs. HPE offers enterprise storage options based on two technologies: hard disk drives (HDDs) or SSDs.

HPE provides server storage solutions using either spinning magnetic media (HDDs) or Solid-State NAND-based technology (SSDs). HPE enterprise HDDs are designed for use in unconstrained (24×7, up to 100% write) I/O workloads.

They are used for mission-critical applications such as large databases, email servers, and back-office applications. They provide maximum reliability, the highest performance, and error management under the most demanding conditions.

HPE also has three enterprise classes of SSD solutions.

- **Enterprise Value SSDs** provide relatively large storage capacities at low costs. They are best suited for high read (for example, boot) environments
- **Enterprise Mainstream SSDs** are suited for high I/O applications with a workload that is balanced between read and write.
- **Enterprise Performance SSDs** have similar capacities as Mainstream SSDs but are suited for mission-critical environments with workloads high in both read and write applications.

SSDs and HDDs use SAS or SATA protocols to communicate with the host system, but SSDs store and retrieve data in flash memory arrays rather than on spinning media. SSDs have no seek or rotational latency time. They address any sector of the NAND flash directly in 0.1 ms. SSD latency includes the time for memory access and transfer combined with controller overhead.

SSDs excel at random read operations, where their performance can be more than 100 times better than that of spinning media drives (Table 3-1). SSDs perform random writes at least 25 times faster than a comparable 15K rpm HDD. This means SSDs provide improved application performance.

Table 3-1 Drive I/O performance in IOPS (system I/O operations per second)

HPE drive	IOPS (70% read/30% write)
15K rpm Enterprise	370
7.2K rpm Midline	137
Enterprise Value/Boot SATA SSD	17000
Enterprise Mainstream SATA SSD	19000
Enterprise Performance SATA SSD	26000

The interconnect type used for accessing drives and enclosures is SAS or SATA with a bandwidth of 3 Gb/s, 6 Gb/s, or 12 Gb/s. Drive I/O performance is usually saturated first; with more drives in the enclosure, interconnect bandwidth can also be saturated.

Unique features of HPE disk enclosures

HPE can provide a complete, end-to-end DAS solution including enclosures, controllers, and servers. Advanced Data Guarding (RAID 6) is the highest level of fault tolerance. It allocates two sets of parity data across drives and allows simultaneous write operations. This level of fault tolerance can withstand two simultaneous drive failures without downtime or data loss. Most of the important components allow hot-plug operations (remove, replace, and add) while the system is running without service disruption. Redundant power and cooling provide increased reliability because the failure of a power supply or fans does not interrupt system functioning.

Drives installed in either the HPE D2600 or the HPE D2700 disk enclosure and monitored using HPE management tools are supported by a Pre-Failure (replacement) Warranty (3 years for SAS drives). The HPE Pre-Failure Warranty allows for the replacement of designated drives in a D2600 or a D2700 disk enclosure before they actually fail.

The D2700 disk enclosure has the unique capability of cascading up to two enclosures behind a single SAS port (depending on the HPE Smart Array Controller). The D2600 disk enclosure can cascade up to four enclosures behind a single SAS port (depending on the Smart Array Controller).

The D2000 family uses the standard, integrated set of Smart Array management and utility software for HPE ProLiant servers. These tools consistently lower the cost of ownership by reducing the amount of training and technical expertise necessary to install and maintain HPE server storage.

Self-Monitoring Analysis and Reporting Technology (S.M.A.R.T), first developed at HPE, detects a possible hard disk failure before it occurs, allowing for replacement of the component before a failure occurs.

HPE D2000 disk enclosures

Figure 3-7 D2700 Disk Enclosure

The D2000 disk enclosures (Figure 3-7) are 6 Gb SAS, low-cost, high-capacity, tiered and modular, external storage systems.

The HPE D2000 disk enclosures give customers the flexibility to mix and match SAS, SATA, and SSDs in the same enclosure, and they extend the benefit of tiered storage to direct attach customers.

With a supported Smart Array Controller, the D2600 and D2700 enclosures can be mixed for expansion, with up to four D2600 and four D2700 (4× large form factor [LFF], 3.5-inch and 4× small form factor [SFF, 2.5-inch]) drives, allowing for expansion up to 148 drives.

The D2000 enclosures support direct attach storage to ProLiant servers with Smart Array controllers. On ProLiant and HPE Integrity servers, dual-domain support is available for the Smart Array P411 and P812 controllers attached to a D2600 or D2700 (single initiator environment only). Blade 6 Gb/s SAS connectivity is enabled using a Smart Array P711m Controller, 6 Gb/s SAS BL switches, and the D2000 disk enclosures.

Dual-domain SAS creates redundant pathways from servers to storage devices. These redundant paths reduce or eliminate single points of failure within the storage network. This increases high availability with redundant paths from the controller to the drives. Dual-domain SAS implementations make it possible to tolerate an external cable failure, an expander failure, or a failure in a spanned disk (JBOD) environment.

Note

One of the key differences between dual-domain and dual-path architectures is the use of SAS or SATA drives. Dual-port SAS drives can provide the advantages of complete pathway redundancy in a dual-domain configuration. Customers seeking lower cost solutions can use single port SATA drives in a dual-path configuration.
A dual-path configuration requires dual-port controllers.

The D2700 disk enclosure is supported when attached to an HPE P2000 G3 MSA family system.

Target customers

The D2000 disk enclosures are ideal for small application environments in small and medium-size businesses, remote offices, departmental locations, tier 2 or 3 storage implementations of enterprise customers, and customers without SANs.

These disk enclosures are designed to provide room to grow as storage demands increase.

With the D2700 SFF, customers can add up to 200 drives, or up to four SFF enclosures can be connected together from a single controller port.

The D2700 has SSD support for up to 25 drives per Smart Array Controller.

D2000 enclosures are the following:

- **Affordable and low cost**—Customers can purchase only what they need today and avoid up-front costs. The modular platform provides investment protection and purchasing only what is needed today simplifies planning and relieves budget pressures.

- **Flexible and scalable**—Storage capacity can grow easily by attaching enclosures as storage demands increase.
- **Easy to manage**—Little IT expertise is required to expand. These disk enclosures are compatible with the latest HPE Smart Array SAS controllers and HBAs. The enclosures can be configured using familiar ProLiant management tools provided through the Smart Controller.

Specifications and performance limits

HPE D2600 Disk Enclosure

- 12 LFF drives
- 6 Gb SAS and 3 Gb SATA
- 15K or 7.2K
- Capacity from 600 GB to 6 TB
    - Up to 7.2 TB with 600 GB or 72 TB with 6 TB MDL drives
- Disk performance from 130 to 370 IOPS

HPE D2700 Disk Enclosure

- 25 SFF drives
- 6 Gb SAS and 3 Gb SATA
- Supports 12 Gb drives
    - 10K, 15K, or 7.2K
- Capacity from 146 GB to 1.2 TB
    - Up to 30 TB with 1.2 TB
- SSDs
    - From 200 GB to 1.6 TB
- Disk performance from 130 to 26000 IOPS

D2220sb Storage Blade

The D2220sb Storage Blade (Figure 3-8) installs in the D2000 disk enclosures and delivers DAS for ProLiant Gen8 and Gen9 blade servers installed in HPE BladeSystem c-Class enclosures. It supports up to 12 hot-plug SFF SAS or SATA HDDs or SAS/SATA SSDs. The enclosure backplane provides a PCI Express connection to the adjacent c-Class server blade and enables high-performance storage access without any additional cables. The D2220sb Storage Blade features an onboard Smart Array

Figure 3-8 D2220sb storage blade

P420i Controller, with 2 GB flash-backed write cache, for increased performance and data protection. Up to eight D2220sb storage, devices can be supported from a single BladeSystem c7000 enclosure for up to 192 TB of capacity.

Use HPE StoreVirtual VSA software to turn the D2220sb Storage Blade into an iSCSI SAN for use by all servers in the enclosure and any server on the network. HPE StoreVirtual VSA software is installed in a virtual machine on a VMware ESX host server adjacent to the D2220sb. HPE StoreVirtual VSA turns the D2220sb into a scalable and robust iSCSI SAN, featuring storage clustering for scalability, network RAID for storage failover, thin provisioning, snapshots, remote replication, and cloning.

You can expand capacity within the same enclosure or to other BladeSystem enclosures by adding D2220sb Storage Blades and HPE VSA licenses. A cost-effective bundle of the D2220sb Storage Blade and a HPE StoreVirtual VSA license makes purchasing convenient. If your storage needs increase, supported external systems can be added and everything can be managed through a single view.

HPE D3000 enclosures

The D3000 disk enclosure is a next-generation 12 Gb SAS, low-cost, flexible, tiered, external storage system. The new 12 Gb SAS enclosures—the LFF D3600 with 12 drive bays and the SFF D3700 (Figure 3-9) with 25 drive bays—offer modular solutions to simplify capacity expansion of HPE ProLiant server environments to external storage without having to make the full move to SAN or NAS.

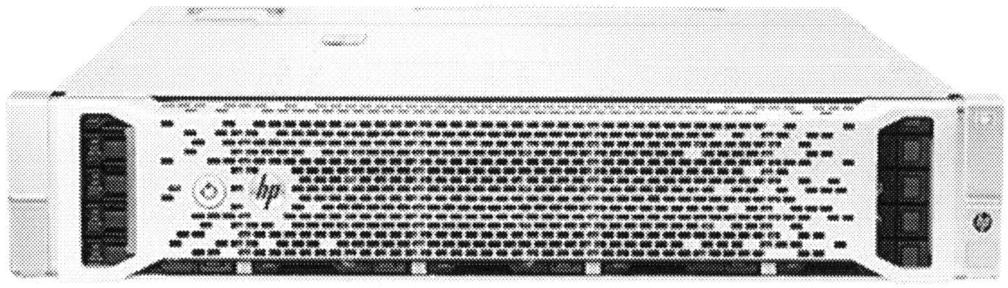

Figure 3-9 D3700 Disk Enclosure

This disk enclosure enables you to buy what is needed today and purchase additional capacity as your data storage needs grow. With the addition of HPE StoreVirtual VSA, you can create a virtual array within the host server. A 3-year limited license for HPE StoreVirtual VSA software with 1 TB of capacity is included at no extra cost.

The HPE D3000 enclosures support HPE SmartDrive Carrier (HPE ProLiant Gen8) drives. Total support can grow as needed to up to 96 LFF drives or 200 SFF drives.

Target customers

The D3000 disk enclosure is ideal for small application environments in small and medium-size businesses, remote offices, departmental locations, tier 2 or 3 storage implementations for enterprise customers, and customers without SANs.

The D3000 disk enclosures are the following:

- **Affordable and low cost**—Customers can purchase only what they need today and avoid up-front costs. The modular platform provides investment protection, and purchasing only what is needed today simplifies planning and relieves budget pressures.
- **Flexible and scalable**—Storage capacity can grow easily by attaching enclosures as storage demands increase.
- **Easy to manage**—Little IT expertise is required to expand. These disk enclosures are compatible with the latest HPE Smart Array SAS controllers and HBAs. The enclosures can be configured using familiar ProLiant management tools provided through the Smart Controller.
- **High-performance 12 Gb host connectivity**—12 Gb SAS host connectivity enables a higher data transfer rate. HPE offers end-to-end 12 Gb SAS connectivity to ensure a high-performance storage solution using a P421, P431, P441, P822, P841, P721m, P731m, or P741m Smart Array Controller and the D3000 enclosures. D3600 and D3700 disk enclosures are also supported with the HPE H221, H222, and H241 HBAs.

Specifications and performance limits

HPE D3600 Disk Enclosure

- 12 LFF drives
- 12 Gb SAS, 6 Gb SAS, and 6 Gb SATA
- 15K or 7.2K
- Capacity from 300 GB to 8 TB
    - Up to 7.2 TB with 600 GB or 96 TB with 8 TB MDL (midline) drives
- SSDs
    - From 80 GB to 1.6 TB
- Disk performance from 130 to 26000 IOPS

HPE D3700 Disk Enclosure

- 25 SFF drives
- 12 Gb SAS, 6 Gb SAS, and 6 Gb SATA
- 10K, 15K, or 7.2K
- Capacity from 146 GB to 2 TB
    - Up to 50 TB with 2 TB
- SSDs
    - From 120 GB to 1.92 TB
- Disk performance from 130 to 26000 IOPS

HPE D6000 Disk Enclosure

With the new 6 Gb/s SAS HPE D6000 Disk Enclosure, HPE extends and redefines DAS for BladeSystem servers (Figure 3-10). It combines the simplicity and cost effectiveness of DAS without sacrificing flexibility or performance.

The 5U high D6000 Disk Enclosure can also be directly connected to selected ProLiant Gen8 servers using the Smart Array P421, P431, P441, P822, and P841 controllers and a standard 2M SAS cable in single- or dual-domain (P822 or P841) environments. The HPE D6000 Disk Enclosure also supports the HPE H221, H222, and H241 HBAs.

Use the D6000 Disk Enclosure as part of a straightforward, in-rack 6 Gb/s SAS implementation that delivers high-density, low-cost, externally zoned DAS for HPE BladeSystem servers. You can easily configure storage on a live system with easy-to-use management software.

Figure 3-10 HPE D6000 Disk Enclosure

Up to six D6000 storage devices can be supported from a single BladeSystem enclosure for up to 1260 TB of capacity. End-to-end 6 Gb/s SAS connectivity enables a high-performance storage solution using an HPE Smart Array P721m, P731m, or P741m controller, 6 Gb/s SAS BL switches, and the D6000. The 6 Gb/s host connectivity enables higher performance, eases configuration and deployment, and broadens the supported features of a D6000 Disk Enclosure, using the HPE Smart Array P721m, P731m, or P741m controller and 6 Gb/s SAS BL switches.

Specifications and performance limits

Features of a D6000 Disk Enclosure using a Smart Array P721m Controller and 6 Gb/s SAS BL switches include the following:

- 5U rackmount form factor
- Two pull-out drive drawers supporting hot-plug LFF dual-ported SAS or SAS midline drives in 5U of rack space (35 hot-plug drives per drawer)
- 6 Gb/s front-end SAS connectivity using 6 Gb/s SAS BL switches in the interconnect bays and HPE Smart Array P721m controllers in each BladeSystem server
- Direct connection to selected ProLiant servers
- 3/0/0 warranty
- Drives:
    - 12 Gb SAS and 6 Gb SAS
    - 7.2K and 15K
    - Capacity from 300 to 8 TB
    - Disk performance from 130 to 370 IOPS

Using the Virtual SAS Manager (VSM) in the HPE SAS BL switch, you can assign or zone D6000 drive bays directly to the BladeSystem server bay.

You can create a dual-domain environment for SAS drives by adding an option kit with two dual port I/O modules. A single BladeSystem enclosure supports up to six D6000 disk enclosures externally.

Sizing disk enclosures

Disk enclosures can be ordered in various configurations so that the solution will fit customer needs.

HPE provides a variety of tools and resources to help size and configure a customer solution. Two of the best places to find online information about HPE products and solutions are the HPE website and HPE Product Bulletin.

Sizing considerations

To correctly select a disk enclosure and connected components, evaluate the following considerations:

- The type of server and the controller storage it is connected to
    - Determine if there are rack-based servers or server blades and the type of the controller (number of ports, type of interface, and bandwidth).
- The total capacity and room for growth
    - Enclosures have a different number of total slots for hard drives and different forms (SFF or LFF).
- The required availability
    - Is dual domain needed? Determine the RAID level.
- The performance requirements
    - Identify the protocols, the rotational speed of the drives, and the size of the cache in the controller.
- The requirements for installation support and enhancements to warranty services
    - Various support options are available.

Reference documents and sizing tools

Sizing tools are described in more detail in Chapter 8 of this course (Figure 3-11). The following reference documents and sizing tools can be used to select and configure the correct disk enclosure:

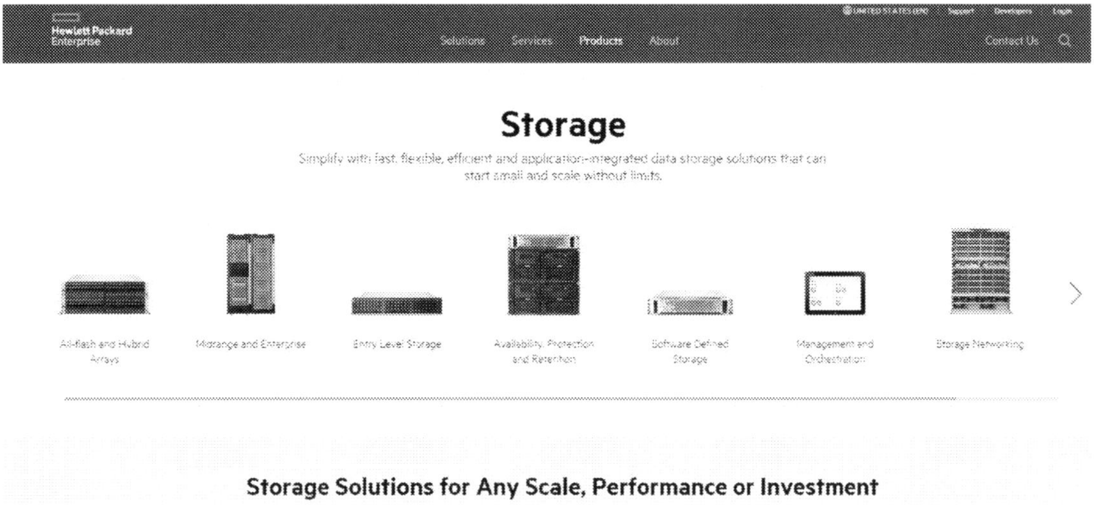

Figure 3-11 HPE website

- HPE QuickSpecs
- HPE Product Bulletin
- HPE website: https://www.hpe.com/us/en/storage.html

Note

There is no publicly accessible sizing tool for disk enclosures.

Learning check

The following questions will help you to measure your understanding of the material presented in this chapter. Read all the choices carefully because there might be more than one correct answer. Select or write the correct answer for each question.

1. List the typical components of a DAS solution.

   ....

   ....

   ....

   ....

   ....

2. Would you suggest the HPE D2700 Disk Enclosure over the D2600 to a customer requiring maximum capacity?

   ☐ Yes

   ☐ No

   Why or why not?

   ....

   ....

   ....

   ....

3. You can use SSDs in both the HPE D3600 and HPE D3700 disk enclosures.

   ☐ True

   ☐ False

4. Which interface can be used to access data on the HPE D6000 Disk Enclosure?

   a. 6 Gb/s SAS
   b. 8 Gb/s FC
   c. 3 Gb/s SATA
   d. 6 Gb/s SATA

5. Where will you find supported drives for the HPE D2700 disk enclosure and its maximum capacity?

   ....

   ....

   ....

   ....

   ....

For answers, see Chapter 3 Answers on page 438.

4 Network Attached Storage (NAS)

OBJECTIVES

In this chapter, you learn to:

- ✓ Describe network-attached storage (NAS) and the HPE NAS portfolio
- ✓ Explain the unique features of the HPE StoreEasy 1000 Storage products
- ✓ Explain the unique features of the HPE StoreEasy 3000 Gateway Storage products
- ✓ Explain how to size NAS solutions and how to locate HPE sizing tools and reference information

INTRODUCTION

This chapter provides an overview of NAS. It begins with the basics about HPE products in this category and then describes the unique features, capacity limits, performance figures, and target customers for the HPE StoreEasy systems. It then identifies some reference documents and tools you can use for sizing and configuration.

Network Attached Storage

Figure 4-1 HPE StoreEasy 1000 and 3000 NAS devices

NAS devices are primarily used for unstructured data and file sharing. NAS adds storage management functionality, increased data protection, performance enhancements, and the ability to share data across an Ethernet network (Figure 4-1). NAS comprises dedicated storage devices connected directly to the network, and it provides file access services over industry-standard protocols such as Common Internet File System/Server Message Block (CIFS/SMB) and Network File System (NFS).

What is NAS?

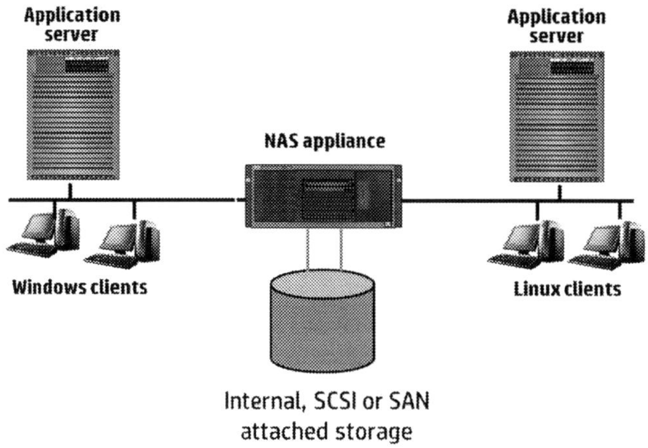

Figure 4-2 NAS solution example

NAS solutions were introduced in Chapter 2 (Figure 4-2). NAS servers provide a file-level interface to storage subsystems. Because NAS devices are server independent, they complement and help ease the burden on overworked file servers by off-loading storage to a single, purpose-dedicated storage device. NAS devices have an operating system that is optimized for file sharing and does not run general server applications, eliminating the major cause of downtime.

A NAS device is a server that is dedicated only to file sharing. NAS devices do not provide any of the typical server activities such as email, authentication, or file management. Instead, the server still handles all of the processing of data, but the NAS device delivers the data. A NAS device can be located anywhere in a LAN.

A NAS solution provides a flexible, intelligent, simple-to-manage solution for file-and-print and application-storage consolidation.

HPE Storage product portfolio

HPE has products to satisfy small and medium-sized customers with a range of direct-attached storage (DAS), NAS, and storage area network (SAN) solutions (Figure 4-3). Ideal for company-wide

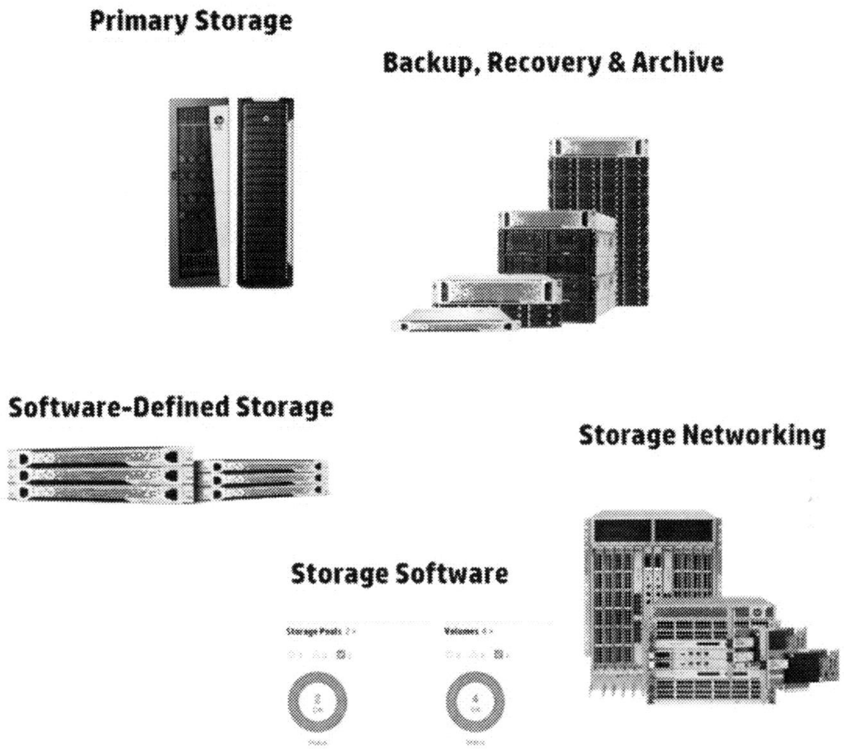

Figure 4-3 HPE Storage product portfolio

deployment and mission-critical applications, HPE Storage solutions offer maximum scalability, industry-leading performance, a fully integrated suite of centralized management tools, and unmatched data protection and disaster tolerance features.

HPE Storage products can be divided into the following categories:

- Primary Storage
- Backup, Recovery, and Archive
- Software-Defined Storage
- Storage Software
- Storage Networking

Primary Storage

Tier-1 Storage is for mission-critical applications, virtualization, and cloud—with high-end and mid-range storage arrays designed for IT-as-a-Service and leveraging a single flash-optimized architecture with support for block, file, and object access.

HPE Primary Storage products include

- HPE 3PAR StoreServ
- HPE StoreVirtual
- HPE MSA
- HPE StoreEasy
- HPE XP
- Disk enclosures

HPE StoreEasy products are typical NAS products belonging to the Primary Storage products category.

What is HPE StoreEasy?

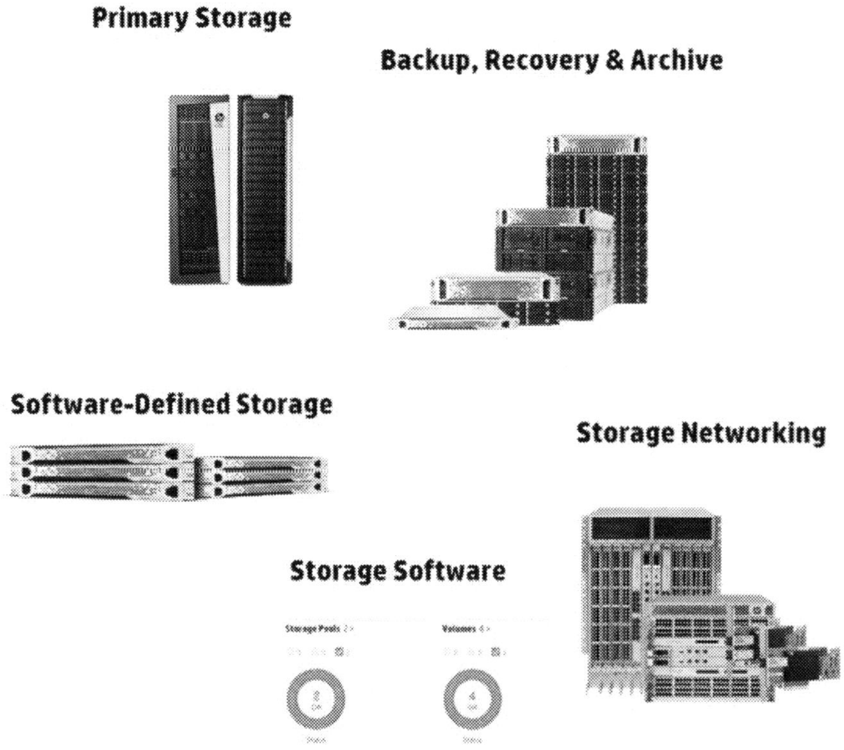

Figure 4-4 HPE Storage product portfolio

HPE StoreEasy products are typical NAS products belonging to the Primary Storage category of HPE product portfolio (Figure 4-4).

StoreEasy is a single, consolidated platform that delivers efficient, secure, and highly available application and file storage. It improves customers' total cost of ownership (TCO) with standard functionalities typically found only in high-end systems such as file and Internet Small Computer System Interface (iSCSI) access, granular deduplication, secure data encryption, and continuous availability. As a Microsoft Windows-based solution, HPE StoreEasy also eliminates the learning curve for customers and enables them to maintain productivity levels without disruption based on their existing familiarity with the operating system.

HPE StoreEasy is an efficient, secure, and highly available storage solution to address file and application storage challenges for large and small organizations. Based on a NAS platform, HPE StoreEasy features Windows Storage Server 2012 R2 Standard Edition with HPE customizations and is built on HPE ProLiant hardware. StoreEasy offers a number of features, including NAS share, iSCSI target, deduplication, clustering, and gateway functionality.

The HPE StoreEasy portfolio consists of two product lines, the HPE StoreEasy 1000 storage devices and the HPE StoreEasy 3000 gateway solutions.

Figure 4-5 HPE StoreEasy 3850 Gateway Storage

Figure 4-6 HPE StoreEasy 1550 Storage

The HPE 3PAR StoreServ File Controller is also part of HPE NAS offerings; however, it is discussed in the SAN chapter, within the 3PAR portfolio (Figures 4-5 and 4-6).

StoreEasy provides the best of NAS and SAN

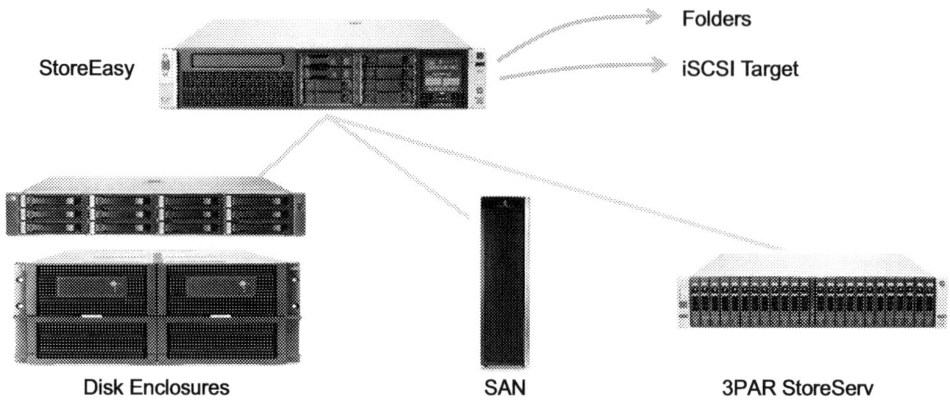

Figure 4-7 StoreEasy provides the best of NAS and SAN

The major difference between SAN and NAS is that a SAN is a separate network away from a company LAN, and it is configured to enable servers to communicate with storage arrays, typically using Fibre Channel (FC). NAS requires a dedicated storage device, typically an optimized server with a number of RAID storage drives that are attached directly to the network (Figure 4-7).

Both options have their strengths and weaknesses, with the primary advantages of a SAN being the major weaknesses of a NAS solution, and vice versa. The benefits of SANs include network speed, reliability, centralization, and data protection. The main strengths of NAS are interoperability, lower TCO, and relative simplicity. In a SAN architecture, administrators present storage to the server as a logical unit number (LUN) that appears to the server as a local disk that can be partitioned, formatted with a file system, and used just like any other disk. It can also be used as raw storage.

HPE StoreEasy can make direct-attached disks (internal or external) as well as storage from SANs available to client systems as either file shares or iSCSI targets. The StoreEasy administrator can use the same set of tools for working with the folders and LUNs.

HPE StoreEasy meets evolving needs

HPE StoreEasy helps you to regain control over file data. HPE StoreEasy products have these benefits (Figure 4-8):

- **Optimized**—Is tailored for file storage
- **Efficient**—Saves money and time
- **Secure**—Protects data
- **Highly available**—Prevents business and user disruption

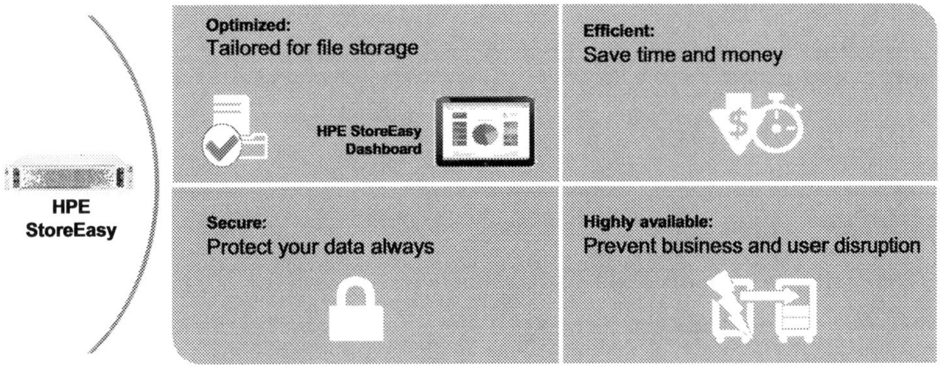

Figure 4-8 HPE StoreEasy meets evolving needs

Optimized—Tailored for file storage

The HPE StoreEasy Storage comes with preconfigured hardware with the latest generation storage operating system for purpose-built file storage. StoreEasy installation and network configuration wizards simplify installation and prevent configuration errors for faster deployment.

The HPE StoreEasy Dashboard consolidates all key status information such as capacity utilization, performance, and health in a single interface and reduces the need to navigate multiple tools and interfaces for system management. Low-capacity warning email alerts are now available to minimize disruption.

The best practices-driven storage provisioning tool discovers system storage and recommends the most efficient setup for optimum performance, optimum capacity, or a balanced configuration based on the available storage.

HDD bundles make it easier for you to buy, deploy, and provision additional storage capacity over the life of the product.

Efficient—Saves money and time

You can achieve an average of 50%–60% space savings with granular data deduplication.

The file sync and share feature enables mobile workers to access their files from anywhere. It provides corporate data controls including data encryption, data wipe, user lock screen password enforcement, file classification, and storage quotas.

You can use the advanced data management to automatically improve use of resources, comply with data retention policies, and enhance the protection of sensitive files.

You can consolidate multiple workloads and IT services in a single platform with included support for a broad range of file access protocols, Microsoft Hyper-V and SQL Server application storage, iSCSI block access, and IT infrastructure services such as print, Domain Name System (DNS), and Dynamic Host Configuration Protocol (DHCP).

You can also reduce branch office WAN usage with Hosted BranchCache, automatically caching frequently accessed corporate data center files.

Secure—Protects your data

HPE StoreEasy Storage protects data while at rest with file system encryption and BitLocker Drive Encryption and while it is being transferred with SMB encryption and signing.

The File Classification Infrastructure helps you to dynamically identify files based on sensitivity and to implement sophisticated access controls using Active Directory Rights Management Services.

You can install and run end-point protection such as antivirus software on the system itself—reducing the cost and complexity of connecting to an external end-point protection server. You can also protect against data loss with Volume Shadow Copy Service online snapshots and support for agent-based backup software.

As an option, you can use the power of HPE LiveVault for easy data backup to the cloud with 7-year retention. In addition, the iTernity Compliant Archive Solution (iCAS), which enables simplified and cost-efficient compliance archiving, has been validated to meet the US Securities and Exchange Commission (SEC) 17a–4(f) requirements.

Highly available—Prevents business and user disruption

HPE StoreEasy Storage increases uptime with standard features such as near continuous health monitoring with HPE Active Health System, mirrored OS drives, redundant fans (not available on all models) and power, and advanced memory protection.

You can confidently deploy large file systems that leverage online self-healing, online verification, and precise and rapid repair functions to reduce the downtime that results from file corruption.

The Vision Solutions Double-Take Availability software provides critical data protection with real-time, byte-level replication. In the event of a failure, you can recover the data—with a minimum of loss—from the replicated data repository.

HPE support options include 3-year parts, 3-year labor, and 3-year on-site support with next-business-day response for the StoreEasy 1650 and 1850 models or 3-year parts, 1-year labor, and 1-year on-site support with next-business-day response for the StoreEasy 1450 and 1550 models. All StoreEasy products also include one year of 24×7 software telephone support for the Windows Storage Server operating system.

HPE StoreEasy use cases

StoreEasy has an incredible number of features that can be implemented. Some typical use cases are

- **Heterogeneous file sharing**—File sharing across multiple client platforms
- **Home directory consolidation**—Consolidated end-user data with greater security, mobility, and efficiency
- **Remote office and branch office (ROBO)**—Simplified branch office deployment and management
- **Applications over SMB**—SAN-like capabilities for Hyper-V and SQL Server at lower cost with simplified management

Heterogeneous file sharing

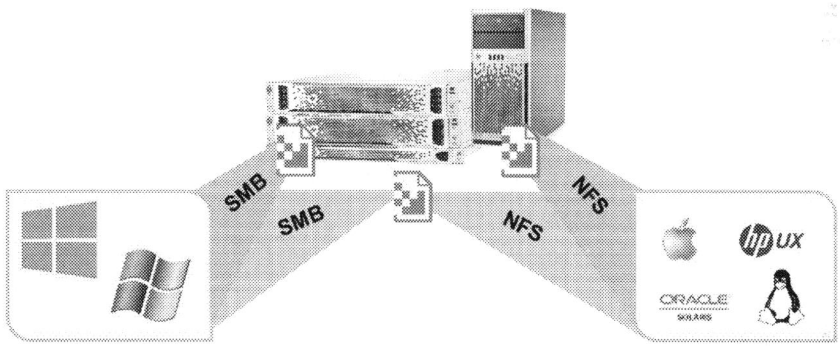

Figure 4-9 Continuously available file sharing across multiple client platforms

HPE StoreEasy products support heterogeneous file sharing through multiprotocol access for (Figure 4-9):

- Windows clients
- Non-Windows clients

Windows clients

To provide continuously available file sharing to Windows clients, the following requirements must be met:

- A failover cluster running Windows Server 2012 with at least two nodes and the configuration of servers, storage, and networking must pass the all tests performed in the Validate a Configuration wizard.
- The file server role must be installed on all cluster nodes.

- The clustered file server must be configured with one or more file shares that were created with the continuously available property.
- The SMB client computers must be running Windows 8, Windows Server 2012, or later.

SMB transparent failover enables administrators to configure Windows file shares, in Windows Failover Clustering configurations, to be continuously available. In the case of a hardware or software failure, the server application nodes will transparently reconnect to another cluster node without interrupting the server applications.

For active-active file access, SMB 3.0 provides active-active file sharing by allowing access to the same folders from multiple systems concurrently. This is achieved through Cluster Shared Volumes (CSV).

SMB encryption ensures that data transfers are secure by encrypting data as it transfers. The biggest benefit of using SMB encryption over more general solutions (such as IPSec) is that there are no deployment requirements or costs beyond changing the SMB server settings. SMB encryption is based on Advanced Encryption Standard (AES).

Non-Windows clients

The NFS 4.1 protocol is a significant evolution of the NFS protocol, and Microsoft delivers a standards-compliant, server-side implementation in Windows Server 2012.

Some of the Windows Server 2012 NFS 4.1 server features include a flexible, single-server namespace for easier share management, full Kerberos v5 support (including authentication, integrity, and privacy) for enhanced security, VSS snapshot integration for backup, and Unmapped UNIX User Access to enable easier user account integration.

Windows Server 2012 supports simultaneous SMB 3.0 and NFS access to the same share, identity mapping using stores based on RFC-2307 for easier and more secure identity integration, and highly available cluster deployments.

Windows Server 2012 includes a new, flat file-based identity mapping store. Windows PowerShell commandlets (cmdlets) also replace cumbersome manual steps to provision Active Directory Lightweight Directory Services (AD LDS) as an identity mapping store and to manage mapped identities.

Note

See these white papers for more information:

Deploying HPE StoreEasy Storage in a UNIX environment

Provisioning storage using SMB, NFS, and iSCSI protocols on HPE StoreEasy Storage

Heterogeneous File Serving on HPE StoreEasy Storage using SMB and NFS protocols

Heterogeneous File Sharing with HPE 3PAR StoreServ File Controller

Deploying HPE 3PAR StoreServ File Controller in a Linux/UNIX environment

---

Home directory consolidation

Users in many organizations store data, such as application data files or documents, locally on individual laptop or desktop PCs. This practice has several disadvantages:

- **Reduced mobility**—Users cannot access the files or the documents from other network locations or other PCs in the organization.
- **Risk of data loss**—Malfunction of the local disk could result in data loss.
- **Inefficient storage usage**—Often, different versions of the same file are stored in several places, which increases total storage-related costs.
- **Difficulty in performing regular data backups**—Individual backups for every PC are cumbersome.
- **Time-consuming maintenance or upgrades**—PC maintenance or upgrade activity requires more time to back up and restore the data.
- **Interferes with the common user experience**—Even if users' data is configured to be stored on a centralized network file share, the desktop, operating system, and application settings are not the same between different desktop and laptop PCs. This requires additional learning time to understand the PC settings and to locate the data, thus decreasing productivity.
- **Network file accessibility**—User data stored on a network file share becomes inaccessible in the event of a network or server outage.

HPE StoreEasy systems can address these challenges, enabling an organization to store user data with greater efficiency and consistency in centralized storage, instead of on individual user PCs, through the following key features:

- **Folder Redirection**—Enables administrators to redirect the path of a known folder to a new location. The new location is a file share on an HPE StoreEasy system. Users interact with files in the redirected folder as if it still existed on the local drive. For example, the administrator can redirect the Documents folder, which is usually stored on a local drive, to a shared location of a StoreEasy system.
- **Offline Files**—Makes network files available, regardless of a slow or unavailable connection to the server. When working online, file access performance is at the speed of the network and the server. When working offline, files are retrieved from the Offline Files folder at local access speeds. For example, users can continue to access their data in the event of a corporate network or server outage. It is even possible to take laptops to a remote location such as a home office and access data without connectivity to the corporate network.
- **Roaming User Profile**—Redirects user profiles to a file share so that users receive the same operating system and application settings on multiple computers.

- **Work Folder**—Provides users with the flexibility to work on files online or offline, and their data will auto-sync with a centralized file server when their system is connected to the Internet. The user should have no worries about the Internet connection, login, or password to access their corporate network, and their files are current whenever they are accessed.

Simplify remote office–branch office

Remote and branch offices can face many deployment challenges, including the following:

- **Business continuity**—Users need access to data at all times. Downtime in a branch office can have major impacts on productivity if access to the corporate data center is not available. Employee mobility adds to the data access challenges. Disaster recovery can be key to the survival of the company.
- **Remote data access**—Branch offices need to access the corporate data center, but often their remote location and their low-bandwidth network connections limit their ability to access the needed data with acceptable performance.
- **Limited IT resources**—IT infrastructure in branch offices is often managed centrally from the corporate data center because there are minimal or no IT resources at the branch offices.
- **Data protection**—Backups must be completed in smaller backup windows. Backup from branch offices requires low-bandwidth connections, and recovery expectations are high.
- **Security**—ROBOs might not need access to all the data in corporate data center, or they might have limited (such as read-only) permissions for data access. In addition, data security guidelines in the ROBOs might not be enforced as strictly as at the corporate data center.
- **Data consolidation**—Critical business data must be available centrally for management and reporting purposes.

HPE StoreEasy products help with these challenges:

- **Minimize WAN utilization with BranchCache**—Stores frequently accessed files in local cache, and deduplication is enabled. Two modes of operation can be used (Figure 4-10):
    - Distributed BranchCache for branch offices with no local StoreEasy
    - Hosted BranchCache for branch offices with local StoreEasy

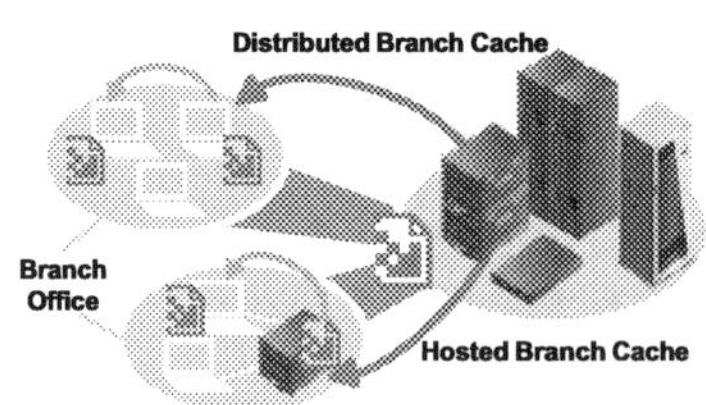

Figure 4-10 Hosted BranchCache with local HPE StoreEasy

- **Enable business continuity with replication**—Enables data and user failover with active-active data access on both sites, data failover with Distributed File System Replication (DFS-R), and user access failover with Distributed File System Namespaces (DFS-N; Figure 4-11).

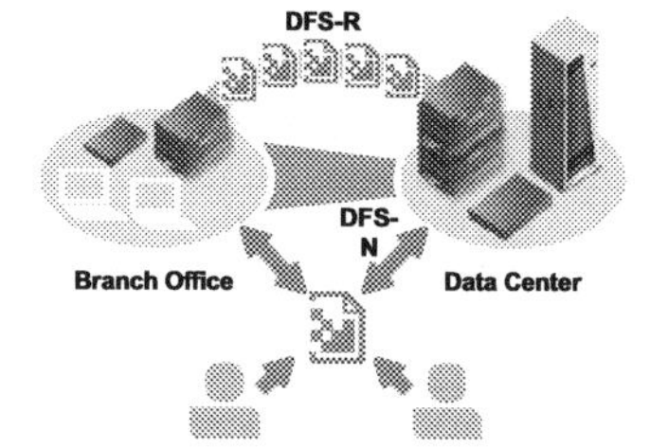

Figure 4-11 HPE StoreEasy used for data replication

DFS-R and DFS-N are role services in the File and Storage Services role in Windows Server 2012 and Windows Server 2012 R2.

  - **DFS Namespaces**—Enables the grouping of shared folders that are located on different servers into one or more logically structured namespaces. Each namespace appears to users as a single shared folder with a series of subfolders. However, the underlying structure of the namespace can consist of numerous file shares that are located on different servers and at multiple sites.
  - **DFS Replication**—Enables efficient replication of folders (including those referred to by a DFS namespace path) across multiple servers and sites. DFS Replication uses a compression algorithm known as remote differential compression (RDC). RDC detects changes to the data in a file, and it enables DFS Replication to replicate only the changed file blocks instead of the entire file.

- **Data compliance and consolidation**—Provides compliance-based data protection and consolidation with HPE StoreOnce (Figure 4-12):
  - Enable fast local backup and recovery
  - Replicate changed blocks to the central data center

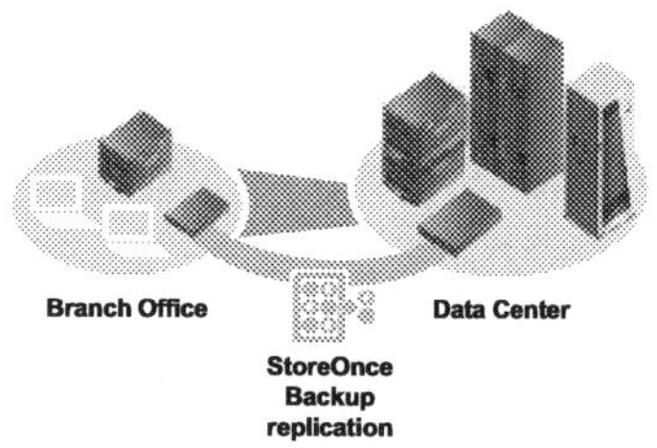

Figure 4-12 Consolidating backups

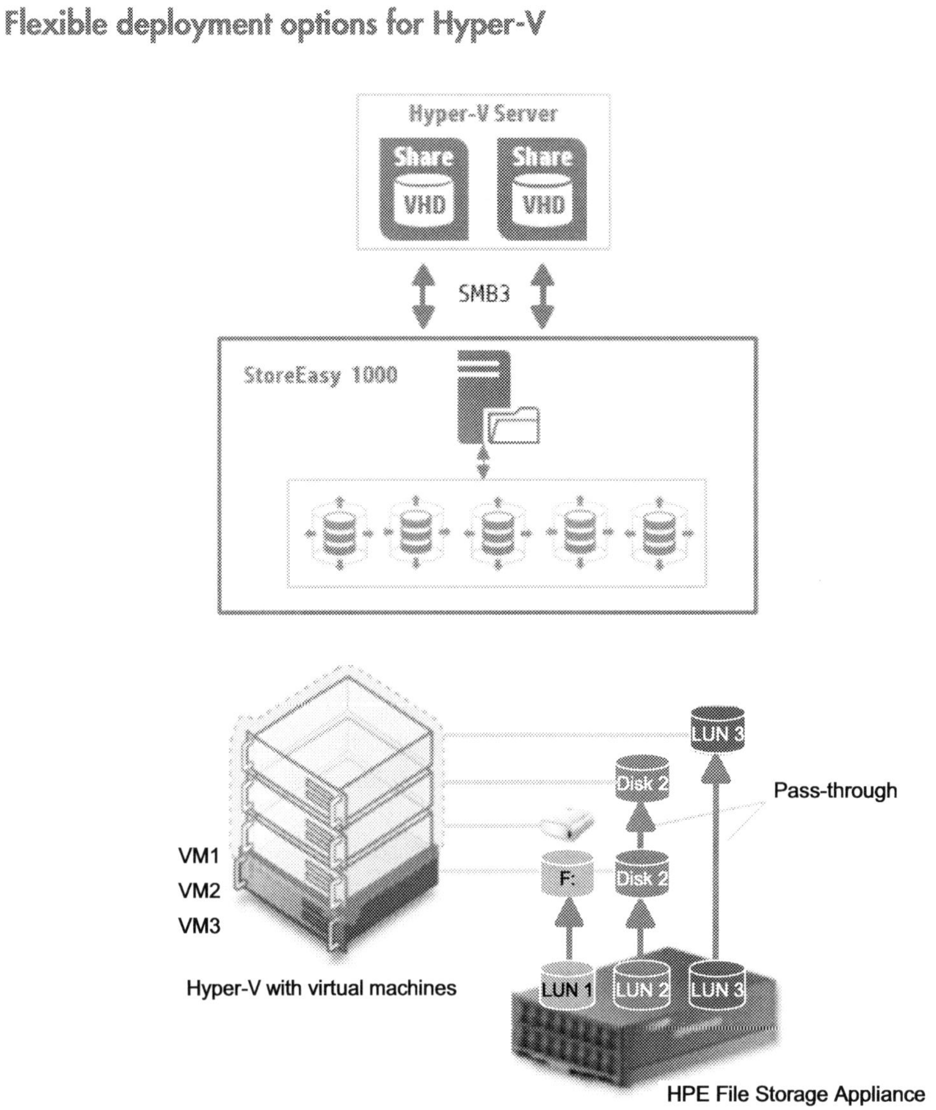

Figure 4-13 Flexible deployment options for Hyper-V

Before Windows Server 2012, remote storage options for Hyper-V were limited to SAN solutions that could be difficult to provision for Hyper-V guests or other more inexpensive options that did not offer many features (Figure 4-13). By enabling Hyper-V to use SMB file shares for virtual storage, administrators have a new option that is simple to provision, with support for CSV v2, and inexpensive to deploy, but also offers performance capabilities and features that rival those available with FC SANs.

- **Fast data transfers and network fault tolerance with SMB Multichannel**—With Windows Server 2012, customers can store application data (such as Hyper-V and Microsoft SQL Server) on remote SMB file shares. SMB Multichannel provides better throughput and multiple redundant paths from the server—for example, Hyper-V or Microsoft SQL Server—to the storage on a remote SMB share.

- **Transparent failover and node fault tolerance with SMB**—Supporting business-critical server application workloads requires the connection to the storage back end to be continuously available. The new SMB server and client cooperate to provide transparent failover to an alternative cluster node for all SMB operations for planned moves and unplanned failures.
- **Traditional SAN capabilities with iSCSI**—The administrator can host Hyper-V virtual machines on HPE StoreEasy, either as a Virtual Hard Disk (VHD) on the parent partition volume or as a pass-through parent partition disk. Another use of iSCSI is to add it as additional storage to the guest OS, with the iSCSI initiator in the child partition directly connected to the LUN.

File storage optimized for SQL workloads

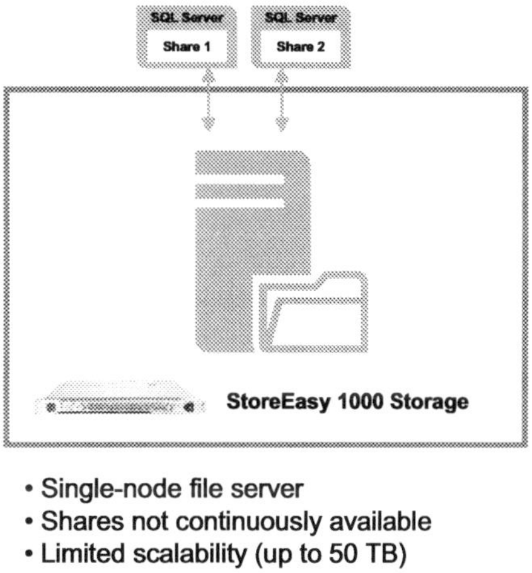

Figure 4-14 SMB 3-based shared data store for SQL

The SMB client and SMB server are optimized for small random read/write I/O to increase performance for common server application workloads such as Microsoft SQL Server online transaction processing (OLTP; Figure 4-14). SMB also uses a large maximum transmission unit (MTU) feature, enabled by default, to significantly improve performance in large sequential transfers such as those for Microsoft SQL Server data warehouse, in addition to database backup and restore operations.

HPE StoreEasy brings fast data transfer and network fault tolerance with SMB Multi-Channel for SQL workloads.

StoreEasy use model

All StoreEasy servers can act as a NAS device to provide storage for other clients. They also support all the same protocols. Where that storage actually comes from, though, depends on the StoreEasy model.

A gateway enables you to present LUNs from the SAN or array to HPE StoreEasy, and you can then offer LUNs as SMB and NFS file shares to clients. This also enables other Windows-based features such as snapshots (Volume Shadow copies), data deduplication, quotas, and much more.

- **StoreEasy 1000**—Uses internal or attached storage
- **StoreEasy 3000**—Acts as a gateway for LUNs from a SAN array
- **3PAR StoreServ File Controller**
  - Acts as a gateway for LUNs from a 3PAR SAN array
  - Based on the StoreEasy 3000 series
  - Includes additional intelligence and integration for 3PAR arrays

Deploying Work Folders in HPE StoreEasy 1000 and 3000 models

Providing a single point of access to the user files, anytime and anywhere, is a major challenge for any organization. Users would be connected to their work or personal computer and to devices in which the commonly used data is duplicated. The bigger challenge is porting the data with the latest revision from one computer or device to another. This issue becomes a concern when there is no Internet connectivity or no login or password access to their corporate network, or when there is no way to know whether users are joined properly to the domain. In all this, there is a certain amount of security risk to the user's files moving from one computer or device to another.

HPE StoreEasy brings an enterprise-class file sync and share solution, with included Microsoft Work Folders and optional Citrix ShareFile, that enables users to securely access their data stored on-premises under IT control, from anywhere at any time. The role Work Folders introduced in Windows 2012 R2 is to give users the flexibility to work on files offline or online, and their data will auto-sync with the centralized file server when it is connected to the Internet.

Consider the following prerequisites for implementing Work Folders on HPE StoreEasy:

- HPE StoreEasy with Quick Restore (QR) version 4.00.0a or later
- Any client machine with Windows 8.1 operating system installed
- Active Directory Domain Service (AD DS) and DNS running with Windows Server 2012
- Users who are familiar with the configuration of AD DS, DNS, and Certificate Authority
- An email address associated with the user profile in AD (required for Work Folders to operate properly)

- Secure Sockets Layer (SSL) configured in both the server and the client machine
- Enough free space on a share configured on the HPE StoreEasy 1000 or 3000 device to store all the user's files in Work Folders

HPE and iTernity Compliant Archive Software

iTernity is one of the leading enterprise software companies exclusively focused on managing and protecting enterprise data and compliance-relevant information. It helps businesses of all sizes to realize a future-proof and hardware-independent "enterprise archive" where information can be accessed and migrated freely without hardware boundaries. iTernity customers gain a competitive edge by securing business continuity and data integrity for long-term relevant information in a most flexible and cost-efficient way.

Together, HPE and iTernity deliver unmatched price: performance in data management, archiving, and protection solutions for crucial enterprise data. It delivers the lowest TCO for legally compliant archiving solutions.

iTernity was the AllianceOne Partner of the Year in 2014 in the HPE Storage category. For many years, HPE and iTernity have worked together to design and deliver advanced archiving and data protection solutions to customers. The complementary technologies and strategies work together to deliver comprehensive, cost-effective solutions that

- Ensure regulatory compliance and integrity of records.
- Maximize flexibility so you can adapt as business needs change.
- Increase security with constant integrity checks of important information.
- Drive down cost.

iTernity Compliant Archive Software (iCAS) is an enterprise archive and data protection software solution that helps organizations store data flexibly and securely. Built on industry standards and Windows platforms and supporting SAN and NAS storage devices, iCAS offers the adaptability that IT departments need.

Note

Although it can be used in the small and medium-size business segment, iCAS is primarily an enterprise archive and data protection software solution.

iTernity customers benefit from these regulatory compliance solutions that can mitigate risk, simplify data management, and improve overall business success. iCAS is the proven solution in healthcare, finance, and compliance-driven markets, and it is certified for more than 80 leading business applications, including Enterprise Content Management (ECM) applications, Enterprise Resource

Planning (ERP) applications, email, and picture archiving and communication systems (PACS). iCAS complements these applications and disk-based storage platforms with compliant archiving and data management features.

iTernity maintains headquarters in Freiburg, Germany, with offices throughout Europe and in North America.

Double-Take Availability

Disaster recovery is an important part of a storage management strategy. It is often important to maintain access to critical data even in the event of site failure or other problems. HPE StoreEasy Storage, when combined with the Double-Take Availability software from Vision Solutions, provides asynchronous replication to copy data from an HPE StoreEasy device in one location to an HPE StoreEasy device in another location in real time as the data changes. This allows access to the replicated data if the master copy becomes unavailable and provides higher levels of availability on HPE StoreEasy storage products.

Double-Take Availability overview

Double-Take Availability protects critical data using real-time byte-level replication. In the event of a failure, the replicated data repository can be accessed with a minimum of data loss.

Customers can specify the data to protect. For example, on a production system, known as the "source" of the replicated data, the administrator identifies the files and directories to replicate. Any changed data is immediately replicated to another system known as the "target." The target can reside on the local network or at a remote location. After making an initial baseline copy, Double-Take Availability monitors the specified data and sends any new changes to the target system. By sending only changes, Double-Take Availability minimizes its use of network and system resources.

Double-Take Availability completes its job with three primary phases:

- **Mirroring**—The system makes an initial baseline copy that creates a secondary copy of the protected data on the target system.
- **Replication**—Double-Take Availability continuously monitors all protected data and immediately transfers any changes to the target system. These changes are merged with the target copy so the target copy stays in sync with the source copy.
- **Failover**—The target system monitors the source system to ensure it is operating properly. In the event of a failure on the source system, the target automatically assumes the identity of the failed source system. By taking on the identity of the source system, and because it holds a current copy of the important data from that system, the replication target can take over the functions of the source system with minimal interruption and data loss.

Note
This is a brief introduction to a few key capabilities of the Double-Take Availability product. For more detailed information, see the "Double-Take Availability" documentation.

HPE StoreEasy 1000 Storage

The HPE StoreEasy 1000 Storage family is designed for small businesses, branch offices, and workgroup environments. The HPE StoreEasy 1000 Storage family integrates easily into new and existing environments, offering storage administrators and IT generalists a straightforward, consistent management experience.

The built-in security means that data remains well-protected and available to users when and where they want to access it. It also enables installation and running end-point protection such as antivirus on the system itself, reducing the cost and complexity of connecting an external, end-point protection server. With features including constant health monitoring, mirrored OS disks, advanced memory protection, and reliable software support, users can rely on a highly available solution that keeps their data ready.

These are the available models:

- HPE StoreEasy 1450 Storage
- HPE StoreEasy 1550 Storage
- HPE StoreEasy 1650 Storage
- HPE StoreEasy 1850 Storage

Differentiators

While HPE StoreEasy 1000 products have a common set of features, many aspects can differ between models.

- Capacity
    - None (the product comes with no drives, so additional drives are needed—4 TB, 8 TB, 12 TB, and 16 TB)
- Form factor
    - Rack
    - Tower

- Drive type
    - Serial Attached SCSI (SAS)
    - Serial ATA (SATA)
    - Large form factor (LFF)
    - Small form factor (SFF)

HPE StoreEasy 1450 Storage

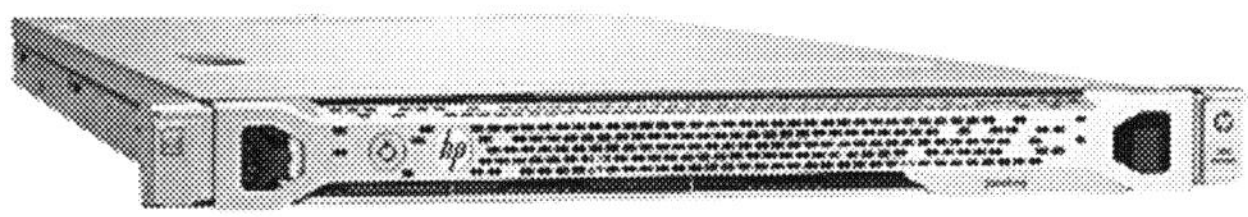

Figure 4-15 HPE StoreEasy 1450 Storage

HPE StoreEasy 1450 Storage solutions deliver multiprotocol file serving and application storage in a compact and affordable 1U rack-mount form factor (Figure 4-15). They are easy to install and manage.

All StoreEasy 1450 models have Windows Storage Server 2012 R2, Standard Edition, preinstalled from the factory on the four internal LFF drives that are configured as RAID 6. The StoreEasy 1450 device is built on an HPE ProLiant DL160 Gen9 server and includes a rail kit and cable management arm.

Sizing guidelines

- Number of users: 50–1300
- Maximum internal capacity supported: 24 TB raw internal SAS or SATA
- Form factor: rack, 1U
- Processor: Intel Xeon E5-2603v3/6-cores
- Memory: 8 GB (maximum 16 DIMM sockets)
- Connectivity: 2-port 361i 1Gb Ethernet
- Warranty: 3/1/1

Models

- HPE StoreEasy 1450 Storage
    - No drives are preconfigured
    - At least four drives must be ordered

- HPE StoreEasy 1450 4TB SATA Storage
- HPE StoreEasy 1450 8TB SATA Storage
- HPE StoreEasy 1450 16TB SATA Storage

HPE StoreEasy 1550 Storage

Figure 4-16 HPE StoreEasy 1550 Storage

HPE StoreEasy 1550 Storage solutions deliver multiprotocol file serving and application storage in a compact and affordable tower form factor (Figure 4-16). They are easy to install and manage.

All StoreEasy 1550 models have Windows Storage Server 2012 R2, Standard Edition, preinstalled from the factory on the four internal LFF drives that are configured as RAID 6. The StoreEasy 1550 is built on an HPE ProLiant ML110 Gen9 Server.

Sizing guidelines

- Number of users: 50–1300
- Maximum internal capacity supported: 48 TB raw internal SAS or SATA

- Form factor: 4.5U tower
- Processor: Intel Xeon E5-2603v3/6-cores
- Memory: 8 GB (maximum 8 DIMM sockets)
- Connectivity: 2-port Broadcom 5717 1 Gb Ethernet
- Warranty: 3/1/1

Models

- HPE StoreEasy 1550 Storage
    - No drives are preconfigured
    - At least four drives must be ordered
- HPE StoreEasy 1550 4TB SATA Storage
- HPE StoreEasy 1550 8TB SATA Storage
- HPE StoreEasy 1550 16TB SATA Storage

HPE StoreEasy 1650 Storage

Figure 4-17 HPE StoreEasy 1650 Storage

HPE StoreEasy 1650 Storage products deliver high capacity and drive down cost per GB (Figure 4-17). Six-core processing, up to 96 TB internal storage capacity, and external expandability enhance this ultra-dense 2U shared storage solution for small, medium, or large IT environments.

All StoreEasy 1650 models have Windows Storage Server 2012 R2, Standard Edition, preinstalled from the factory on two rear internal 120B 6G Value Endurance SFF Solid State Drives (SSDs) configured as RAID 1. The StoreEasy 1650 is built on an HPE ProLiant DL380 Gen9 Server and includes a rail kit and cable management arm.

Sizing guidelines

- Number of users: 100–3400
- Maximum internal capacity supported: 96 TB raw internal SAS or SATA

- Form factor: 2U rack
- Processor: Intel Xeon E5-2609v3/6-cores
- Memory: 16 GB (maximum 24 DIMM sockets)
- Connectivity: 4-port 331i Adapter plus optional HPE Flexible LOM
- Warranty: 3/3/3

Models

- HPE StoreEasy 1650 Storage
- HPE StoreEasy 1650 16TB SAS Storage
- HPE StoreEasy 1650 32TB SAS Storage
- HPE StoreEasy 1650 48TB SAS Storage
- HPE StoreEasy 1650 90TB SATA Storage

HPE StoreEasy 1850 Storage

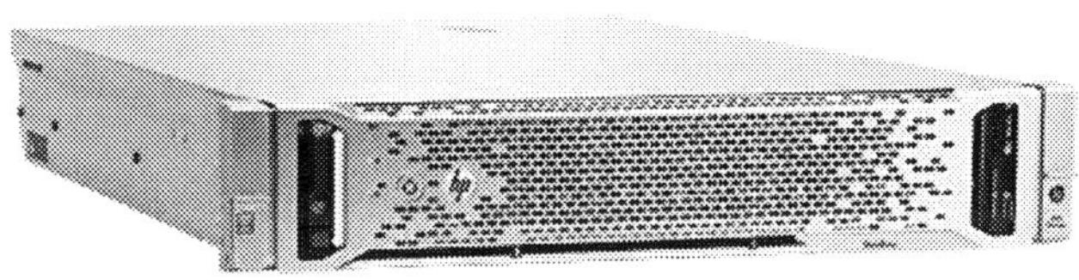

Figure 4-18 HPE StoreEasy 1850 Storage

HPE StoreEasy 1850 Storage products deliver enterprise-class performance, expandability, and flexibility (Figure 4-18). Six-core processing, up to 24 internal SFF data drives for performance, and external expandability make this ultra-dense 2U shared storage solution ready for small, medium, or large IT environments.

All StoreEasy 1850 models have Windows Storage Server 2012 R2, Standard Edition, preinstalled from the factory on two rear internal 120B 6G Value Endurance SFF SSDs configured as RAID 1. The StoreEasy 1850 is built on an HPE ProLiant DL380 Gen9 Server and includes a rail kit and cable management arm.

Sizing guidelines

- Number of users: 100–3400
- Maximum internal capacity supported: 43.2 TB raw internal SAS or SATA
- Form factor: 2U rack

- Processor: Intel Xeon E5-2609v3/6-cores
- Memory: 16 GB (maximum 24 DIMM sockets)
- Connectivity: 4-port 331i Adapter plus optional HPE Flexible LOM
- Warranty: 3/3/3

Models

- HPE StoreEasy 1850 Storage
- HPE StoreEasy 1850 9.6TB SAS Storage
- HPE StoreEasy 1850 14.4TB SAS Storage

HPE StoreEasy 3000 Gateway Storage

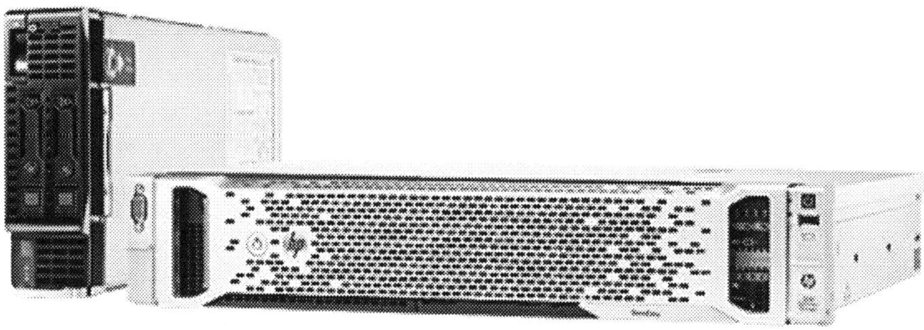

Figure 4-19 HPE StoreEasy 3000 Gateway Storage

HPE StoreEasy 3000 Gateway Storage solutions are an optimized, efficient, secure, and highly available storage to address the file storage challenges of medium-to-large business and branch office SAN environments (Figure 4-19). These devices are built on industry-leading HPE ProLiant DNA and Microsoft Windows Storage Server, and they integrate easily into new and existing SAN environments with a straightforward, consistent management experience for IT generalists or storage administrators.

These are the available models:

- HPE StoreEasy 3850 Gateway System
- HPE StoreEasy 3850 Gateway Blade Storage

HPE StoreEasy 3850 Gateway System

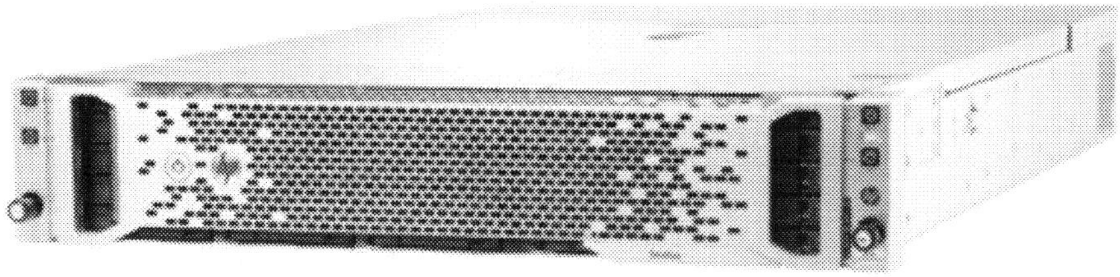

Figure 4-20 HPE StoreEasy 3850 Gateway System

The HPE StoreEasy 3850 Gateway System sets the standard for performance and expandability in gateway services for an array or FC SAN (Figure 4-20). This system consists of the StoreEasy 3850 Gateway Storage 2U chassis and one or two StoreEasy 3850 Gateway Single Nodes.

A StoreEasy 3850 Gateway System with two nodes provides high availability by clustering the nodes together. Up to four StoreEasy 3850 Gateway Systems can be configured together as an eight-node cluster (when using two StoreEasy 3850 Gateway Nodes per chassis).

It comes with Windows Storage Server 2012 R2, Standard Edition preinstalled.

Sizing guidelines

Up to four StoreEasy 3850 Gateway Systems can be configured together as an eight-node cluster (when using two StoreEasy 3850 Gateway Nodes per chassis).

- Number of users: 100–20,000
- Rack form factor: 2U
- Warranty: 3/3/3

It is designed to use external array or SAN storage.

Specification per single node

- Processor: Intel Xeon E5-2609v3/6-cores
- Memory: 32 GB RAM (16 DIMM sockets)
- Connectivity: 2 × 1 GB Ethernet
    - Optional FlexLOM Adapter
- Bootable storage: 2 × 120 GB 6G SATA Value Endurance SSDs
    - RAID 1

HPE StoreEasy 3850 Gateway Blade Storage

Figure 4-21 HPE StoreEasy 3850 Gateway Blade Storage

The HPE StoreEasy 3850 Gateway Blade Storage enables you to add file services from your BladeSystem c3000 or c7000 Enclosure to your array or FC SAN (Figure 4-21). This device is built on a ProLiant BL460c Gen9 Server Blade.

This storage blade comes preconfigured with hardware and the latest generation storage OS for purpose-built file storage: Microsoft Windows Storage Server 2012 R2, Standard Edition.

StoreEasy installation and network configuration wizards simplify installation and prevent configuration errors for faster deployment.

Sizing guidelines

- Number of users: 400–25,000
- Number of servers per c7000 enclosure: Up to 16
- Warranty: 3/3/3

It is designed to use external array or SAN storage.

Specification per single node

- Processor: Intel Xeon E5-2609v3/6-cores
- Memory: 32 GB RAM (16 DIMM sockets)
- Connectivity: 2-port FlexibleLOM
- Bootable storage: 2 × 120 GB 6G SATA Value Endurance SFF 2.5-in SC Enterprise Boot SSDs containing factory-installed OS
  - Configured as a RAID 1 mirrored pair

Sizing NAS products

HPE StoreEasy storage systems can be ordered in various configurations so that the solution will fit customer needs.

HPE provides a variety of tools and resources to help size and configure a customer solution. Two of the best places to find online information about HPE products and solutions are the HPE website and HPE Product Bulletin.

Sizing considerations

To correctly select a StoreEasy system, evaluate the following considerations:

- The total capacity and room for growth
  - NAS products have a different number of total slots for hard drives and different forms (SFF/LFF).
- The number of users accessing the storage
  - NAS products are designed to support a specific amount of users.
- Required availability
  - This includes RAID level selection and cluster configuration.
- Performance requirements
  - These include the protocol, rotational speed of drives, and the size of the cache in the controller.
- Requirements for installation support and enhancements to warranty services
  - Various support options are available.

Performance best practices for SMB 3

HPE StoreEasy 1000 and 3000 Storage systems are built on the Windows Storage Server 2012 R2 operating system. Windows Storage Server 2012 and 2012 R2 include major features in the SMB 3 network protocol stack, which can improve the performance and reliability of SQL Server with databases stored on the SMB 3 File Share on a StoreEasy Storage product.

SMB multichannel uses multiple network interfaces if possible. (This applies to all StoreEasy models.)

You can increase the performance and reliability of the system as follows:

- Improve performance by providing the combined throughput of all network adapters and cluster nodes together.
- Improve reliability by transparently failing back to the remaining network adapters or cluster nodes if one component fails.

SMB 3 protocol compatibility

SMB 3 is compatible with clients using version 1 or 2 of the SMB protocol, but new features are available only to client operating systems that are also using SMB version 3. As of March 2015, only Windows 8 and 8.1 and Windows Server 2012 and 2012 R2 operating system variants support the SMB 3.0+ protocol.

Using SMB 3 multichannel

SMB 3 multichannel is a feature that enables SMB clients to automatically identify all the network interfaces on a target server and to simultaneously use all the network interfaces on the client that can connect to those on the server. This is similar in concept to network teaming, except that it requires no configuration in the Network and Sharing Center. All available network devices are automatically used. If one of the links fails, SMB automatically reroutes requests through the remaining links. The application will experience only a slight delay on completion of one of its I/Os.

To use SMB 3 multichannel, connect multiple network interfaces, both on the Windows 2012 or 2012 R2 server hosting the SQL Server application and on the StoreEasy 1000 or 3000 device.

Using SMB 3 scale-out

SMB 3 scale-out enables multiple clustered servers to share the same file system. A client accesses that share through all clustered servers simultaneously, provided the file share is on a Cluster Shared Volume (CSV).

CSVs are suitable for SQL data files or for similar applications making many I/Os into a few large files that are kept open for a long time.

They are not suitable for workloads generating a large number of metadata operations, such as opening files, closing files, creating new files, or renaming existing files.

To use SMB 3 scale-out, follow these steps:

1. Create a CSV on a StoreEasy 3000 system.
2. Create a clustered share of that volume on both nodes of the StoreEasy 3000 system.
3. Configure SQL Server to store its data files on the clustered share.

Configuring SQL Server to use an SMB share

Configure the default database locations using UNC paths (\\server\share) instead of a local drive (D:).

Virtualization best practices using Hyper-V on SMB or iSCSI with HPE StoreEasy products

When planning for the hosting Hyper-V on an HPE StoreEasy 1000 or 3000 system, you need to ensure that the storage will be configured to give both the level of reliability that the customer expects and the performance the customer needs. Table 4-1 lists the high-level performance that is possible with the different RAID configurations.

Table 4-1 High-level performance possible with the different RAID configurations

Features	RAID 1	RAID 10	RAID 5	RAID 50	RAID 60
Minimum number of drives	Two	Four	Three	Six	Eight
Data protection	Single-drive failure	Up to one disk failure in each pair	Single-drive failure	Up to one disk failure for every three disks	Up to two disk failures for every four disks
Read performance	High	High	High	High	High
Write performance	Medium	Medium	Low	Medium	Medium
Read performance (while in a degraded state)	Medium	High	Low	Medium	Medium
Write performance (while in a degraded state)	High	High	Low	Medium	Low

With a low number of disks and a high workload, customers and clients might experience slow virtual machine performance. Adding more disks to a RAID drive set might increase the performance of the virtual machines, but other factors might be hindering the performance of those virtual machines. See the Microsoft article, "Measuring Performance on Hyper-V," for more information.

Note

If you are experiencing problems with performance when using Hyper-V on HPE StoreEasy systems, use the following Microsoft article to help diagnose where the bottlenecks are on the system: "Measuring Performance on Hyper-V" (Microsoft TechNet) http://technet.microsoft.com/en-us/library/cc768535 (v=bts.10). aspx.

When provisioning storage for use by Hyper-V, the actual procedure might differ for the different HPE StoreEasy products. For more information on how to connect, provision, and monitor storage using the HPE StoreEasy system, see the HPE StoreEasy user guide or external SAN array user guide.

After you have provisioned storage from your array, you can then create an iSCSI target using either the System Manager Console or PowerShell commandlets. To set up iSCSI targets using the Windows Storage Server 2012 R2 System Manager, refer to the following article: http://blogs.technet.com/b/filecab/archive/2012/05/21/introduction-of-iscsi-target-in-windows-server-2012.aspx.

To set up iSCSI targets using PowerShell commandlets in Windows Storage Server 2012 R2, refer to the following article: http://blogs.technet.com/b/keithmayer/archive/2013/03/12/speaking-iscsi-with-windows-server-2012-and-hyperv.aspx#.UbYW8iTnZaQ.

With Windows Storage Server 2012 R2, you can use SMB 3 file shares as shared storage for Hyper-V. After you have provisioned storage from your internal array or external array, configure your HPE StoreEasy system to use SMB 3 for Hyper-V. For more information, see the following article: http://technet.microsoft.com/en-us/library/jj134187.aspx.

Reference documents and sizing tools

Sizing tools are described in more detail in Chapter 8. You can use the following reference documents and sizing tools to select and configure the correct NAS storage system:

- HPE QuickSpecs
- HPE Product Bulletin

- HPE website: http://www8.hp.com/us/en/products/data-storage/storeeasy.html.

- HPE Storage Sizing Tool
- StoreEasy simulators
- Links to solution descriptions in the "Key StoreEasy solutions" attachment at the end of this module

HPE Storage Sizing Tool

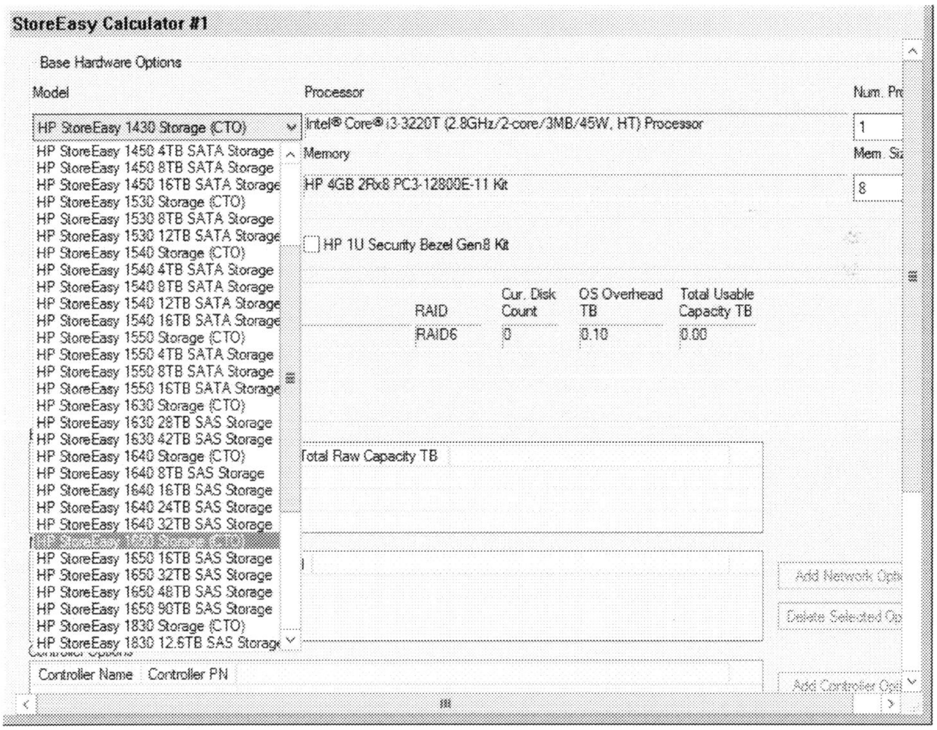

Figure 4-22 HPE Storage Sizing Tool

The HPE Storage Sizing Tool can be used to size both empty and prepopulated models (Figure 4-22). Results can be exported to Excel with part numbers and prices. The sizer provides several features to make sizing easier, such as simplified selection of drives by size or amount.

Example—Using the HPE Storage Sizing Tool to configure a StoreEasy solution

This example will show you how to configure a StoreEasy solution using the HPE Storage Sizing Tool. It is based on the following customer description:

- Multiple ProLiant DL380 Gen9 servers
- 12U of free space in the HPE 11642 Rack
- 2 TB of storage required by two Windows-based ProLiant DL380 Gen9 servers over SMB 3
- Deduplication and easy management required
- Additional 3 TB on a separate LUN for users accessible over CIFS
- Hard drive must not be a single point of failure
- Customer requires enterprise and high-performance SAS drives with 15K rpm

---

Note

You can install and configure the HPE Storage Sizing Tool and create a similar configuration on your own laptop.

Following are some key points for using the sizing tool:

- You can start the HPE Storage Sizing Tool from the icon on the desktop.

Figure 4-23 Storage Sizing Tool icon

- The **Storage Calculators** section in the left-hand menu contains the Storage Sizing Tool calculators (Figures 4-23 and 4-24).

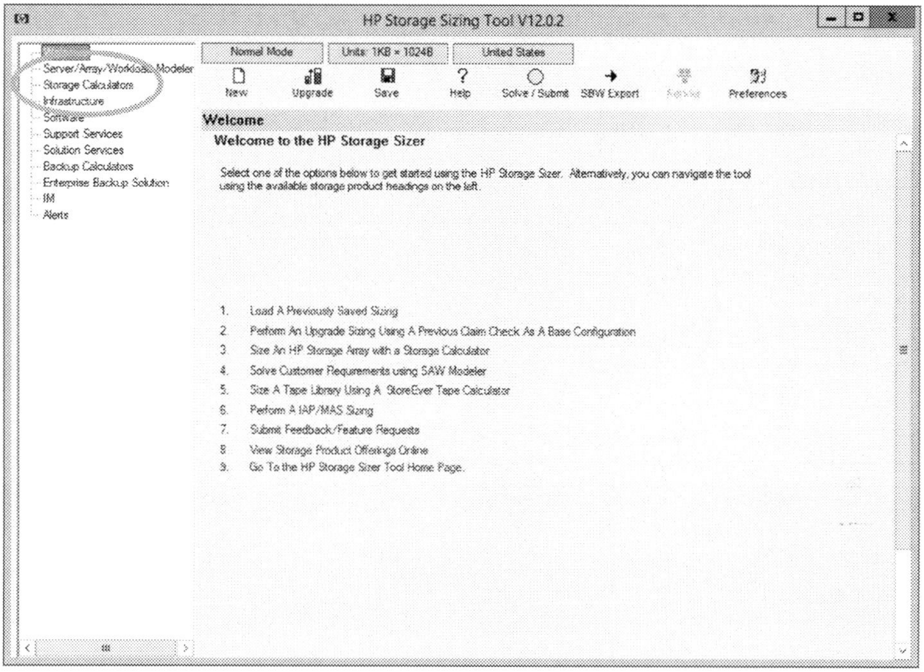

Figure 4-24 Storage Sizing Tool

4. You can start a calculator by clicking a link such as the **HPE StoreEasy 14X0/15X0/16X0/18X0/5530 Calculator** link (Figure 4-25).

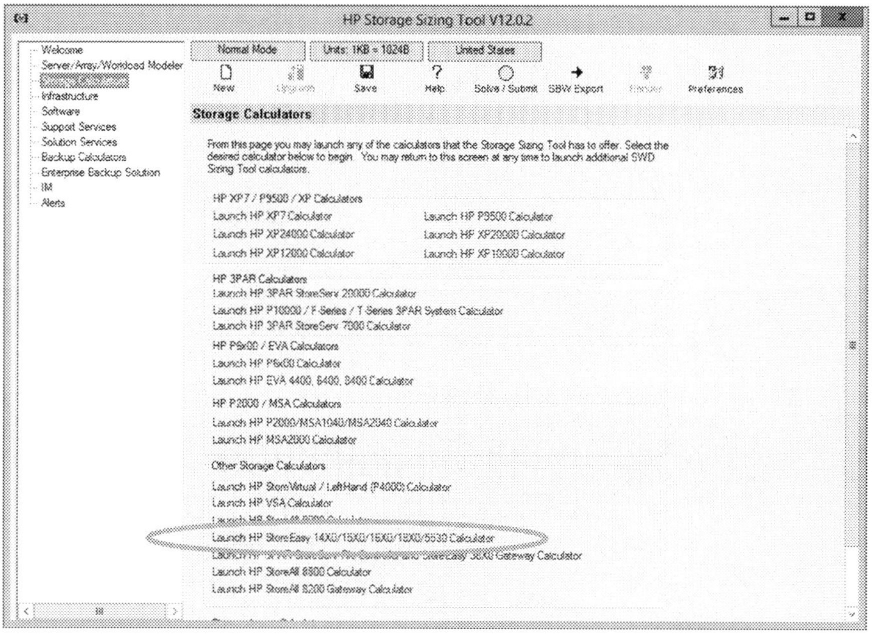

Figure 4-25 HPE StoreEasy Calculator link

- You can select a particular model from the Model section (Figure 4-26).

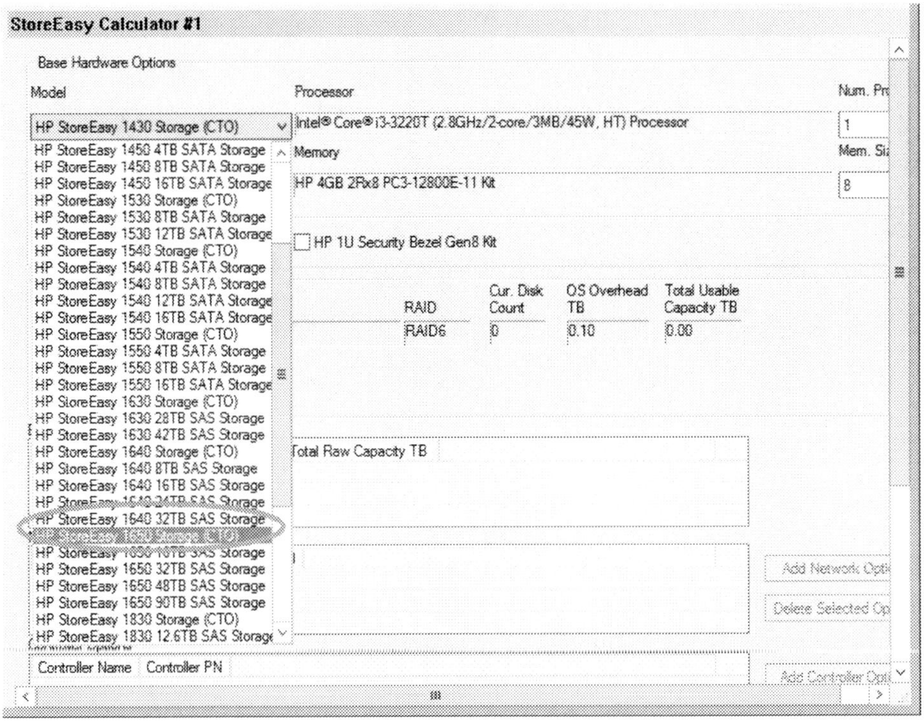

Figure 4-26 HPE StoreEasy Calculator Model section

- You can select disks from the Disk Type field (Figure 4-27).

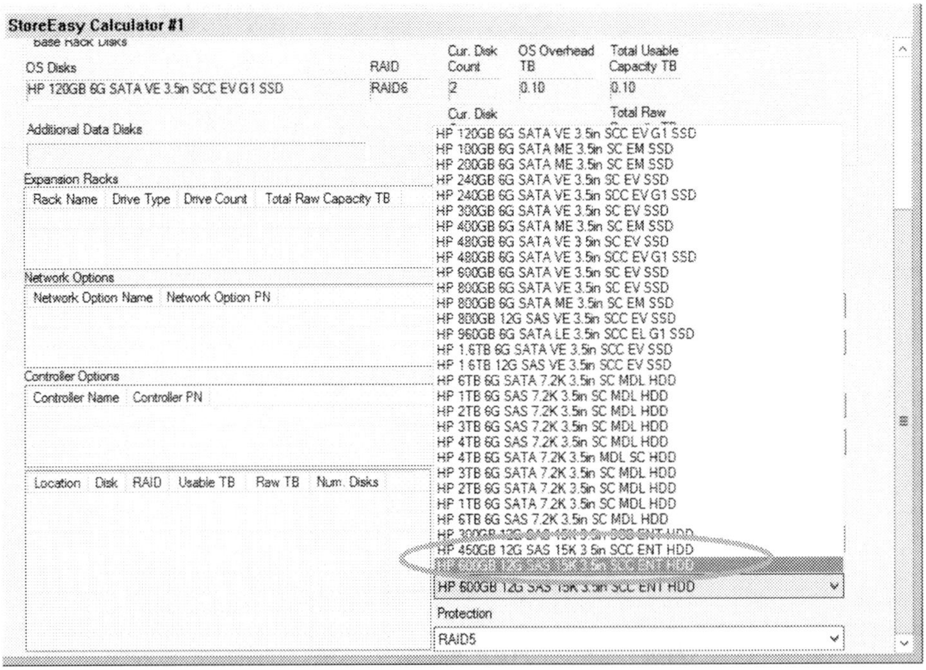

Figure 4-27 HPE StoreEasy Calculator Disk Type field

- You can select a RAID type from the Protection field to achieve the required redundancy (Figure 4-28).

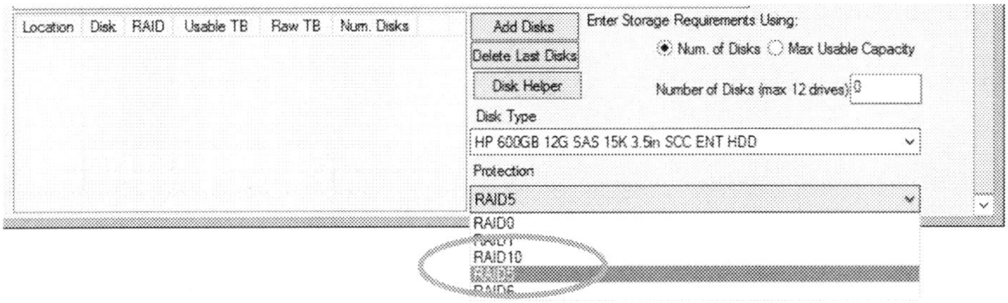

Figure 4-28 HPE StoreEasy Calculator Protection field

- You can add a group of disks by clicking the **Add Disks** button (Figure 4-29). You can also add multiple disk groups if, for example, a customer requests a second LUN for users.

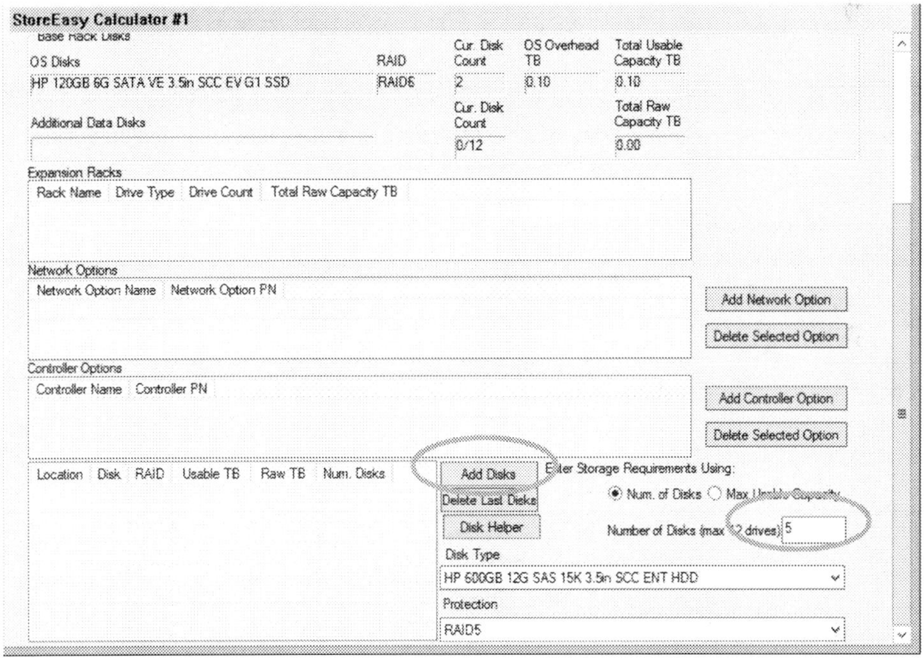

Figure 4-29 Add a group of disks by clicking the **Add Disks** button

- An alternative to selecting an amount of disks is to select **Max Usable Capacity** (Figure re 4-30).

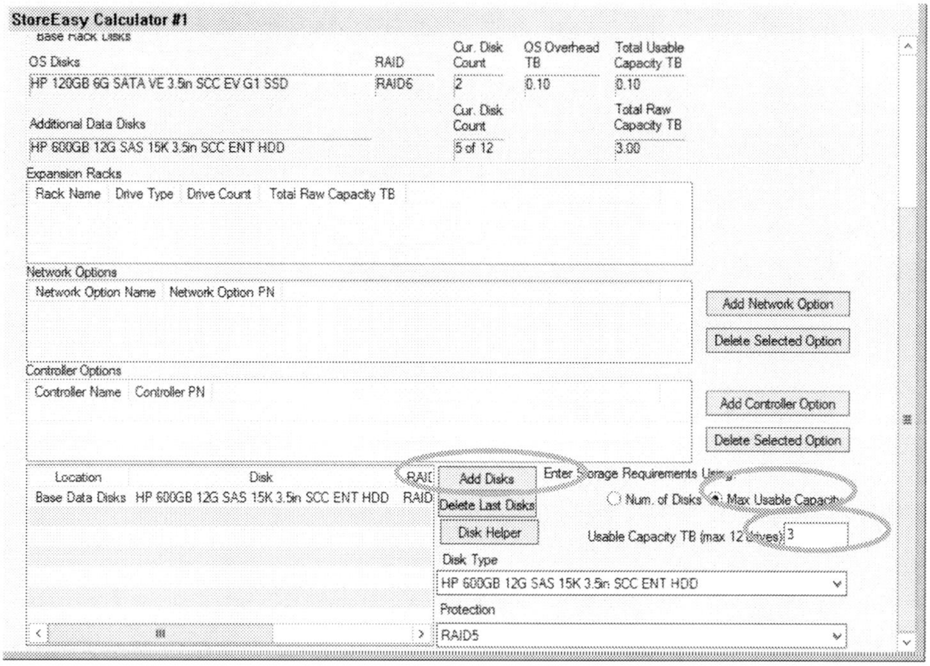

Figure 4-30 Alternative is to select **Max Usable Capacity**

- The **Solve/Submit** button is used to create a configuration based on user input (Figure 4-31).

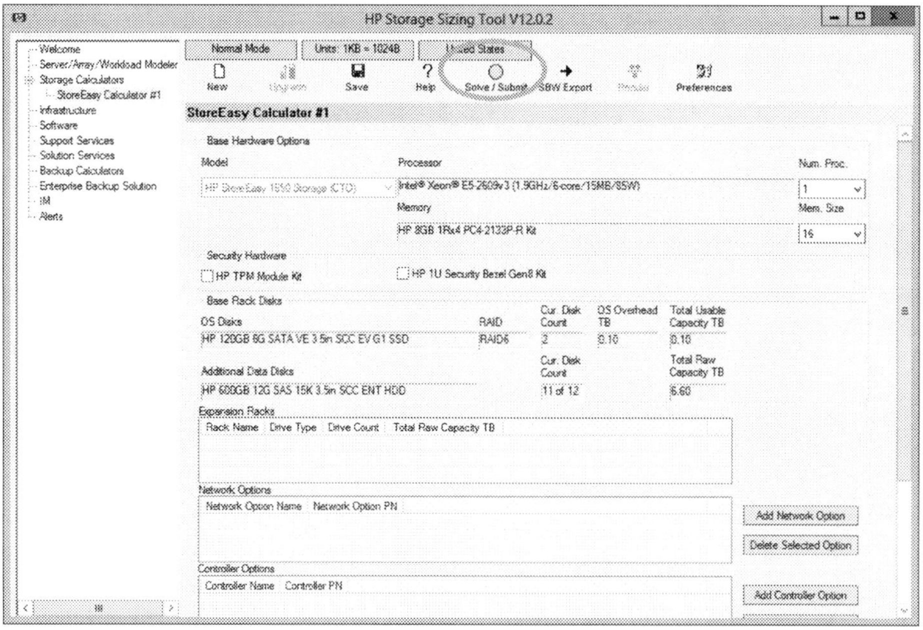

Figure 4-31 **Solve/Submit** button used to create a configuration

- You can export the resulting file to Microsoft Excel (Figure 4-32).

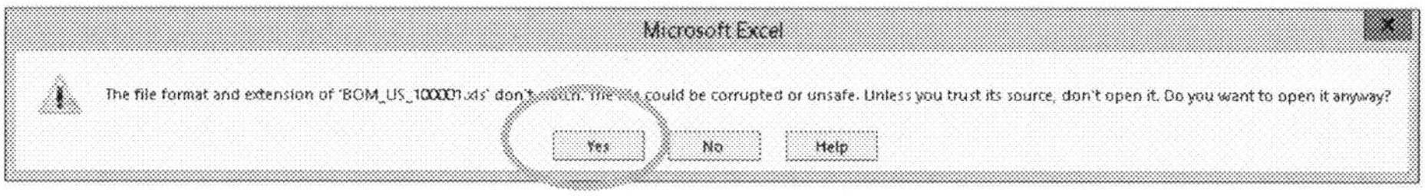

Figure 4-32 Export the resulting file to Microsoft Excel

The exported file will look similar to the graphic in Figure 4-33.

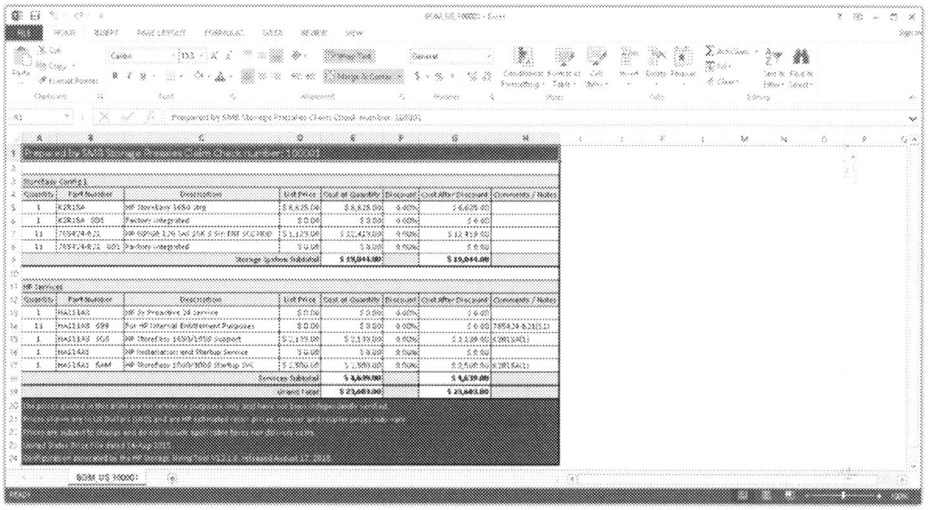

Figure 4-33 The exported file

Key StoreEasy solutions

Solution	Description
Enterprise File Sync and Share	Gain control over your employees' data. HPE StoreEasy with included Microsoft® Work Folders and optional Citrix ShareFile enables users to securely access their data stored on-premises under IT control from anywhere at any time. Deploying Work Folders in HPE StoreEasy 1000 and 3000 (PDF, 884 KB): http://www8.hp.com/h20195/V2/GetDocument.aspx?docname=4AA5-1651ENW&cc=us&lc=en.

Solution	Description
User File Sharing	Facilitate user collaboration behind your firewall. HPE StoreEasy delivers simple yet sophisticated multiprotocol file access over SMB and NFS so that users can share files regardless of their computer's operating system. Heterogeneous file sharing on HPE StoreEasy Storage White Paper (PDF, 694 KB): http://www8.hp.com/h20195/v2/GetDocument.aspx?docname=4AA4-7478ENW&cc=us&lc=en.
Home Directory Consolidation	Empower users while increasing security and control by consolidating home directories. With rich features such as folder redirection, offline files, and roaming user profiles, HPE StoreEasy makes storing user data in a centralized location rather than on user controlled devices easy. Data Consolidation Best Practices White Paper (PDF, 397 KB): http://www8.hp.com/h20195/v2/GetDocument.aspx?docname=4AA4-7459ENW&cc=us&lc=en.
Compliance Archiving	Preserve your data with iTernity Compliant Archive Software (iCAS). Sitting between more than 70 different business applications and HPE StoreEasy, iCAS prevents your data from being manipulated or deleted. Your key to legally compliant data archiving (PDF, 590 KB): http://h22168.www2.hp.com/docs/iternity/HP-and-CAS_Flyer_EN.PDF.
Microsoft Hyper-V	Get the storage reliability and performance you need for your virtualized environment with Microsoft Hyper-V and HPE StoreEasy Storage. Using an optimized and integrated combination of software capabilities, this solution lets you adapt storage to your rapidly changing needs. Virtualization best practices using Hyper-V and HP StoreEasy Storage (PDF, 153 KB): http://www8.hp.com/h20195/v2/GetDocument.aspx?docname=4AA4-7392ENW&cc=us&lc=en.

Solution	**Description**
Microsoft SQL Server	Improve the performance and reliability of SQL Server with HPE StoreEasy Storage. With your SQL Server database stored on SMB 3.0 File Share on StoreEasy, you benefit from the combined throughput of all network adapters and can transparently fail back to the remaining network adapters should one component fail. HPE StoreEasy Storage Performance Best Practice Guidelines for SQL (PDF, 159 KB): http://www8.hp.com/h20195/v2/GetDocument.aspx?docname=4AA4-7418ENW&cc=us&lc=en.
Hybrid Cloud Backup	Back up data on-premises and in the cloud with HPE LiveVault powered by HPE StoreEasy. Now including TurboRestore Appliance installation software, it is even easier to deploy StoreEasy as an on-site turnkey backup appliance within the LiveVault secure cloud and hybrid cloud data protection solution. ESG: HPE Helps SMBs to Protect Their Data and Keep Their Businesses Safe (PDF, 592 KB): http://www8.hp.com/h20195/V2/GetDocument.aspx?docname=4AA5-7557EAW&cc=us&lc=en.

Learning check

The following questions will help you to measure your understanding of the material presented in this chapter. Read all the choices carefully, because there might be more than one correct answer. Select or write the correct answer for each question.

1. What is a network attached storage (NAS) solution?

   ....

   ....

   ....

   ....

   ....

2. List two typical StoreEasy use cases.

   ....

   ....

   ....

   ....

   ....

3. HPE StoreEasy 1000 Storage is based on HPE ProLiant servers.

   ☐ True

   ☐ False

4. HPE StoreEasy 1850 9.6TB SAS Storage comes with preinstalled Windows Storage Server 2012 R2.

   ☐ True

   ☐ False

5. Up to four StoreEasy 3850 Gateway Systems can be configured together as a four-node cluster (when using two StoreEasy 3850 Gateway Nodes per chassis).

   ☐ True

   ☐ False

6. Is the SMB 3 protocol compatible with previous versions, and are all features available? Please explain your answer.

    ....

    ....

    ....

    ....

For answers, see Chapter 4 Answers on page 439.

5 SAN Storage

OBJECTIVES

In this chapter, you should learn to:

- ✓ Describe storage area networks (SANs)
- ✓ Explain the unique features and target customers for the HPE Storage MSA product family
- ✓ Explain the unique features and target customers for the HPE StoreVirtual product family
- ✓ Explain the unique features and target customers for the HPE 3PAR StoreServ product family
- ✓ Identify HPE sizing and configuration tools and resources
- ✓ Differentiate HPE 3PAR StoreServ 7000 and 8000 products based on a feature comparison

INTRODUCTION

This chapter provides an overview of SAN products. It begins with basics of HPE products in this category and then describes the unique features, capacity limits, performance figures, and target customers for the HPE MSA, HPE StoreVirtual, and HPE 3PAR StoreServ products. It then describes reference documents and tools you can use for sizing and configuration.

Storage Area Networks

Figure 5-1 HPE StoreVirtual 4730 and HPE StoreServ 8000 base chassis with add-on enclosure

HPE has products to satisfy business customers with a range of direct attached storage (DAS), network-attached storage (NAS), and SAN solutions. Ideal for company-wide deployment and mission-critical applications, HPE Storage solutions offer maximum scalability, industry-leading performance, a fully integrated suite of centralized management tools, and unmatched data protection and disaster tolerance features (Figure 5-1).

SANs use storage repositories that are attached to multiple host servers, with centralized storage management and scalability and enterprise features.

What is a SAN?

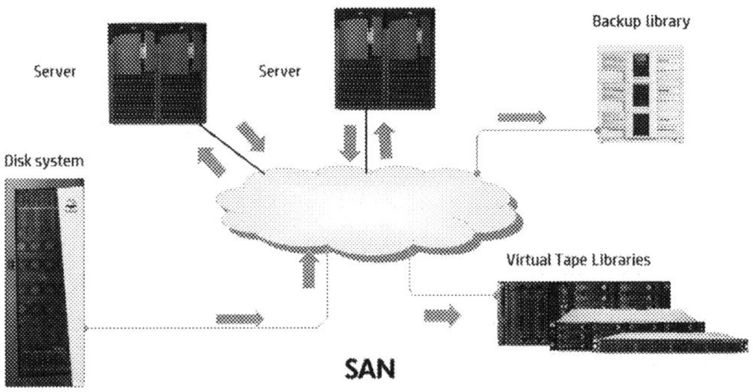

Figure 5-2 SAN architecture example 1

A SAN is an intelligent infrastructure that interconnects heterogeneous servers with shared, heterogeneous storage systems (Figure 5-2). It is a dedicated storage network, designed specifically to connect storage, backup devices, and servers.

SANs are maintained separately from parallel general-purpose networks and are isolated from the messaging network. They are optimized for the movement of data from servers to disk and to tape.

SANs use multiple paths to connect different storage devices with associated servers, and they can provide backup and archival storage for multiple or remote locations. HPE StorageWorks SANs deliver value-added storage applications, management tools, storage arrays, and virtualization technology.

General-purpose networks such as LANs and WANs carry heavy user communications traffic involving printers, email, and so forth. A SAN is the back-end network that carries storage traffic, which provides a clear separation of storage devices from processing and presentation activities and enables the front-end LAN to carry normal TCP/IP traffic.

SANs have become almost synonymous with Fibre Channel (FC). However, FC is not a required component because almost any networking or serial Small Computer System Interface (SCSI) technology can be used to create a SAN. In addition, the FC protocol is designed to carry more than just SCSI traffic; it can also carry TCP/IP traffic and other protocols.

SNIA definition of a SAN

Figure 5-3 SNIA logo

The Storage Network Industry Association (SNIA) provides this statement defining SANs (Figure 5-3):

> *A network whose primary purpose is the transfer of data between computer systems and storage elements, and among storage elements. A SAN consists of a communication infrastructure which provides physical connections, and a management layer which organizes the connections, storage elements, and computer systems so that data transfer is secure and robust.*

Within this definition, there is no mention of FC. SNIA recommends using the term *FC SAN* when the network is based on FC technology.

Although a SAN typically references FC, it can be based on other technologies such as Enterprise System Connection (ESCON) and Internet SCSI (iSCSI), which is a protocol that enables a server to access storage through an Ethernet network adapter.

Note

For more information on the SNIA, see the website at: http://www.snia.org.

Better utilization of backup and restore solutions

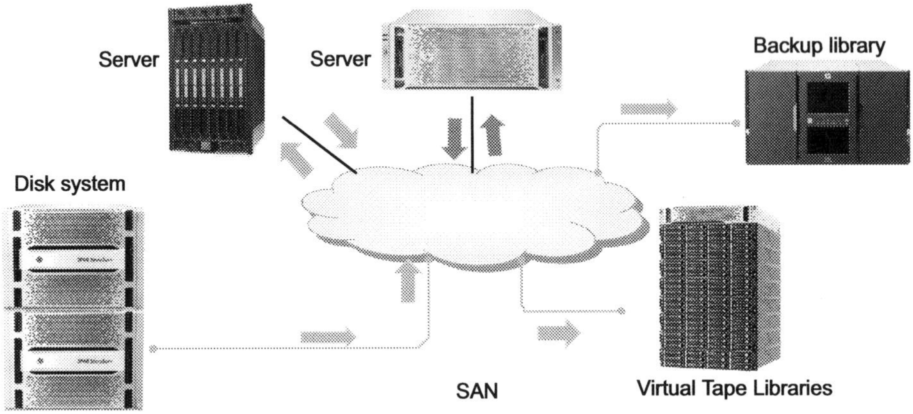

Figure 5-4 SAN architecture example 2

The amount of data being stored, the speed of access to that data, and the amount of time that data is kept online are all growing exponentially (Figure 5-4). Backup and recovery operations are coming under increasing pressure from collapsing backup windows and growing storage requirements.

Because SANs remove backup and recovery traffic from the LAN, congestion is reduced and backup windows are improved. System performance is dramatically increased because data and communications traffic no longer competes for the limited bandwidth on a standard LAN.

Centrally managed, high-performance storage resources such as tape libraries reduce backup time and overhead. SANs also can be configured for serverless (active fabric) backup. Serverless backups use the Network Data Management Protocol (NDMP), the standard for backing up NAS, and SCSI Extended Copy to move the data directly from disk to tape. This enables continuous, uninterrupted access to data and applications during the backup and restore processes.

This technology also eliminates the read/write processes through the application and backup hosts, resulting in less processing power used. By removing the LAN, CPU, and I/O resources from the data path, network bottlenecks are eliminated and high-application performance is maintained. Serverless backup is ideal for large databases and for file servers for which there are no backup windows.

Business continuance

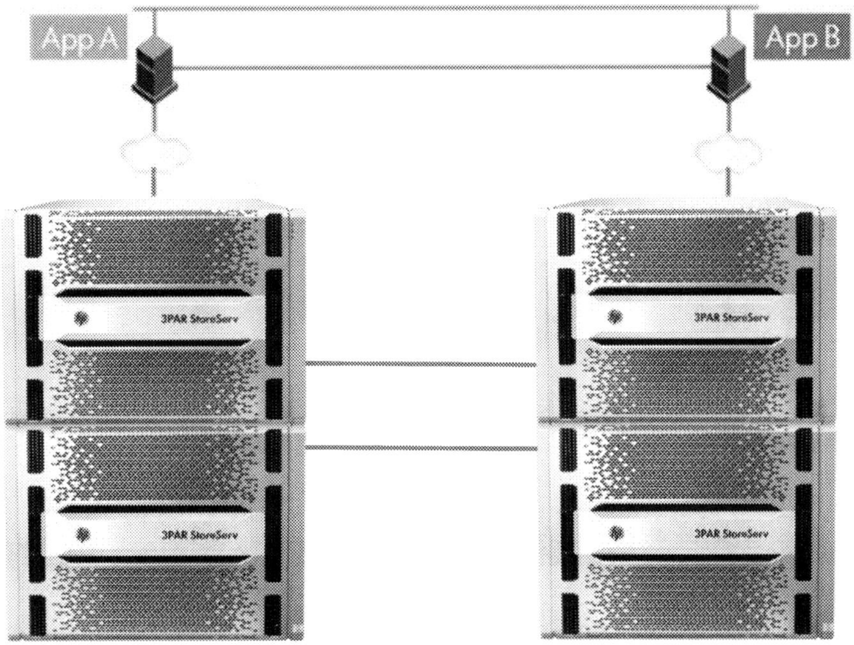

Figure 5-5 Business continuance example

SANs can eliminate single points of failure, incorporate failover software, and support mirroring at geographically dispersed data centers for disaster recovery (Figure 5-5). Businesses can quickly restore productivity after a power failure or a component downtime.

Business continuance (sometimes referred to as business continuity) describes the processes and procedures an organization puts in place to ensure that essential functions can continue during and after a disaster. Business continuance planning seeks to prevent the interruption of mission-critical services and to reestablish full functioning as swiftly and smoothly as possible.

Although business continuance is important for any enterprise, it might not be practical for any but the largest enterprise to maintain full functioning throughout a disaster crisis.

According to many experts, the first steps in business continuity planning are deciding which of the organization's functions are essential and then apportioning the available budget accordingly. After the crucial components are identified, failover mechanisms can be put in place.

New technologies such as disk mirroring over the Internet make it feasible for an organization to maintain current copies of data in geographically dispersed locations so that data access can continue uninterrupted if one location is disabled.

A business continuance plan should include the following:

- A disaster recovery plan, which specifies an organization's planned strategies for post-failure procedures
- A business resumption plan, which specifies a means of maintaining essential services at the crisis location
- A business recovery plan, which specifies a means of recovering business functions at an alternate location
- A contingency plan, which specifies a means of dealing with external events that can seriously impact the organization

High availability (HA)

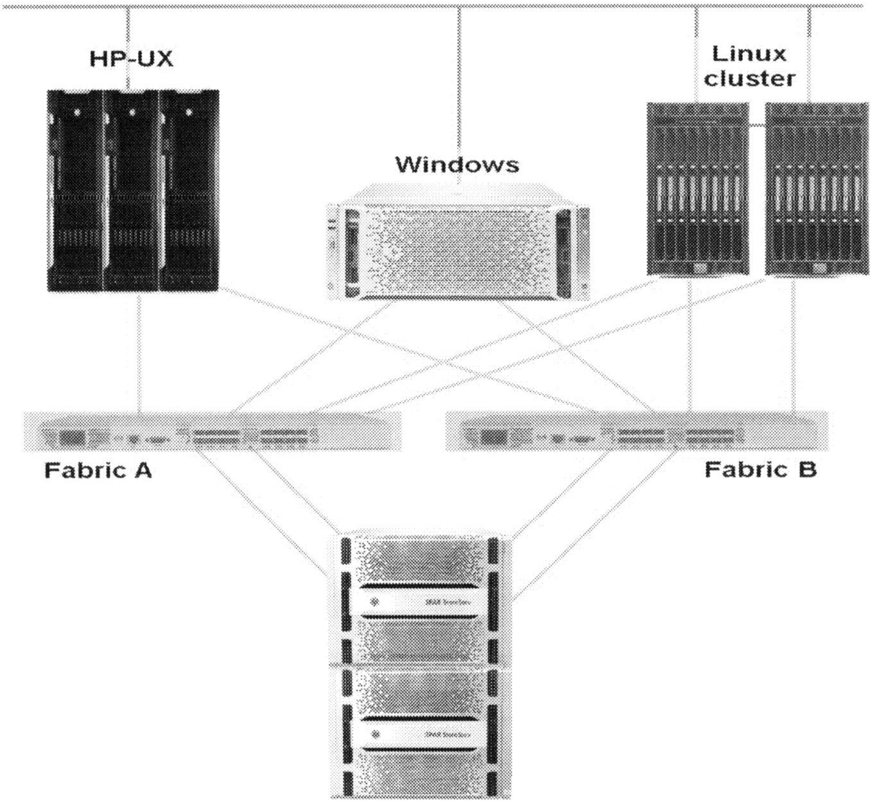

Figure 5-6 HA configuration example

Well-designed SANs are fault tolerant with no single point of failure (Figure 5-6). Redundant fabric designs, server clustering, storage replication, and traffic rerouting provide enterprise-class availability.

Because the long-distance capabilities of FC can be used to mirror data to a remote location, SANs also can help to facilitate disaster recovery.

SANs use fiber optic and copper connections to create dedicated networks for servers and their storage systems.

With a SAN, several servers can access a common data set for server clustering arrangements. This configuration provides several advantages:

- If an application fails, it can be restarted on another server without having to move the data.
- If a server fails, the SAN enables fast failover. This means that operations from the failed server are automatically transferred to another node in the cluster. Again, the data does not have to be moved as part of the failover.
- A SAN provides an automatic method for transferring operations from a failed or down system to a secondary, identical system.

Redundant design

The robust high-availability architecture of a SAN requires redundant components throughout, including servers, host bus adapters (HBAs), switches and hubs, and storage units.

In the extended architecture shown in Figure 5-6, both servers are equipped with dual HBAs so that the failure of one adapter will not disable the server. Both servers are connected to both switches so that the loss of one switch will still leave the servers with an intact path to storage. If an entire storage unit fails, the redundant storage unit will enable the servers to continue to access the application data.

Data mirroring

Data mirroring copies data from one location to a storage device in real time, resulting in an exact copy, or mirror, of the data at that point in time. SAN data mirroring capabilities take advantage of FC distances to enable the replication of key systems to a remote site.

Server and storage consolidation

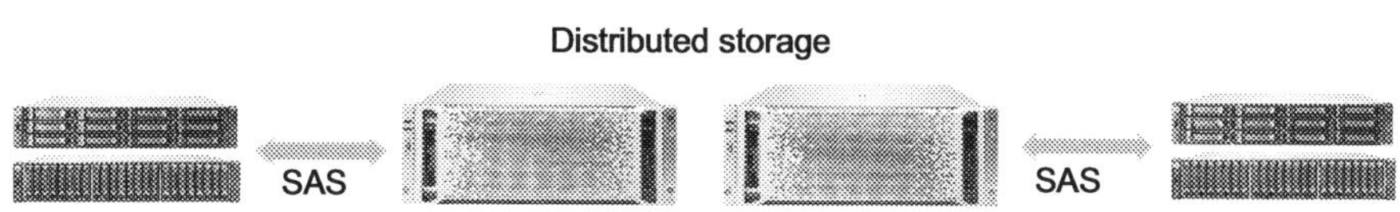

Figure 5-7 Distributed storage example

SAN implementations often begin as storage consolidation efforts and evolve into storage expansion mechanisms as storage needs grow (Figure 5-7).

High connectivity enables storage for many servers to be consolidated on a small number of shared storage devices, reducing costs and easing the management of capital assets.

When storage is managed as a pooled resource that can be allocated to servers in large or small amounts, administrators can deploy the most cost-effective storage products without concern about what fits in the limited number of slots available in the servers.

Adding more storage to a SAN is easy and rarely results in loss of service for the application server, other than what it takes to mount the new storage device. Servers can be added and removed from the SAN, and storage can be reassigned, all while the SAN is online. The number of switches in a fabric can also be increased to provide the number of ports required, or routing technology can be used to connect multiple SAN fabrics (Figure 5-8).

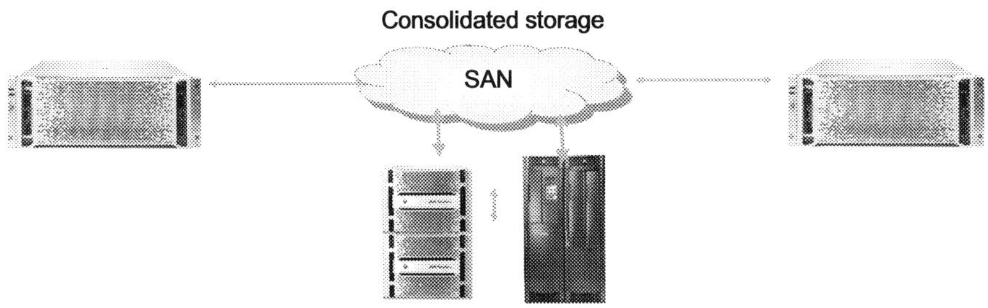

Figure 5-8 Consolidated storage example

SAN components

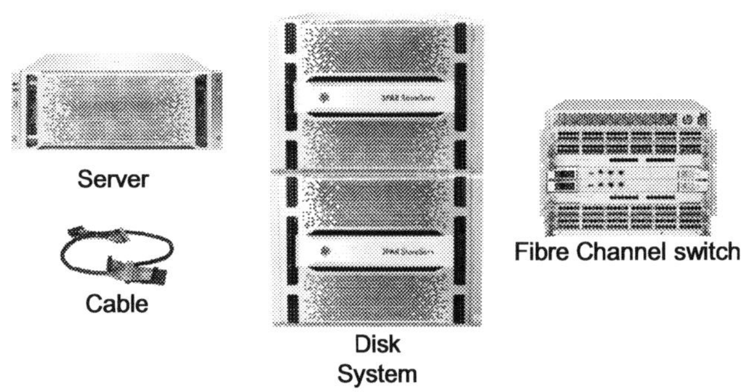

Figure 5-9 SAN components

A SAN consists of both hardware and software components (Figure 5-9):

- Switches:
    - Create the fabric of the SAN

Note

"Fabric" means a network of one or more FC switches that transmit data between any two N_Ports on any of the switches.

-
    - Ernable scalability
- Routers, bridges, and gateways:
    - Enable device sharing, multiprotocol technologies, and fault isolation
    - Extend the SAN over long distances
- Storage devices:
    - Can be integrated among multiple types
    - Can be configured as RAID or JBOD
- Servers:
    - Connect to the SAN with HBAs
    - Can be of any variety
- Cabling and connectors:
    - Can be fiber optic or copper
- SAN management applications:
    - Manage and monitor components
    - Ensure optimal SAN operation

HPE Storage product portfolio

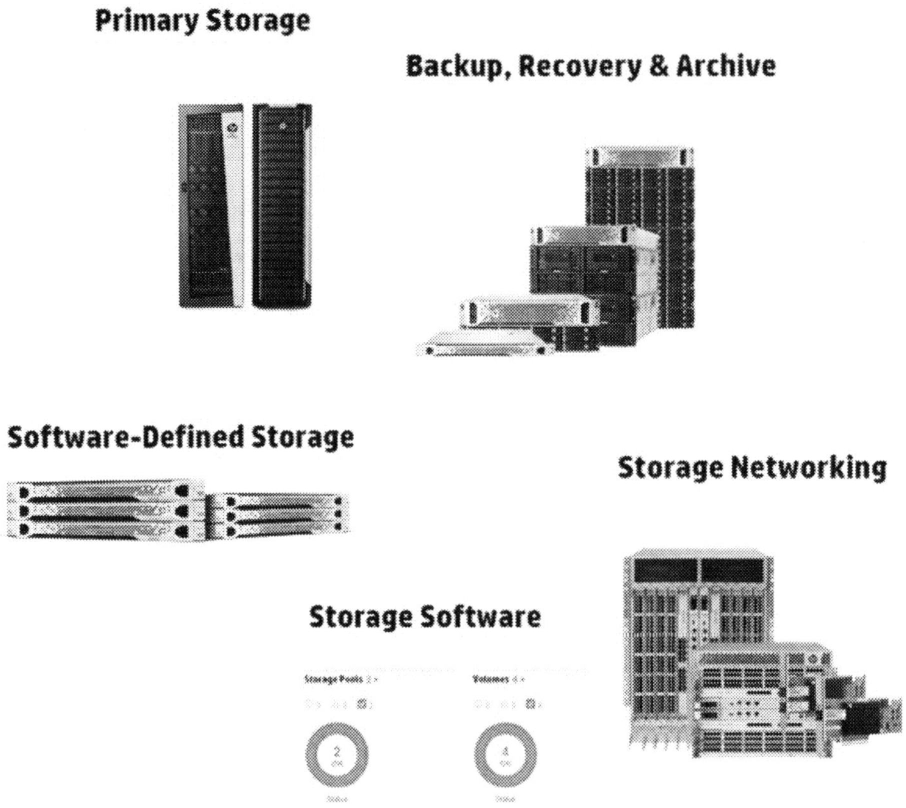

Figure 5-10 HPE Storage product portfolio

HPE Storage products can be divided into the following categories (Figure 5-10):

- Primary Storage
- Backup, Recovery, and Archive
- Software-Defined Storage (SDS)
- Storage Software
- Storage Networking

Primary Storage

Tier-1 Storage is for mission-critical applications, virtualization, and cloud—with high-end and mid-range storage arrays designed for IT-as-a-Service (ITaaS) and leveraging a single flash-optimized architecture with support for block, file, and object access.

HPE Primary Storage products include the following:

- HPE 3PAR StoreServ
- HPE StoreVirtual
- HPE MSA
- HPE StoreEasy
- HPE XP
- Disk enclosures

HPE 3PAR StoreServ, HPE StoreVirtual, and HPE MSA are typical SAN products belonging to the Primary Storage products category.

SAN Storage portfolio

As mentioned, the HPE SAN Storage portfolio consists of these product families:

- HPE 3PAR StoreServ
- HPE StoreVirtual Storage
- HPE MSA Storage

HPE 3PAR StoreServ

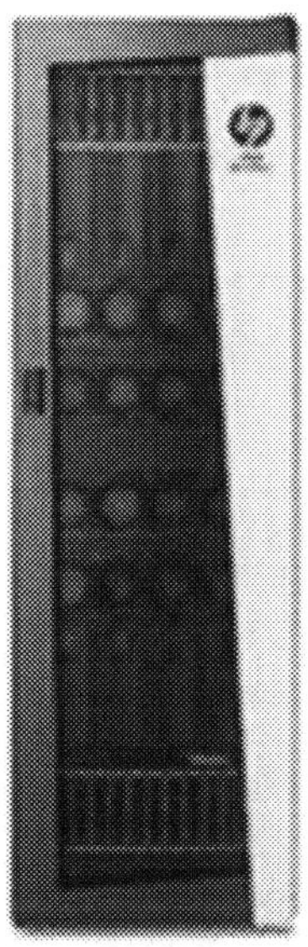

Figure 5-11 HPE 3PAR StoreServ storage

Business environments can start small and grow affordably and nondisruptively with HPE 3PAR StoreServ. HPE 3PAR StoreServ is a Tier 1 storage offering built for virtualization, including midrange, enterprise, and all-flash arrays. It has these advantages (Figure 5-11):

- **Multitenant and federated**—You can manage unpredictable workloads and support huge numbers of applications, mixed workloads, and customers securely and concurrently, without performance or resiliency impacts and with double virtual machine (VM) density. HPE Peer Motion software provides nondisruptive data mobility between federated systems.
- **Efficient**—You can reduce acquisition and operational costs by 50% guaranteed. 3PAR StoreServ enables you to reduce up-front capacity requirements, reduce technology refresh costs, and eliminate stranded capacity by using thin technologies. The subvolume storage tiering helps to reduce cost per GB by up to 30%.
- **Autonomic**—You can save up to 90% of administrator time. 3PAR StoreServ simplifies, automates, and expedites storage management intelligently and without administrator intervention. The system is self-managing, self-healing, and self-configuring.

HPE StoreVirtual Storage

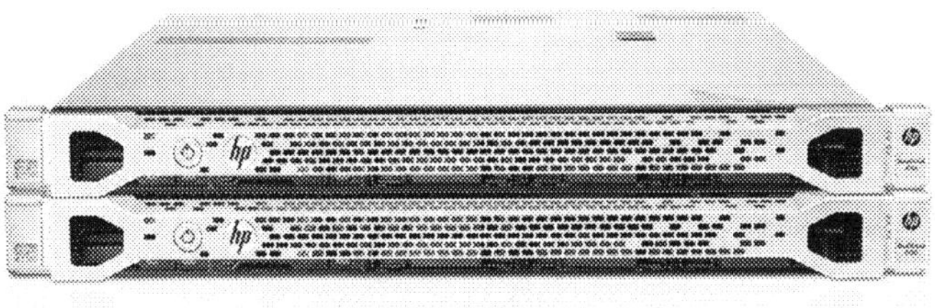

Figure 5-12 HPE StoreVirtual Storage

The versatile HPE StoreVirtual Storage family spans SDS, dedicated storage appliances, and fully hyper-converged virtualization systems to meet the rapidly changing demands of businesses. HPE StoreVirtual Storage scales out to meet business virtualization needs (Figure 5-12).

The all-inclusive enterprise feature set delivers in all aspects—from HA to ease of management. You can scale from a software solution such as a Virtual Storage Appliance (VSA) to rack or blades as the business grows, or you can use a combination of federated platforms that all share the same DNA.

HPE MSA Storage

Figure 5-13 HPE MSA Storage

Whether you need best-in-class performance or the most capacity for your investment, the MSA product family has a fast, affordable solution (Figure 5-13). Every MSA comes with an intuitive setup and management interface that anyone can master.

Note

HPE MSA Storage is an industry-leading entry-level storage platform, according to the IDC Q2 2013 Worldwide Quarterly Disk Storage Systems Tracker for external disk subsystems, entry-level storage category (price bands 1-4), and SAN attached storage (FC, iSCSI, and SAS protocols).

HPE MSA STORAGE

Two product families are available in the HPE MSA Storage portfolio:

- **HPE MSA 1040 Storage**—Affordable entry-level storage

These products provide simple, fast, shared storage with direct SAS connection for up to four servers. No SAN infrastructure is required. They have the lowest entry price of any SAS entry-level arrays in the industry (based on Tier 1 system provider, publicly available US list prices).

The MSA 1040 models have options to support iSCSI and FC connectivity as well as flexibility for both small form factor (SFF) and large form factor (LFF) disk drives.

- **HPE MSA 2040 Storage**—High-performance entry-level storage

You can achieve new levels of performance in entry-level storage with the GL201 MSA Firmware. It delivers up to 45% faster reads and 18% faster writes (based on a comparison of the GL200 code release to the GL210R004 code release on the MSA 2040 FC model with Solid-State Drives (SSDs), with a random 8K block workload and average latency of 30 ms or less).

Energy-Star compliance supports data center energy savings initiatives.

The MSA 2040 storage supports high-performance flash media and flexible connectivity options, and it ships standard with snapshot and volume copy software for increased protection.

The HPE Storage Management Utility (SMU) is a web-based application for configuring, monitoring, and managing the storage system. The MSA 1040 and MSA 2040 storage devices support two versions of SMU:

- Version 3 is the primary web interface for the MSA storage enclosures, providing access to all common management functions for both linear and virtual storage.
- Version 2 is a secondary web interface for the MSA storage enclosures, providing access to traditional linear storage functions. This legacy interface provides certain functionality that is not available in the primary interface.

The SMU CLI enables administrators to interact with the storage system using command syntax entered on the keyboard or through scripting. The CLI preferences can be set to use v3 or v2 terminology in the command output and system messages.

Note

See the *HPE MSA 1040/2040 SMU Reference Guide* to find more information about using SMU v3 and v2.
https://h20566.www2.hpe.com/hpsc/doc/public/display?docId=c04220794

HPE MSA 1040 Storage

Figure 5-14 HPE MSA 1040 Storage arrays

The HPE MSA 1040 Storage products are more affordable MSA arrays using the latest shared storage technologies and delivering entry-level features to business customers. They are simple to deploy and easy to manage (Figure 5-14).

These products put entry-level consolidation and virtualization initiatives within budgetary reach. HPE MSA 1040 Storage offers many of the HPE MSA 2040 features but at substantially lower price points.

HPE ProLiant and HPE BladeSystem administrators as well as IT generalists will find storage management tasks to be simple and intuitive with MSA. The MSA 1040 arrays leverage a new fourth-generation controller architecture with a new processor, two host ports, and 4 GB cache per controller.

Customers can take advantage of the latest hard drive advancements by choosing the form factor and technology that best meets their application and budgetary requirements. Unlike many competing arrays, the MSA 1040 array ships are standard with intuitive management tools and powerful data services such as Snapshots and Volume Copy for increased data protection.

Key features and benefits

Why choose HPE MSA Storage?

- **Simple**—Shared storage without the learning curve
- **Fast**—Proven fourth-generation MSA architecture built for speed with faster processors and 2× cache
- **Affordable**—New levels of affordability in entry storage, whether the goal is to optimize cost or performance
- **Future-proof**—Expandable and upgradable to meet business needs today and tomorrow

Simple: Flexible architecture, easy to set up, and easy to manage

The integrated setup and management web-based interface (WBI) makes MSA products easy to manage for HPE ProLiant and BladeSystem administrators and IT generalists.

These products leverage the latest LFF or SFF hard disk drives (HDDs), with a choice of high-performance enterprise-class SAS or high-capacity SAS midline drives based on application, performance, and budgetary requirements.

Onboard management tools augment OS and hypervisor views, navigate firmware upgrades, and monitor array performance with ease, locally or remotely.

Fast: Up to 50% more performance compared to the HPE MSA P2000 G3 Modular Smart Array System.

The MSA 1040 array leverages the fourth-generation MSA architecture to provide two port controllers with 4 GB cache each, a faster processor, and a performance-optimized I/O engine. The increased performance translates into better application response times, and it facilitates entry-level consolidation and virtualization initiatives.

Affordable: Popular iSCSI/, SAS/, and FC configurations

Dual-controller configurations are standard and available at single-controller price points, bringing more capacity and performance within budget.

Future proof: Built-in expandability along with an MSA 2040 upgrade option

Customers can add disk enclosures as needed to boost the capacity and spindle count. Most low-cost entry arrays do not allow expansion. Data-in-place upgrades eliminate time-consuming and risky data migrations.

The MSA 1040 array allows simple controller swaps to the MSA 2040 array. Other value-priced arrays force customers to compromise.

Virtualization features

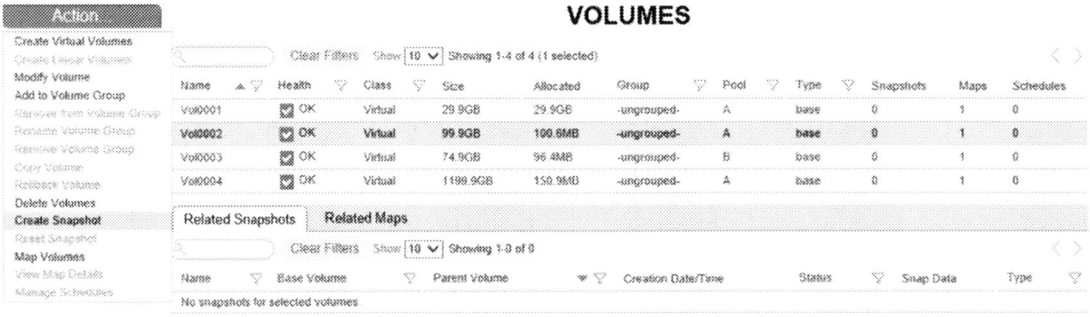

Figure 5-15 MSA user interface

Fourth-generation MSA arrays support a powerful set of data services that are enabled with a new virtualized array system architecture (Figure 5-15). MSA 1040 customers have access to a number of these new features through a firmware upgrade and paid license feature.

- **Thin provisioning**—This feature enables the allocation of physical storage resources only after they are consumed by an application.
- **Automated Tiering (Archive Tiering)**—This functionality uses a real-time I/O engine to intelligently position data on the most appropriate "tier" of storage. MSA 1040 customers leveraging the Archive Tiering functionality will be able to better use storage resources to move data that has not been accessed for long periods of time to more cost-effective midline SAS HDDs.
- **Virtualized Snapshot services**—These services were created specifically for the MSA virtualized storage architecture. They enable the MSA 1040 to use redirect-on-write technology to provide better snapshot performance, and they enable customers to create more complex snapshots with simplified snapshot management.
- **New web user interface**—The powerful, embedded SMU simplifies the setup and management of the MSA 1040 and the new features. Users familiar with the previous MSA web UI will have the option to continue using it or to take advantage of the new UI.

- **Wide striping**—By leveraging virtualized storage pools, MSA 1040 users can also leverage wide striping technology to simplify volume expansion and to facilitate the full utilization of all resources allocated to a specific volume. MSA 2040 storage also supports these features as well as SSD Reach Cache and a Performance Tiering function. Upgrading an MSA 1040 array to an MSA 2040 array using a data-in-place upgrade and a controller swap would bring the full suite of MSA 2040 virtualized features to an MSA 1040 customer.

Note

For more information about the virtualization features and the MSA 2040 platform, see the QuickSpecs or visit http://hp.com/go/msa.

Sizing and performance guidelines

- Capacity
    - 384 TB maximum
    - Maximum including expansion depends on the model
- Drive description
    - 99 SFF SAS/MDL (midline) SAS
    - 48 LFF SAS/MDL SAS
    - Maximum including expansion depends on the model
- Host interface
    - 8 Gb FC, two ports per controller
    - 1 Gb iSCSI, two ports per controller
    - 10 Gb/s iSCSI, two ports per controller
    - 12 Gb/s SAS, two ports per controller
- Storage controller
    - Two MSA 1040 2-port FC controllers
    - Two MSA 1040 2-port 1G iSCSI controllers
    - Two MSA 1040 2-port 10G iSCSI controllers
    - Two MSA 1040 2-port 12G SAS controller

- Storage expansion options
  - HPE MSA 2040 LFF (3.5 in) Disk Enclosure or D2700 SFF (2.5 in) Disk Enclosure
  - Clustering support
  - Windows, Linux, HP-UX, or OpenVMS
- SAN backup support
  - Yes
- HPE Systems Insight Manager support
  - Yes
- Compatible operating systems
  - Microsoft Windows Server 2012
  - Microsoft Windows 2008
  - Microsoft Windows Hyper-V
  - Red Hat Linux
  - SUSE Linux
  - VMware ESXi 5.x
  - VMware ESXi 6.x

Note

Detailed information about compatible operating systems and other supported configurations is available at: http://www.hp.com/storage/spock

---

- Form factor
  - 2U
- Warranty 3/0/0
- Drives
  - High-performance, enterprise-class SAS
  - High-capacity midline SAS

HPE MSA 1040 Storage models

- HPE MSA 1040 2-port FC Dual Controller LFF Storage
- HPE MSA 1040 2-port FC Dual Controller SFF Storage
- HPE MSA 1040 2-port SAS Dual Controller LFF Storage
- HPE MSA 1040 2-port SAS Dual Controller SFF Storage
- HPE MSA 1040 2-port 1GbE iSCSI Dual Controller LFF Storage
- HPE MSA 1040 2-port 1GbE iSCSI Dual Controller SFF Storage
- HPE MSA 1040 2-port 10GbE iSCSI Dual Controller LFF Storage
- HPE MSA 1040 2-port 10GbE iSCSI Dual Controller SFF Storage

Note

Drives are not included. Small form-factor pluggable (SFP) transceivers are installed for FC and iSCSI models.

Performance RAID 10, RAID 5, and RAID 6

The preliminary performance figures provided in Tables 5-1, 5-2, and 5-3 are for reference only because many variables exist between array configurations, workloads, hard drive types, vdisk setup parameters, and host system setups. All performance information is measured using linear storage.

Performance RAID 10

Table 5-1 The preliminary performance figures provided for Performance RAID 10

MSA 1040 Array Performance	HPE MSA 1040 8 Gb FC Controller	HPE MSA 1040 10 GbE ontroller	HPE MSA 1040 1 GbE Controller	HPE MSA 1040 SAS
Protocol (host connect)	8 Gb Fibre Channel	10 GbE iSCSI	1 GbE iSCSI	6 Gb SAS
MSA 1040 RAID 10 Performance Results				
Random Reads IOPS	29,400	29,200	28,900	29,500
Random Writes IOPS	23,000	22,500	22,500	22,000
Random Mix 60/40 IOPS	25,500	25,000	25,000	25,000
Sequential Reads MB/s	3100	3050	440	3100
Sequential Writes MB/s	1650	1650	420	1650

Performance RAID 5

Table 5-2 The performance figures provided for Performance RAID 5

MSA 1040 Array Performance	HPE MSA 1040 8 Gb FC Controller	HPE MSA 1040 10 GbE Controller	HPE MSA 1040 1 GbE Controller	HPE MSA 1040 SAS
Protocol (host connect)	8 Gb Fibre Channel	10 GbE iSCSI	1 GbE iSCSI	6 Gb SAS
MSA 1040 RAID 5 Performance Results				
Random Reads IOPS	29,000	28,800	28,600	29,500
Random Writes IOPS	13,000	12,800	12,500	12,500
Random Mix 60/40 IOPS	16,500	16,500	16,500	16,500
Sequential Reads MB/s	3100	3050	440	3100
Sequential Writes MB/s	2900	2700	430	2800

Performance RAID 6

Table 5-3 The preliminary performance figures provided for Performance RAID 6

MSA 1040 Array Performance	HPE MSA 1040 8 Gb FC Controller	HPE MSA 1040 10 GbE Controller	HPE MSA 1040 1 GbE Controller	HPE MSA 1040 SAS
Protocol (host connect)	8 Gb Fibre Channel	10 GbE iSCSI	1 GbE iSCSI	6 Gb SAS
MSA 1040 RAID 6 Performance Results				
Random Reads IOPS	29,000	28,800	28,500	29,500
Random Writes IOPS	9500	9500	9500	8700
Random Mix 60/40 IOPS	12,500	12,700	12,500	12,500
Sequential Reads MB/s	3100	3050	440	3100
Sequential Writes MB/s	2700	2650	420	2750

Note

For this test configuration, 300 GB 15K SAS drives were used in a dual controller configuration of eight vdisks consisting of 12 disks per vdisk, 3.3 TB volumes, and two volumes per host. Four hosts directly attached to the HPE MSA 1040 arrays were used in this test configuration (results cannot be expected with a single host).

MSA 1040 tests with 1 GbE iSCSI used four hosts directly attached to the HPE MSA 1040 array.

Sequential tests results were achieved with 256K block sizes and random tests were based on 8K block sizes.

For sequential workloads with a queue depth greater than one, each sequential stream is targeted to operate on a separate Logical Block Addressing (LBA) range. Other types of sequential workloads that target specific LBA ranges might achieve higher results.

All FC results were measured using 16 Gb FC HBAs running at 8 Gb. All 10 GbE iSCSI results were measured using 10 Gb iSCSI HBAs. All 1 GbE iSCSI results were measured using 1 Gb NICs. All 6 Gb SAS results were measured using 6 Gb SAS HBAs.

The number and types of applications, the number and types of drives, the operating systems used, and the number of hosts will affect overall performance. Tables 5-1 to 5-3 are provided strictly as test lab comparisons.

These numbers reflect a full array configuration with the maximum number of front-end ports, disks, and controllers. The test results shown for the HPE MSA 1040 are designed to give a conservative reference point for comparisons.

HPE MSA 2040 Storage

Figure 5-16 HPE MSA 2040 Storage arrays

The latest MSA array features two high-performance controllers with support for the industry's latest FC, iSCSI, and SAS host interfaces (Figure 5-16). The MSA 2040 array delivers mid-market features to entry-level SAN customers who want to use the latest spinning or SSD technologies. It is simple to deploy and simple to manage.

Changing storage requirements have raised the bar for entry-level shared storage arrays. The MSA 2040 array uniquely addresses the shared storage and data protection needs of HPE ProLiant and BladeSystem server customers by reducing total cost of ownership (TCO) while dramatically increasing performance and availability. The entry-level MSA 2040 arrays use technologies such as SSDs and self-encrypted drives (SEDs) and several advanced data services that were previously only available on midrange and enterprise disk arrays.

The HPE MSA 2040 arrays are positioned to provide an excellent value for customers needing increased performance to support initiatives such as consolidation and virtualization.

The MSA 2040 delivers this performance by leveraging a fourth-generation controller architecture with more processing resources, 4 GB cache per controller, and four host ports per controller standard. The controller architecture delivers high performance and enables users to extract the full benefits from the latest storage technologies such as SSDs.

Key features and benefits

Customers choose HPE MSA 2040 Storage because it is simple, fast, future-proof, and ENERGY STAR Certified.

Simple: Flexible architecture, easy to set up, easy to manage

- The choice of 8 Gb/16 Gb FC, 1 Gb/10GbE iSCSI, and 12 Gb SAS matches the configuration needs of the infrastructure.
- Customers can deploy single or dual controllers, depending on their high-availability and budgetary requirements.
- HPE offers a choice of disk enclosures with LFF or SFF drives and a choice of high-performance SSDs, enterprise-class SAS or SAS midline drives, or self-encrypting drives (SEDs) to meet application, performance, and budgetary needs.
- The integrated setup and management tools make MSA arrays easy to manage for HPE ProLiant and BladeSystem IT managers.

Fast: New standards for $/IOPS in entry-level SANs, 3x to 4x the competition

- The new high-performance Converged SAN controller and SAS controller offer 4x the performance of other entry-level SAN arrays.

- A four-port Converged SAN controller supports FC and iSCSI, and a SAS controller with 4 GB cache translates into better application response time and support for more virtualized environments.
- SSD support with an integrated "wear gauge" helps to improve application performance and enables customers to reduce their operating costs by reducing footprints and power consumption.
- SEDs are designed to safeguard critical personal and business information and to comply with regulatory mandates.

Future-proof

- Twice the bandwidth and the first entry-level SAN with 16 Gb FC.
- The fourth-generation of the unique MSA data-in-place upgrades provides investment protection.
- Host ports are upgradable by SFPs (8 Gb/16 Gb FC) or (1 GB/10 GBE iSCSI) so customers can reduce their TCO.

MSA 2040 is now ENERGY STAR Certified

ENERGY STAR certified power supplies lead to energy consumption efficiencies that translate into savings or regulatory rebates for customers.

Virtualization features

The fourth generation of MSA arrays provides support for a powerful set of data services, enabled with a new virtualized array system architecture. Existing MSA 2040 customers have access to these new features through a firmware upgrade. These features are supported in the virtualization firmware:

- **Thin provisioning**—This feature enables the allocation of physical storage resources only after they are consumed by an application.
- **SSD Read Cache**—Improves random read performance by allowing an SSD to be used as an extension of the MSA 2040 controller cache.
- **Automated Tiering (with Performance and Archive Tiering)**—This functionality uses a real-time I/O engine to intelligently position data on the most appropriate "tier" of storage. Customers leveraging Performance Tiering capacities will see improved application response times on both read and write activities for frequently accessed data and better utilization of storage resources by being able to move data that has not been accessed for long periods of time to more cost-effective midline SAS HDDs (Archive Tiering).

- **Virtualized Snapshot services**—These services were created specifically for the MSA virtualized storage architecture. The MSA 2040 can use redirect-on-write technology to provide better snapshot performance and more complex snapshots, with simplified snapshot management.
- **New web user interface**—The powerful, embedded SMU simplifies the setup and management of the MSA 2040 array and the new features. Users who are familiar with the previous MSA web UI will have the option to continue using it or to take advantage of the new UI.
- **Wide striping**—By leveraging virtualized storage pools, MSA 2040 users can leverage wide striping technology to simplify volume expansion and to facilitate the full utilization of all resources allocated to a specific volume.

Note
For more information on MSA software, visit http://hp.com/go/MSA.

Sizing and performance guidelines

- Capacity
    - 768 TB maximum
    - Including expansion, depending on the model
- Drive description
    - 199 SFF SAS/MDL SAS/SSD
    - 96 LFF SAS/MDL SAS
    - Maximum including expansion is supported, depending on the model
- Host interface
    - 16 Gb/8 Gb FC, four ports per controller
    - 1 GbE/10 GbE iSCSI, four ports per controller
    - 12 Gb/sec SAS, four ports per controller
- Storage controller
    - Two MSA 2040 SAN controllers
    - Two MSA 2040 SAS controllers

- Storage expansion options
  - HPE MSA 2040 LFF Disk Enclosure or D2700 2.5 in, disk enclosure
- Clustering support
  - Windows, Linux, HP-UX, or OpenVMS
- SAN backup support
  - Yes
- HPE Systems Insight Manager support
  - Yes
- Compatible operating systems
  - Microsoft Windows Server 2012
  - Microsoft Windows 2008
  - Microsoft Windows Hyper-V
  - HP-UX
  - Red Hat Linux
  - SUSE Linux
  - VMware ESXi 5.x
  - VMware ESXi 6.x

Note
Detailed information about compatible operating systems and other supported configurations is available at http://www.hp.com/storage/spock.

- Form factor
  - 2U
- Warranty
  - 3/0/0
- Drives
  - High-performance, enterprise-class SAS

- High-capacity SAS midline
- SSD or SED

MSA software and data services

Several software offerings are available for MSA 2040 arrays.

HPE MSA Snapshot Software

HPE MSA Snapshot Software services enable increased data protection by creating recovery points for data, taking a "picture" of data at a specific point in time. Snapshots are then maintained even as data continues to change. In the event of a failure, recovery to any previous snapshot is possible. Snapshots are a great complement to a tape or a disk backup strategy.

The MSA Snapshot functionality is controller-based so no host resources are used. MSA 2040 Copy Services use copy-on-write capabilities when operating on linear volumes and redirect-on-write capabilities when operating on virtualized volumes.

The MSA 2040 ships standard with 64 snapshots enabled. An optional 512 Snapshot Software License is available.

HPE Volume Copy Software

HPE MSA Volume Copy Software makes a physical copy of the data to another set of disks within the same array. One of the benefits of Snapshot and Volume Copy software is the ability to mount a Snap or a Volume Copy to another server for backup, application testing, or data mining.

HPE Volume Copy Software is controller-based, meaning no host resources are used.

HPE MSA 2040 arrays ship standard with Volume Copy. Volume Copy only works with linear volumes.

Note

Linear Storage is the traditional storage that has been used for the four MSA generations. With Linear Storage, the user specifies which drives make up a RAID group and all storage is fully allocated.

Virtual Storage is an extension of Linear Storage. The data is virtualized across a single disk group, as in the linear implementation, and also across multiple disk groups that have different performance capabilities and use cases.

HPE MSA Array Remote Snap Software

This software is built on the asynchronous replication technology that provides remote replication on the HPE MSA 1040 and 2040 arrays and on the HPE P2000 G3 arrays (FC, iSCSI, and combination FC/iSCSI models). This optional software is based on the core MSA Snapshot technology. It supports both Ethernet (iSCSI) and FC interconnects and provides flexible remote data protection options between two MSA arrays, enabling local and remote recovery. Remote Snapshots only function with linear volumes.

Note
For more information, visit http://www.hp.com/go/RemoteSnap.

MSA 2040 Performance Tiering upgrade

The MSA 2040 Automated Tiering engine provides Performance Tiering functionality as well as Archive Tiering functionality. The Performance Tiering functionality moves data between an enterprise SAS HDD tier and an SSD tier based on real-time access trends placed on the array. Archive Tiering is a standard feature with the virtualized firmware architecture. It moves data between enterprise SAS HDDs and midline SAS HDDs based on real-time I/O patterns.

HPE StoreEasy File Services

An HPE StoreEasy 3000 Gateway device can add file, print, iSCSI, and management hosting services to an MSA 2040 array.

Note
For more information, visit http://www.hp.com/go/StoreEasy.

HPE MSA 2040 Storage models

- HPE MSA 2040 SAN Controller
- HPE MSA 2040 SAS Controller
- HPE MSA 2040 Energy Star SAN Dual Controller LFF Storage
- HPE MSA 2040 Energy Star SAN Dual Controller SFF Storage
- HPE MSA 2040 Energy Star SAS Dual Controller LFF Storage
- HPE MSA 2040 Energy Star SAS Dual Controller SFF Storage
- HPE MSA 2040 Energy Star SAN Dual Controller with 21.6TB Bundle
- HPE MSA 2040 Energy Star SAN Dual Controller with 28.8TB Bundle
- HPE MSA 2040 Energy Star SAS Dual Controller with 28.8TB Bundle
- HPE MSA 2040 Energy Star SAS Dual Controller with 21.6TB Bundle
- HPE MSA 2040 SAN DC with SSD and 10K SAS Auto Tier LTU 6.2TB Bundle
- HPE MSA 2040 SAS DC with SSD and 10K SAS Auto Tier LTU 6.2TB Bundle

Performance RAID 10, RAID 5, and RAID 6—End-to-end performance figures

The performance figures provided here are for reference only because many variables exist between array configurations, workloads, hard drive types, disk group setup parameters, and host system set-ups. All performance information is measured using linear storage.

HPE has traditionally published a set of end-to-end MSA performance specifications that feed into HPE Sizer tools and are based on conservative, real-world configurations. End-to-end performance figures guarantee that performance numbers are a guideline as established by tests using RAW I/O in an operating system–agnostic test lab environment. These numbers are preliminary and are subject to change without notice.

Performance RAID 10

Table 5-4 Performance RAID 10

MSA 2040 Array Performance	HPE MSA 2040 Converged SAN Controller with HDD	HPE MSA 2040 Converged SAN Controller with SSD	HPE MSA 2040 Converged SAN Controller with HDD	HPE MSA 2040 Converged SAN Controller with SSD	HPE MSA 2040 Converged SAN Controller with HDD	HPE MSA 2040 Converged SAN Controller with SSD	HPE MSA 2040 Converged SAS Controller with HDD	HPE MSA 2040 Converged SAS Controller with SSD
Protocol (host connect)	16 Gb FC	16 Gb FC	10 GbE iSCSI	10 GbE iSCSI	1 GbE iSCSI	1 GbE iSCSI	12 Gb SAS	12 Gb SAS
MSA 2040 RAID 10 Performance Results **NOTE: RAID 1 was used for SSD testing								
Random Reads IOPS	57,000	112,500	56,500	102,000	56,500	93,000	56,500	112,500
Random Writes IOPS	32,000	31,500	30,500	31,500	30,500	31,500	31,000	32,500
Random Mix 60/40 IOPS	45,000	57,500	44,500	54,500	44,500	54,500	44,500	58,000
Sequential Reads MB/s	5000		4700		860		4720	
Sequential Writes MB/s	2400		2300		850		2300	

Performance RAID 5

Table 5-5 Performance RAID 5

MSA 2040 Array Performance	HPE MSA 2040 Converged SAN Controller with HDD	HPE MSA 2040 Converged SAN Controller with SSD	HPE MSA 2040 Converged SAN Controller with HDD	HPE MSA 2040 Converged SAN Controller with SSD	HPE MSA 2040 Converged SAN Controller with HDD	HPE MSA 2040 Converged SAN Controller with SSD	HPE MSA 2040 Converged SAS Controller with HDD	HPE MSA 2040 Converged SAS Controller with SSD
Protocol (host connect)	16 Gb FC	16 Gb FC	10 GbE iSCSI	10 GbE iSCSI	1 GbE iSCSI	1 GbE iSCSI	12 Gb SAS	12 Gb SAS
MSA 2040 RAID 5 Performance Results **NOTE: RAID 1 was used for SSD testing								
Random Reads IOPS	57,000	106,500	55,500	100,000	55,500	87,000	55,500	108,000
Random Writes IOPS	18,000	20,500	17,500	20,500	17,500	20,000	18,000	20,500
Random Mix 60/40 IOPS	30,000	37,500	29,500	37,500	29,500	36,500	29,500	37,000
Sequential Reads MB/s	4900		4700		860		4700	
Sequential Writes MB/s	4000		3600		850		4100	

Performance RAID 6

Table 5-6 Performance RAID 6

MSA 2040 Array Performance	**HPE MSA 2040 Converged SAN Controller with HDD**	**HPE MSA 2040 Converged SAN Controller with SSD**	**HPE MSA 2040 Converged SAN Controller with HDD**	**HPE MSA 2040 Converged SAN Controller with SSD**	**HPE MSA 2040 Converged SAN Controller with HDD**	**HPE MSA 2040 Converged SAN Controller with SSD**	**HPE MSA 2040 Converged SAS Controller with HDD**	**HPE MSA 2040 Converged SAS Controller with SSD**
Protocol (host connect)	16 Gb Fibre Channel	16 Gb Fibre Channel	10 GbE iSCSI	10 GbE iSCSI	1 GbE iSCSI	1 GbE iSCSI	12 Gb SAS	12 Gb SAS
MSA 2040 RAID 6 Performance Results **NOTE: RAID 1 was used for SSD testing								
Random Reads IOPS	57,000	106,500	54,500	97,500	54,500	87,000	55,500	108,000
Random Writes IOPS	12,500	16,500	12,000	16,000	12,000	16,000	12,500	16,500
Random Mix 60/40 IOPS	23,000	31,500	22,500	31,000	22,500	30,500	23,000	32,000
Sequential Reads MB/s	4900		4600		860		4500	
Sequential Writes MB/s	3900		3500		850		3800	

Note

You can find additional details about testing in the HPE Product Bulletin.

For MSA 2040 HDD results, 300 GB 15K SAS drives were used in a dual-controller configuration of 16 vdisks consisting of 12 disks per vdisk, 3.3 TB volumes, and four volumes per host. Four hosts directly attached to the HPE MSA 2040 array were used in this test configuration (results cannot be expected with a single host).

MSA 2040 tests with 1 GbE iSCSI used eight hosts directly attached to the HPE MSA 2040 array.

For MSA 2040 SSD results, 200 GB and 400 GB Enterprise Mainstream SSDs were used in a dual-controller configuration of four vdisks consisting of two disks per vdisk, 200 GB and 400 GB volumes, and one volume per host. Four hosts directly attached to the HPE MSA 2040 array were used in this test configuration (results cannot be expected with a single host).

MSA 2040 tests with 1 GbE iSCSI used eight hosts directly attached to the HPE MSA 2040 array.

Sequential tests results were achieved with 256K block sizes, and random tests were based on 8K block sizes.

For sequential workloads with a queue depth greater than one, each sequential stream was targeted to operate on a separate LBA range. Other types of sequential workloads that target specific LBA ranges might achieve higher results.

All SAS results were measured using 6 Gb SAS HBAs. All configurations were tested with GL210 firmware.

All FC results were measured using 16 Gb FC HBAs. All SAS results were measured using 6 Gb SAS HBAs. All 10 GbE iSCSI results were measured using 10 GbE iSCSI HBAs. All 1 GbE iSCSI results were measured using 1 GbE network interface controllers (NICs).

The number and types of applications, the number and types of drives, the operating systems used, and the number of hosts will affect overall performance. The Table 5-6 Performance RAID 6 is provided strictly as a test lab comparison.

These numbers reflect a full array configuration with the maximum number of front-end ports, disks, and controllers. The test results shown for the HPE MSA 2040 are designed to give a conservative reference point for comparisons.

HPE StoreVirtual

HPE StoreVirtual storage provides a virtualized pool of storage resources to deliver enterprise-level storage functionality that enhances virtual environments, simplifies management, and reduces costs. It is easy to deploy, grow, and maintain. HPE StoreVirtual storage ensures that critical business data remains available.

This innovative approach to storage provides a unique data protection level across the entire array that reduces vulnerability without increasing costs the way traditional SANs can. HPE StoreVirtual storage is affordable storage designed for a virtualized infrastructure. It supports continuous data growth and keeps businesses up and running.

Introducing HPE StoreVirtual

HPE StoreVirtual, which is based on the LeftHand operating system, is a scale-out storage platform that is designed to meet the dynamic needs of virtualized environments. Intuitive, common management, and storage federation provide for simplicity and flexibility in virtual data centers. HPE StoreVirtual storage enables data mobility across tiers, locations, and between physical and virtual storage.

The all-inclusive enterprise feature set makes HPE StoreVirtual the most versatile storage platform on the market. Its SDS VSA software and ProLiant rack- and BladeSystem-based hardware models provide options to fit any infrastructure and budget. Enterprise-class storage software functionality and leading virtualization software integration are built in. This makes HPE StoreVirtual storage the ideal platform for supporting virtualization growth at all stages.

HPE StoreVirtual redefined shared network storage in 2002 with the introduction of the first full-featured Internet Protocol (IP) SAN. By building a SAN using existing Ethernet infrastructure, the HPE StoreVirtual solution eliminates the learning curve, the expensive FC components, and the complexity associated with FC SANs. Version 10.0 introduces FC capability to allow for a mixed protocol environment for customers who need more flexibility.

An HPE StoreVirtual SAN is ideal for use with Microsoft Exchange, Microsoft SQL Server, server consolidation, local or geographic disaster recovery, and disk-to-disk backup projects.

The LeftHand OS v12.x software (formerly, SAN/iQ) brings VMware multipathing, Space reclamation, REST APIs, Adaptive Optimization (VSA and 4335), Standard Management Initiative Specification (SMI-S) support for Microsoft System Center VM Manager (SCVMM), Optimized Snapshot Manager, Online Upgrade enhancements, and HPE StoreVirtual VSA for HPE Helion OpenStack®.

Models

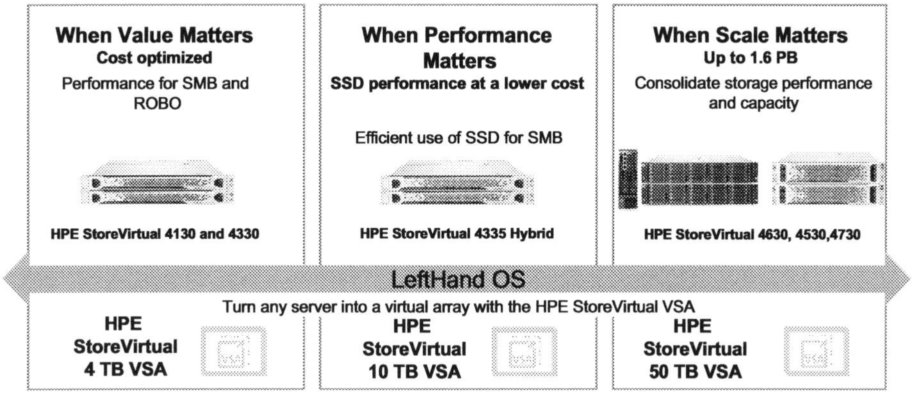

Figure 5-17 HPE StoreVirtual portfolio

HPE StoreVirtual VSA

A proven enterprise feature set delivers the capabilities of a dedicated array without the cost. The HPE StoreVirtual VSA appliance works with all major hypervisors and any x86 server. No required hardware means lower costs and less complexity (Figure 5-17). HPE StoreVirtual VSA expands the value of the server by colocating data services and applications. It is the number one SDS solution with more than 1 million VSAs shipped to date.

HPE Hyper Converged 250 System

This is the fastest and easiest way to deploy and manage virtual infrastructure. You can simplify, optimize, and virtualize with the new HPE hyper-converged platform that combines server, storage, networking, and management tools in a single networked appliance that deploys in minutes.

HPE StoreVirtual 4335 Hybrid Storage

You can optimize performance with powerful, easy-to-use SSD tiering. Businesses can match the benefits of SSD performance with their budget and management needs.

HPE StoreVirtual 4530/4630/4730

HPE StoreVirtual products enable you to simplify and protect virtual infrastructure with built-in application integration for leading virtualization applications. You can elevate storage performance with a solution that supports rapid deployment across multiple sites, delivers enterprise-class features, and provides mission-critical availability.

HPE StoreVirtual 4330/4130

HPE StoreVirtual 4330/4130 devices expand storage seamlessly so customers can respond to the unpredictable demands of today's business environment. They can achieve enterprise-class availability within their budgets with simple management and ease-of-maintenance that saves time and money.

Differentiators

Differentiators between HPE StoreVirtual models:

- Form factor—Blade or rack
- Capacity
- Drive type
    - SAS or MDL SAS
    - SSD
- Performance
- Host connectivity
    - iSCSI, FC
    - Common Internet File System/Server Message Block (CIFS/SMB), Network File System (NFS), HTTP, and FTP with the HPE StoreEasy 3830 Storage Gateway

Managing storage pools

With the HPE IP SAN, clustered storage nodes are managed as one storage pool, and additional capacity can be added to the cluster with no downtime. Adding a storage node is as simple as adding any other Ethernet device; the intelligence within the IP SAN automates the process of adding the capacity to the storage pool.

The storage pool is carved into volumes using a simple GUI. Creating or resizing a volume takes only minutes with no downtime, so storage can be added and managed during normal business hours.

An HPE IP SAN supports a true "pay as you grow" architecture that eliminates the need for capacity planning and complicated budgeting.

The capacity of the IP SAN storage pool can be assigned to application servers as needed, completely eliminating the issue of inefficient provisioning. As a particular application requires more storage, administrators simply resize the volume associated with that application. Similarly, underutilized capacity can be reassigned, allowing double the capacity utilization over a typical DAS solution.

HPE StoreVirtual SAN provides

- High-availability, high-reliability hardware
    - Redundant, hot-swappable power supplies
    - Hot-swappable drives
    - Dual NIC connectivity
    - FC connectivity (available with certain models)
    - Robust power and cooling diagnostics
- Modular hardware for granular scalability
- Pooled storage with centralized management

HPE StoreVirtual components

HPE StoreVirtual solutions usually consist of the following components:

- Storage system—An Ethernet-based storage device with the LeftHand OS
    - Provides a complete IP SAN solution for Windows environments
    - Can be based on ProLiant servers, BladeSystem components, or a VSA
- LeftHand OS—The operating system that runs on the storage system
    - Manages all network traffic and storage operations
    - Virtualizes multiple storage modules
- Central Management Console (CMC)
    - Provides the intuitive, GUI-based administrative interface to LeftHand OS
- iSCSI initiator
    - Is required to allow a client to access an HPE IP SAN volume

HPE StoreVirtual all-inclusive feature set

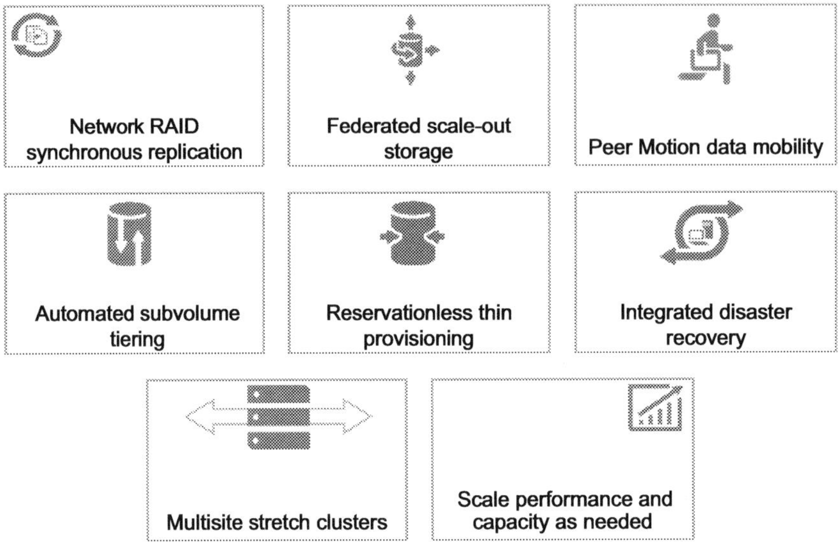

Figure 5-18 HPE StoreVirtual all-inclusive feature set

The all-inclusive license of HPE StoreVirtual storage covers enterprise-class storage features, management, and value-add tools (Figure 5-18). There is no need to purchase additional software; customers buy only the storage needed today and grow nondisruptively in the future.

Scalable storage

Scale linearly in both capacity and performance with the scale-out architecture of HPE StoreVirtual. Scaling storage is as simple as adding nodes to the cluster so that customers can seamlessly and nondisruptively grow as the business grows.

Network RAID

Network RAID technology aggregates all resources and capacity into a single storage pool, so every volume in the cluster can access all available capacity. This also means that all cluster resources—CPUs, RAID controllers, NICs, caches, and disk spindles—can contribute to performance.

To remove any single point of failure, the Network RAID functionality stripes and protects multiple copies of data across a storage cluster. Redundancy is managed on a per-volume basis to improve storage utilization and to match the data protection of the volume to the application requirements.

iSCSI and FC connectivity

With iSCSI and optional FC connectivity, HPE StoreVirtual storage works great even in mixed-protocol environments. A single, flexible storage solution can be deployed for all locations. StoreVirtual customers are prepared for what the future might bring—they can migrate their storage systems from one protocol to another without having to redesign or replace them.

High data availability and superior disaster recovery

With proven 99.999% HA and reliability, a StoreVirtual storage cluster can sustain multiple concurrent failures and still keep data online and accessible to applications. This provides customers with worry-free storage for centralized data centers and for remote and branch offices.

VMware-certified multisite disaster recovery delivers business continuity with failover that is transparent to users and applications. The multisite configuration maintains data availability beyond a single physical or logical site and validates full compatibility with VMware HA features.

LeftHand Remote Copy enables you to asynchronously replicate thin provisioned, reservationless snapshots between primary and remote locations. Backup and disaster recovery can be centralized on a per-volume basis while using application-integrated snapshots for faster recovery.

LeftHand Peer Motion

With LeftHand Peer Motion, an online data mobility technology, volumes can be moved across all deployed HPE StoreVirtual nodes regardless of the platform or drive type. HPE StoreVirtual storage is the only product that offers data mobility across tiers and locations and between virtual and physical storage.

Adaptive Optimization

Adaptive Optimization is an auto-tiering technology to optimize the cost and performance of HPE StoreVirtual storage clusters. It automatically balances data on volumes between tiers.

Scale and mix and match

With HPE StoreVirtual storage, you can scale from SDS VSA software to ProLiant rack- and BladeSystem-based hardware models. You can mix and match product models that best fit the infrastructure. The result is simplified management and feature consistency.

HPE StoreVirtual DSM for Microsoft MPIO

The HPE StoreVirtual Device Specific Module (DSM) for the Microsoft Multipath I/O (MPIO) framework enhances MPIO functionality as follows:

- Enables the automatic creation of an I/O path to each storage system in the cluster on which the volume resides, plus a path for the administrative connection.
- Provides an improved performance architecture over native MPIO solutions.
  - Read I/Os are always serviced by a storage system that holds a copy of the data being requested.
  - Write I/Os are always serviced by a storage system that receives a copy of the data. Remaining copies (replicas) of the data are forwarded to the appropriate storage system based on the data protection level of the volume.
- Provides the ability to build a robust, fault-tolerant solution because an I/O path is built to every storage system in the cluster. For example, in a cluster with five storage systems, DSM for MPIO-connected volumes have six iSCSI MPIO connections to the SAN, one for each storage system and one for the administrative connection. Five of the six connections could go offline and the I/O would still be serviced. The native MPIO provided by the operating system does not open iSCSI connections to all HPE StoreVirtual nodes.

REST API interface

REST API simplifies HPE StoreVirtual management and allows scripting. This feature provides a new, fast, and platform-independent way of managing and provisioning storage. It uses a RESTful interface that is ideal for scripting and achieving high levels of automation.

It also enables the new Cinder driver in the OpenStack Icehouse release that simplifies the management and orchestration of cloud infrastructures built on HPE StoreVirtual Storage. Customers running this OpenStack release will automatically use this new interface on HPE StoreVirtual solutions as soon as the storage pool is upgraded to LeftHand OS 11.5.

In this first version, the REST API provides access to information about storage pools and provisioning operations for volumes, snapshots, and servers (including settings for Network RAID and Adaptive Optimization). It is significantly faster than the equivalent operation in the CLI because of a re-engineered management stack. Four operations are available through the REST API: Create, Read, Update, and Delete.

Space reclamation

Figure 5-19 Space reclamation

The space reclamation feature in HPE StoreVirtual solutions reclaims unused storage from volumes in VMware and Microsoft environments to free up space after VMs and other files are deleted (Figure 5-19). Unused storage is returned to the storage pool to maximize data storage.

HPE StoreVirtual Recovery Manager

The easiest way to recover files and folders on Microsoft Windows Server from snapshots on HPE StoreVirtual storage is to use HPE StoreVirtual Recovery Manager. HPE StoreVirtual Recovery Manager provides smart search capabilities for discovery and recovery to any location (the original location, a file share, or a local file system). It can also be used to automatically access up to five of the most recent snapshots of a volume.

Now servers with FC connectivity to HPE StoreVirtual storage can be used to recover files and folders from snapshots. Servers running HPE StoreVirtual Recovery Manager for Windows require IP connectivity to the HPE StoreVirtual solution for the discovery of snapshots and for assigning snapshots to the recovery server.

In addition, the auto-mount policies to mount snapshots can also include Remote Snapshots (when using Remote Copy) to allow for automated processing of replicated volumes to a remote site.

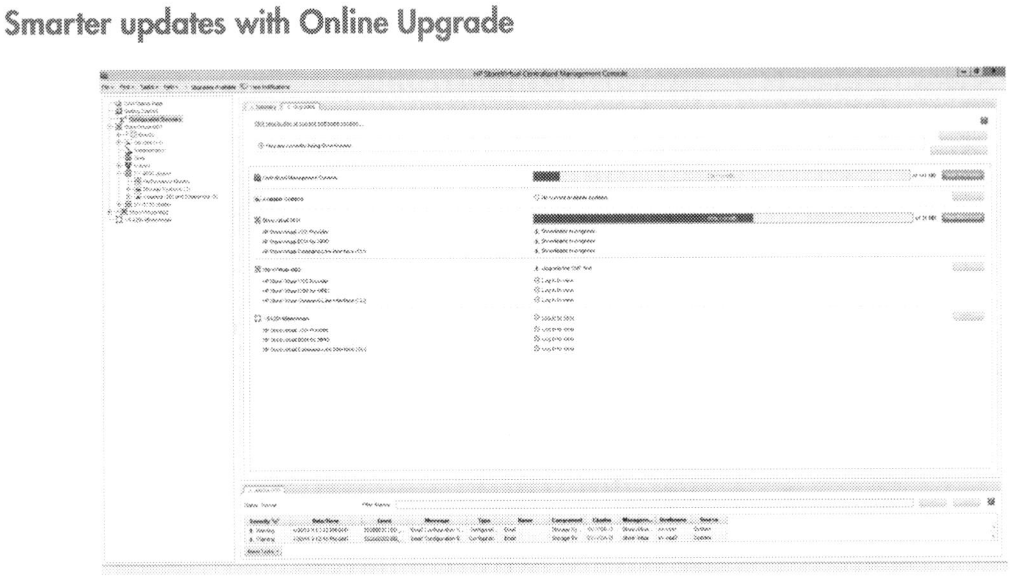

Figure 5-20 HPE StoreVirtual CMC Online Upgrade screen

Online Upgrade in the HPE StoreVirtual Centralized Management Console helps to keep an HPE StoreVirtual installation secure and current with the latest enhancements (Figure 5-20). Applicable upgrades for the environment are identified as soon as the administrator logs in to the management group and can be downloaded and installed when ready. Automatic Online Upgrades allow for the immediate download of upgraded components whenever new upgrades are available from HPE.

Full support for HTTP and SOCKS proxies, including proxy authentication, is available for Online Upgrades in the new version of the Centralized Management Console.

Sites without connectivity to the HPE website can be upgraded by taking the Centralized Management Console outside of the dark site to download all upgrade components.

HPE StoreVirtual 4130

Figure 5-21 HPE StoreVirtual 4130

The HPE StoreVirtual 4130 system has these features (Figure 5-21):

- Proven platform built on ProLiant DL360 Gen8
- One Intel E5-2620 2.0 GHz six-core processor, 12 threads

- 8 GB RAM – DDR3 1333 MHz
- Four 600 GB 10K SFF SAS drives
- Support for RAID 5 only
- Smart Array P420i controller with 2 GB Flash-backed write cache (FBWC)
- Four 1 Gb Ethernet iSCSI ports
- HPE integrated Lights-Out (iLO) 4 remote management
- LeftHand OS 12.0

HPE StoreVirtual 4330

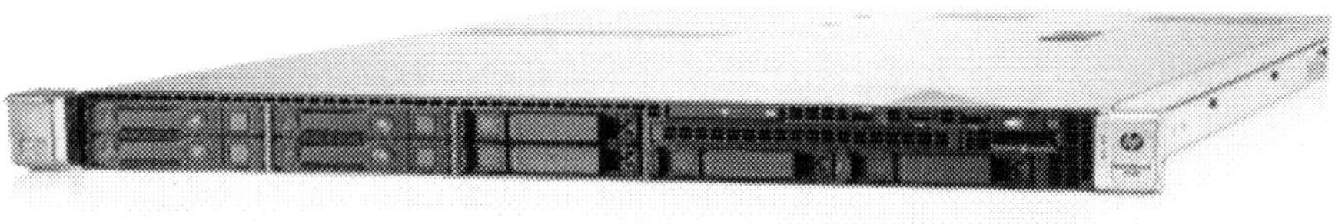

Figure 5-22 HPE StoreVirtual 4330

The HPE StoreVirtual 4330 system has these features (Figure 5-22):

- One Intel E5-2620 2.0 GHz six-core processor, 12 threads
- 32 GB RAM—DDR3 1333 MHz
- Drive types
    - Eight 450 GB 10K SFF SAS
    - Eight 900 GB 10K SFF SAS
    - Eight 1 TB 7.2K SFF MDL SAS
- Support for RAID 5, 6, and 10
- Smart Array P420i controller with 2 GB FBWC
- Four 1 Gb Ethernet iSCSI ports
- 10 Gb upgrade option
- ILO 4 remote management
- LeftHand OS 12.0

HPE StoreVirtual 4330FC

Figure 5-23 HPE StoreVirtual 4330FC

The HPE StoreVirtual 4330FC system is the same as the HPE StoreVirtual 4330 system with the following differences (Figure 5-23):

- Offers only 900 GB SAS drives (eight)
- Requires the 10 Gb option
- Includes a dual-port Emulex 8Gb FC card

Other features include the following:

- One Intel E5-2620 2.0 GHz six-core processor, 12 threads
- 32 GB RAM—DDR3 1333 MHz
- Support for RAID 5, 6, and 10
- Smart Array P420i controller with 2 GB FBWC
- Four 1 Gb Ethernet iSCSI ports
- 10 Gb required (NC552SFP)
- Dual-port Emulex 8Gb FC card (AJ763A)
- ILO 4 remote management
- LeftHand OS 12.0

Starting with LeftHand OS version 10.0 and supported models, FC can be incorporated into the HPE StoreVirtual environment.

In a traditional iSCSI configuration, the storage nodes are all connected by means of a 1 GbE or 10 GbE IP network. Any application servers or clients that need to access the storage cluster do so using the iSCSI protocol over the same IP network on which the storage nodes reside.

The IP network in this architecture is used to handle iSCSI storage traffic between the servers and the storage nodes and all internode communications. This includes Network RAID replication, data redistribution that occurs when nodes are added or removed to a cluster, and administrative traffic that, among other things, serves to ensure consistency between all the nodes in a storage cluster.

In the HPE StoreVirtual FC solution, the communication path between the nodes remains in place, but the host-to-cluster communication occurs over a separate FC network for FC hosts.

HPE StoreVirtual 4335 hybrid storage solution

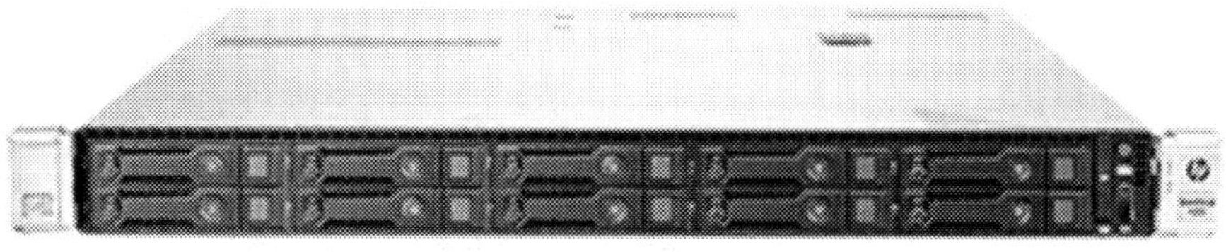

Figure 5-24 HPE StoreVirtual 4335 hybrid storage solution

The HPE StoreVirtual 4335 hybrid storage solution has these features (Figure 5-24):

- 7.5 TB capacity; 240 TB maximum
- Expansion options
    - Storage clustering capability
    - Incremental HPE StoreVirtual storage nodes in the array
- Four 1 GbE iSCSI ports
- Two 10 GbE iSCSI ports
- Drive types
    - Three 400 GB SFF SSD
    - Seven 900 GB 6G 10K SFF dual-port ENT SAS
- Support for Adaptive Optimization

The HPE StoreVirtual 4335 hybrid and HPE StoreVirtual VSA leverage the Adaptive Optimization feature of the LeftHand OS to automate the dynamic movement of frequently accessed data onto SSDs, while placing less frequently accessed data on more economical HDD capacity. Adaptive Optimization provides intelligent, automatic, and transparent performance optimization by monitoring data access patterns at a granular level based on a page size of 256 KB.

With the use of Adaptive Optimization, the HPE StoreVirtual 4335 hybrid identifies pages that contain the most frequently accessed, or "hot," data on the system. Adaptive Optimization automatically moves these pages to higher performing, Tier 0 storage comprising SSD capacity. The system maintains a data "heat map" to continually track data access frequency and accelerate the hot pages dynamically for the most efficient use of SSD capacity.

To balance cost and performance, Adaptive Optimization enforces extremely efficient use of valuable SSD capacity by keeping the SSD tier full and applying change thresholds to prevent data thrashing. In addition, the system is designed so that, regardless of which Network RAID protection level is implemented, only a single copy of data is maintained on Tier 0 storage, so the most valuable capacity real estate can be reserved for application data.

Unlike caching, Adaptive Optimization does not require capacity to be set aside, which leads to accelerated application performance and greater overall capacity efficiency. When Adaptive Optimization is enabled, the HPE StoreVirtual 4335 hybrid can maintain nearly 100% utilization of SSD capacity.

HPE StoreVirtual 4530 Storage

Figure 5-25 HPE StoreVirtual 4530 Storage

The HPE StoreVirtual 4530 Storage has these features (Figure 5-25):

- Twelve LFF dual-port SAS drives
    - 450 GB, 600 GB, 2 TB MDL SAS, 3 TB MDL SAS, or 4 TB MDL SAS
- Four 1 GbE iSCSI ports; to two 10 GbE iSCSI ports
- Storage clustering capability
- HPE StoreVirtual storage nodes; hyper-redundant clustered storage
- Management features
    - CMC
    - iLO 4 remote management
    - Baseboard management controller (BMC) Intelligent Platform Management Interface (IPMI) 2.0
- LeftHand OS 12.0 software

HPE StoreVirtual 4630 Storage

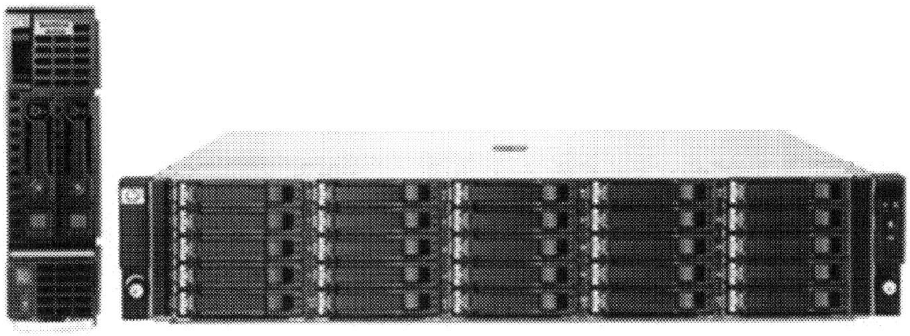

Figure 5-26 HPE StoreVirtual 4630 Storage

The HPE StoreVirtual 4630 Storage has these features (Figure 5-26):

- HPE StoreVirtual 4630sb Storage Blade
- HPE D2700 storage enclosure
- 22.5 TB starting capacity; 720 TB maximum
- 25x 900 GB 6G 10K SFF dual-port SAS
- Two 10 GbE iSCSI ports
- Management features
  - CMC
  - iLO 4 remote management
  - BMC IPMI 2.0
- LeftHand OS 12.0
- Availability features

HPE StoreVirtual 4730 Storage

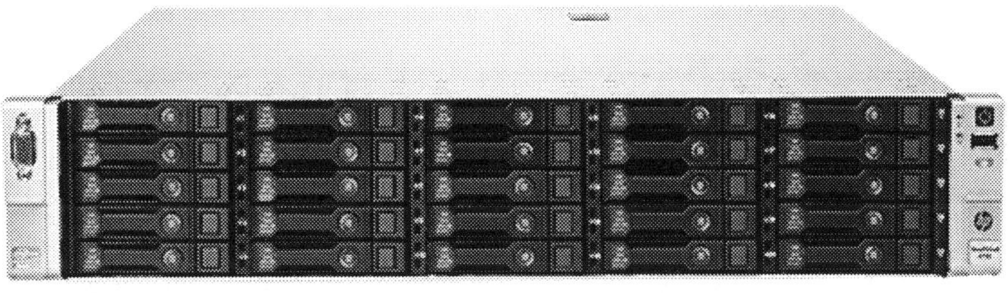

Figure 5-27 HPE StoreVirtual 4730 Storage

The HPE StoreVirtual 4730 Storage (Figure 5-27) has these features:

- 25× 600 GB or 900 GB 6G 10K SFF dual-port SAS
- Four 1 GbE iSCSI ports and two 10 GbE iSCSI ports
- Management features
    - CMC
    - iLO 4 remote management
    - BMC IPMI 2.0
- LeftHand OS 12.0
- Availability features

HPE StoreVirtual VSA

Figure 5-28 HPE StoreVirtual VSA implementation

VSAs are an example of SDS and offer a flexible, cost-effective way to provide advanced data services to virtual environments (Figure 5-28).

The HPE StoreVirtual VSA appliance runs on any x86 hardware platform and spans vSphere and Hyper-V environments. It integrates with Veeam software for enhanced backup and recovery, and it offers licensing options based on capacity levels.

HPE StoreVirtual VSA enables administrators to

- Use internal and direct-attached disks for shared storage
- Turn heterogeneous and disconnected physical drives into a single pool of logical storage

The hardware requirements of an HPE StoreVirtual VSA are

- 3 GB of RAM reserved
- One virtual CPU with 2 GHz reserved
- 5 GB to 2 TB of disk space per virtual disk, up to 10 TB total per VSA
- A dedicated gigabit virtual switch

Advanced data services can include data protection and replication, disaster recovery with multisite and remote copy functionality, centralized management, and multiple points of hypervisor and management integration.

The nature of VSAs as SDS is to be hardware agnostic and hypervisor independent, which enables you to create an agile infrastructure with data mobility across platforms, locations, and even hardware generations.

As of February 2015, HPE StoreVirtual VSA supports up to 50 TB of virtualized storage per appliance.

Unlock server capacity

HPE StoreVirtual VSA transforms the server's internal or direct-attached storage into a full-featured, shared storage array without the cost and complexity associated with dedicated storage. This VSA is optimized for VMware vSphere and Microsoft Hyper-V environments. It creates a virtual array within the application server and scales as storage needs evolve.

The ability to use internal storage within an environment greatly increases storage utilization and eliminates the costs and complexity associated with dedicated external storage. The unique scale-out architecture enables you to add storage capacity while the system is running without compromising performance. It is built-in HA and disaster recovery features ensure business continuity for the entire virtual environment.

Administrators can use the straightforward and simple management console to manage all HPE StoreVirtual products, physical and virtual, regardless of their physical location.

Transform internal storage into a shared resource

HPE StoreVirtual VSA brings all the benefits of traditional SAN storage without a physical SAN. These appliances enable you to build enterprise-level SAN functionality into server infrastructure to deliver lower cost of ownership and superior ease of management.

You can give new life to existing servers by virtualizing the environment and using HPE StoreVirtual VSA to convert disks in servers into a new tier of storage.

Using HPE StoreVirtual VSA with SSDs provides a high-performance storage solution in the environment. You can create an all-flash tier with VSAs for maximum performance or use HPE StoreVirtual Adaptive Optimization to automatically migrate to SSD only the data that needs the best performance.

Another use for the HPE StoreVirtual VSA appliance is to provide the dedicated performance required for a primary site and to deliver full-featured, shared storage and disaster recovery for remote sites.

HPE StoreVirtual VSA runs on all HPE ProLiant or third-party servers and HPE BladeSystem c-Class server blades certified with VMware vSphere or Microsoft Hyper-V.

With tens of thousands of systems deployed in production, HPE StoreVirtual VSA is the solution trusted by Fortune 500 companies and small businesses alike to deliver enterprise storage features without a SAN.

Consolidate storage

HPE StoreVirtual VSA solutions can consolidate storage VMs onto the same servers where application VMs reside, providing fault-tolerant server and storage environments.

- By removing a physical storage array from the picture, HDD capacity can be added to servers, along with HPE VSA software, to reduce up-front investment in new technologies.
- Converging server and storage technology into the same system cuts the physical footprint in half for each solution.
- You can regain 60% savings in energy costs by taking advantage of the efficiency of VSAs in servers, instead of having servers and storage arrays. These savings are linear—the more solutions based on VSA, the more can be saved.

Centralized Management Console

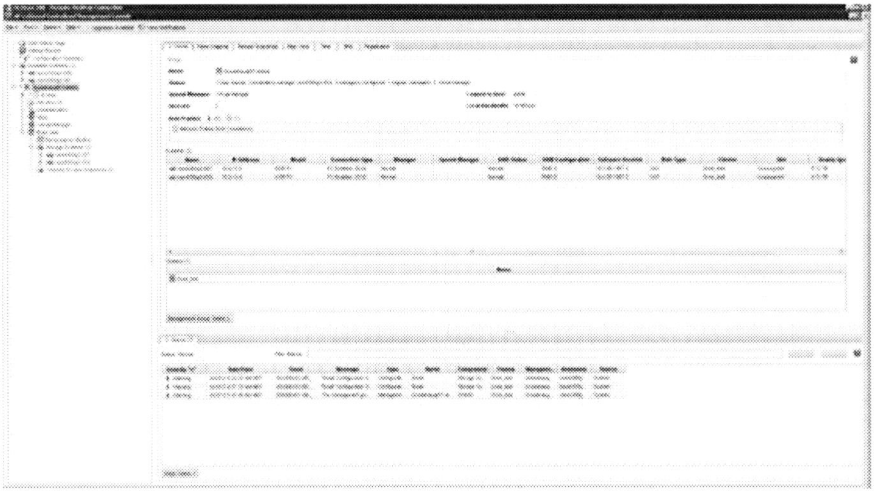

Figure 5-29 Centralized Management Console screen

The Centralized Management Console is the intuitive, GUI-based administrative interface for the LeftHand OS (Figure 5-29). It is used to configure and manage storage volumes spanning clustered storage nodes, and it provides a single graphical layout of the storage environment. The tab view provides details of the selected object.

Example—Using the Centralized Management Console

The following example shows key points for using the Centralized Management Console to create a new volume on an HPE StoreVirtual system (Figure 5-30).

- You can start the Centralized Management Console from the icon on the desktop.

Figure 5-30 HPE StoreVirtual Centralized Management Console icon

- To log in to the StoreVirtual system (Figure 5-31), it must first be selected in the left-hand pane.

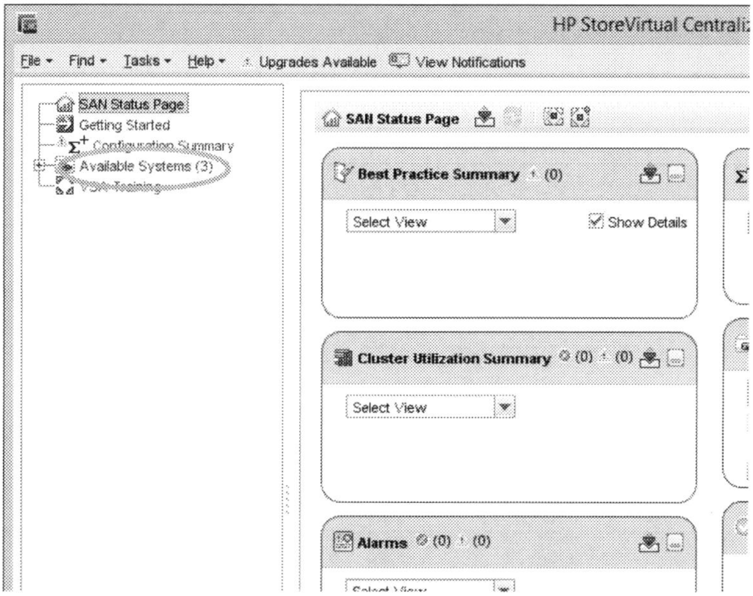

Figure 5-31 Logging in to the StoreVirtual system

- You need to enter login credentials to access the system (Figure 5-32).

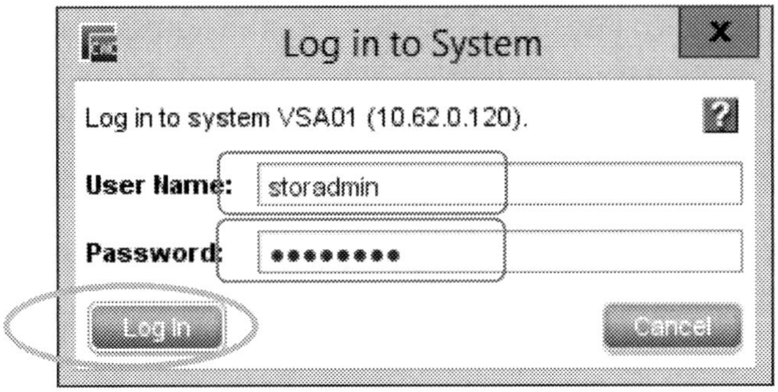

Figure 5-32 Login credentials

- Configured clusters and Storage Systems display (Figure 5-33) in the console together with their health information.

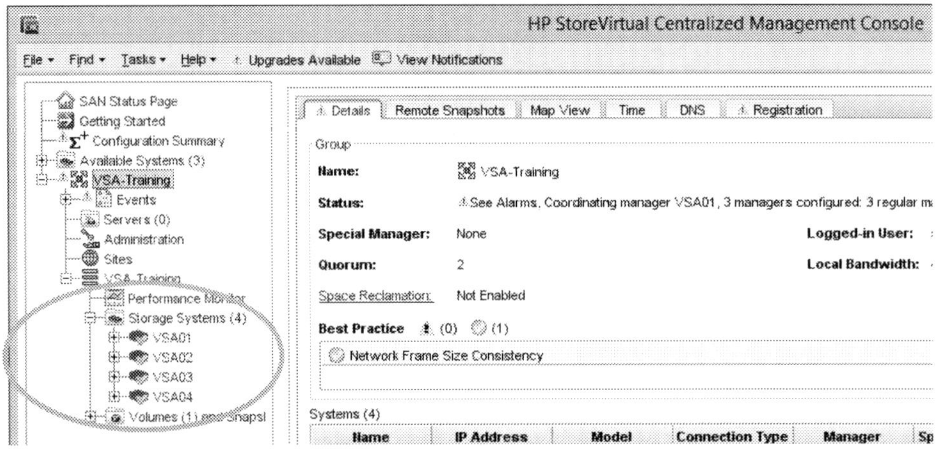

Figure 5-33 Configured clusters and Storage Systems

- You can expand the volumes section to see the configured volumes (Figure 5-34), their size, the provisioning type, and additional details.

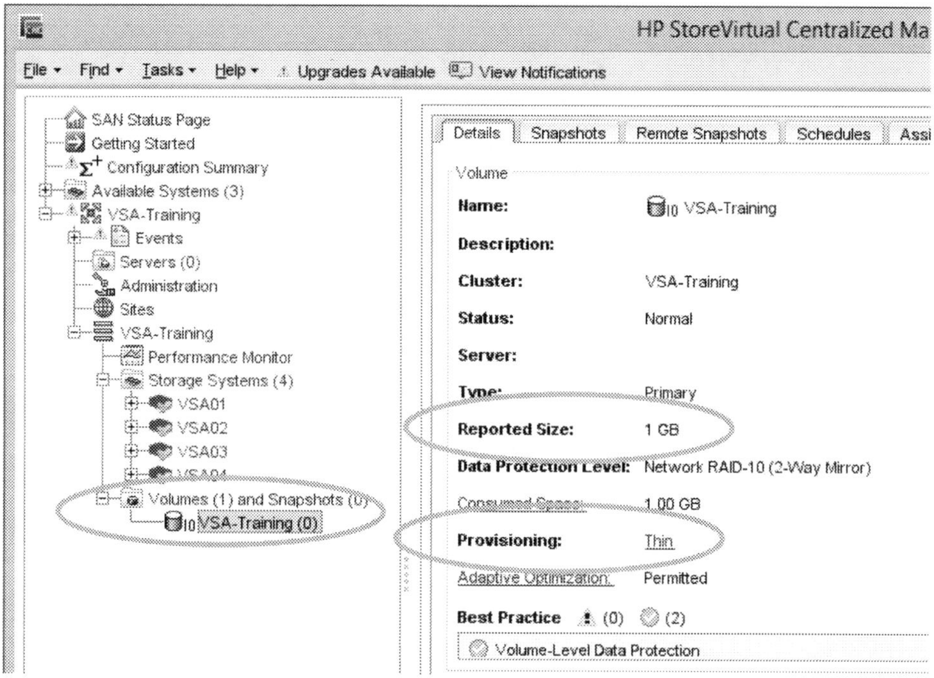

Figure 5-34 The volumes section to see the configured volumes

- The **Servers** section enables you to create a new iSCSI Initiator (Figures 5-35 and 5-36).

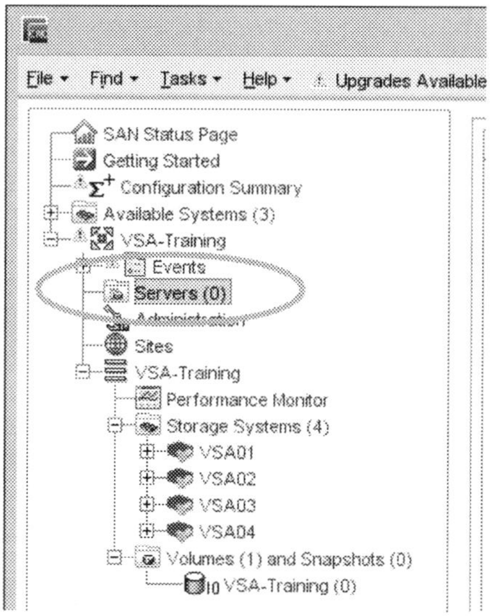

Figure 5-35 The **Servers** section

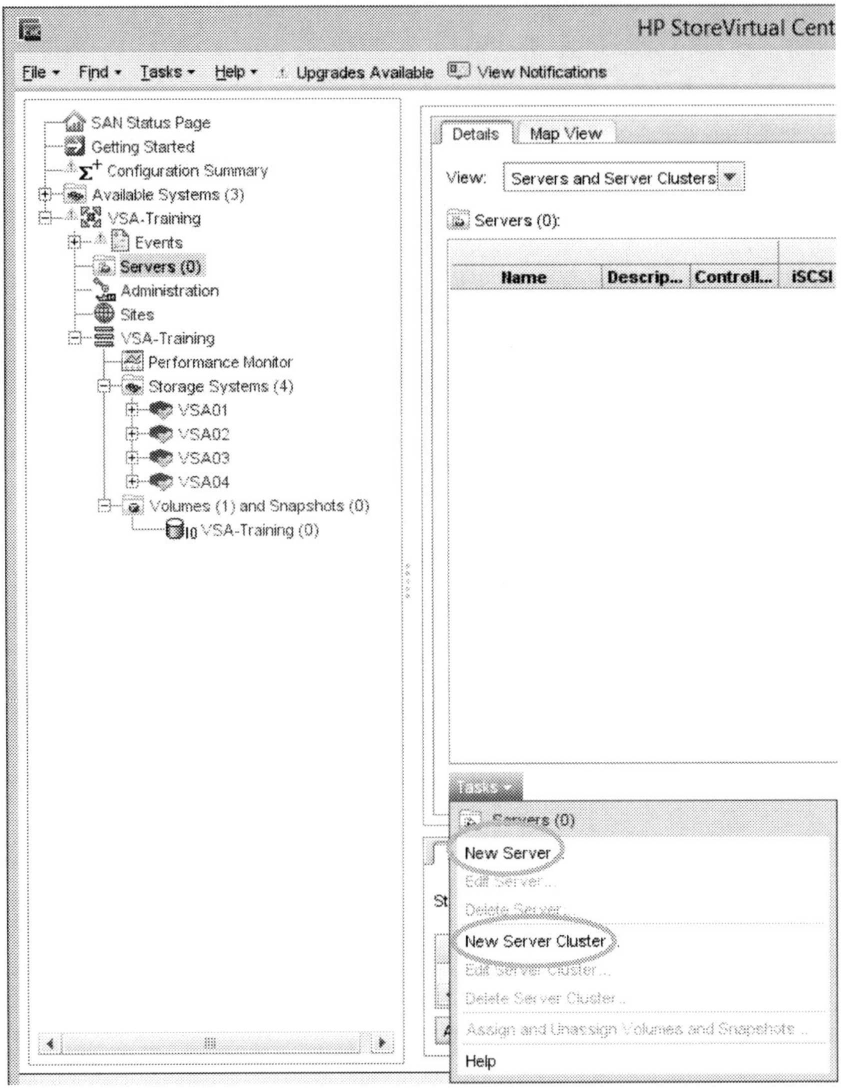

Figure 5-36 Server and New Server Cluster

- Enter information to describe the new server in the form.

Figure 5-37 IP address to use

- To attach a volume to your newly created server, go to the **Server Tasks** section and click the **Assign and Unassign Volumes and Snapshots** link (Figure 5-38).

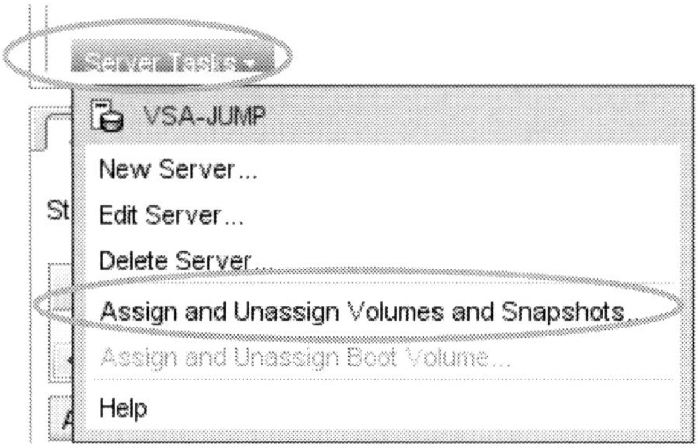

Figure 5-38 **Server Tasks**

- Then you can assign the appropriate volumes or snapshots to the server Figure 5-39 and 5-40.

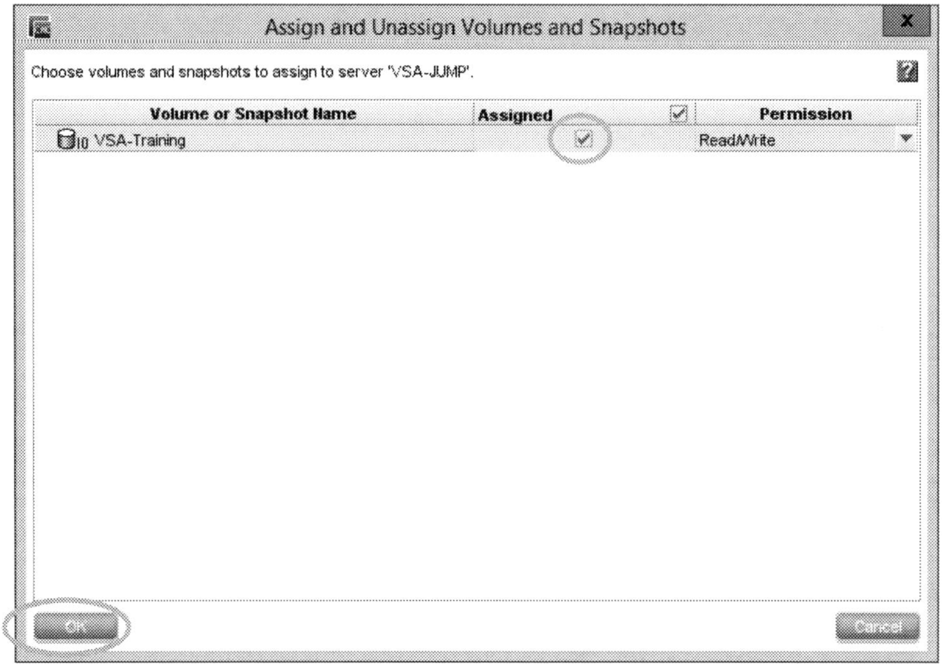

Figure 5-39 Assign and Unassign Volumes and Snapshots

Figure 5-40 Volumes and Snapshots

iSCSI Initiator

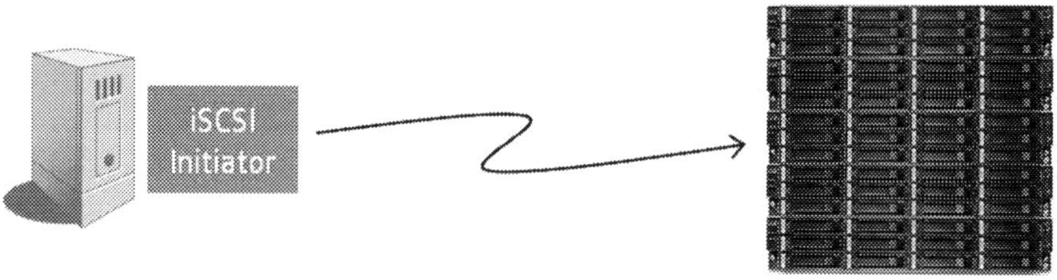

Figure 5-41 iSCSI Initiator

The iSCSI Initiator connects a server through Ethernet hardware to the volumes created on the SAN and simulates a virtual SCSI device. For Windows or another OS to communicate with the remote iSCSI SAN device, an iSCSI client driver is needed. This iSCSI Initiator appears to Windows to be a standard device driver, and as far as any application can tell, it is talking to a normal, locally connected SCSI drive.

Behind the scenes, however, the iSCSI Initiator converts the SCSI commands into iSCSI commands and relays them across the network. The remote iSCSI device receives the commands and performs the desired operation as if it were connected directly to the SCSI bus on the Windows system.

The iSCSI Initiator can use the Challenge Handshake Authentication Protocol (CHAP) to improve the security of the connection. Often, however, the client and the SAN are on the same isolated and secure LAN, so network intrusion is not a concern.

Standard best practices architecture overview

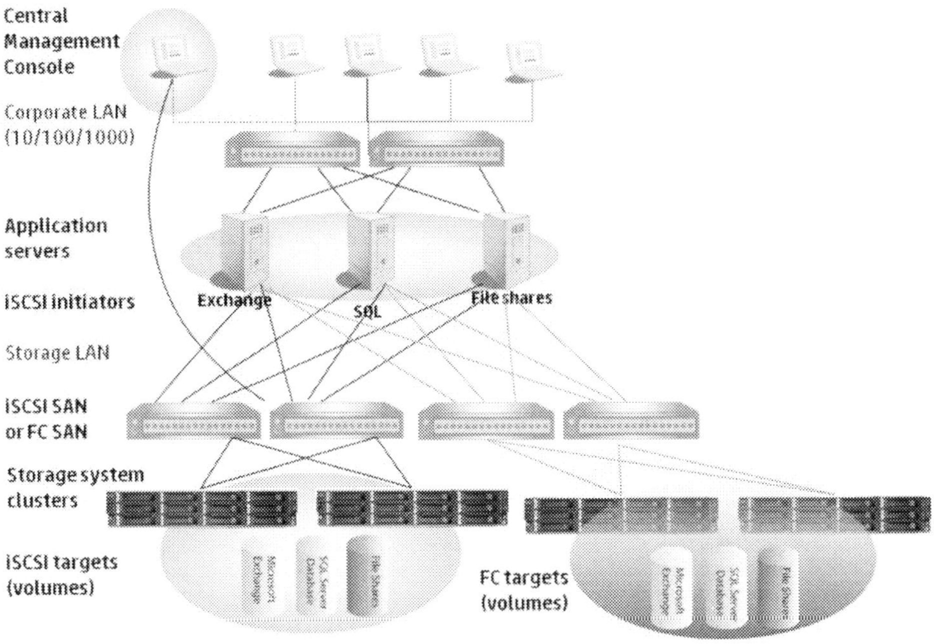

Figure 5-42 IP SAN architecture overview

There are many ways to configure a network with HPE StoreVirtual solutions, but one rule should always be followed: Do **not** run SAN traffic on a normal business LAN (Figure 5-42). There is a lot of traffic between iSCSI Initiators and the storage systems, and in particular, there is an enormous amount of traffic between the individual storage systems. This network traffic would overwhelm a business LAN. So as shown Figure 5-42, business traffic should be placed on one or more networks, and the SAN traffic should be isolated (between the iSCSI Initiators and the storage systems) on its own network. The CMC must have a route to the storage systems.

In addition to that basic rule, several options can improve the reliability of network. Multiple network paths should be provided to remove any single points of failure; for example, parallel switches (as shown in the Figure 5-42) should provide multiple independent paths from the user systems to the data servers and from the servers' initiators to the SAN cluster. Then, when the network is designed to maximize availability and reliability, the SAN cluster can be designed to do the same.

Note that in Figure 5-42 , the CMC station is located on the corporate LAN, and thus would have no access to the storage systems it must manage. Two solutions are available for this situation:

- Dedicate a management system for CMC access and add a connection from the CMC system to the storage LAN, as shown in Figure 5-42.
- Install the CMC on one of the servers, and access it using Remote Desktop or a similar tool.

What is hyper-converged?

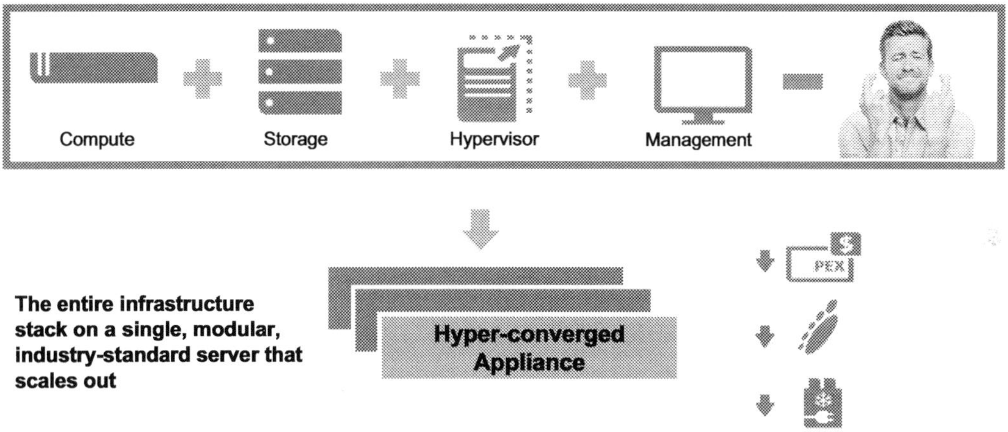

Figure 5-43a Hyper-converged system

Hyper-converged systems integrate server, storage, networking, and management in a small footprint, with simplified installation and administration (Figure 5-43a). The easy-to-use and compact systems are smaller and simpler to use than buying and configuring individual components (servers, storage, networking, and management tools) separately. These systems can be deployed and expanded quickly, and they can be managed from a single interface.

HPE ConvergedSystem 250-HC StoreVirtual system

Figure 5-43b HPE ConvergedSystem 250-HC StoreVirtual system

For customers who are looking for a simple, agile, and highly available virtualization system, the new HPE Hyper Converged 250 (HC 250) product family eliminates complex solution stacks

(Figure 5-43b). It combines a powerful HPE Apollo server platform and mature HPE StoreVirtual data services into one building block to simplify the delivery of virtualized servers and desktops. The Hyper Converged 250 family now supports VMware and Microsoft virtualization technologies and continues to provide a turnkey virtualization solution for medium-sized businesses, enterprises, and Infrastructure-as-a-Service (IaaS) providers.

Designed from the ground up for the software-defined data center, the HC 250 system enables a standardized, appliance approach to virtual server deployment, regardless of whether it is used as a primary virtualization platform in medium-sized businesses or as a dedicated resource pool for specific applications in the enterprise. Unlike other hyper-converged systems on the market, the HC 250 system can be customized and ready for virtualized workloads with a few clicks.

All hardware and software components are preinstalled and preintegrated by Hewlett Packard Enterprise. Quick customization using the HPE OneView InstantOn software enables faster time to value that is unique to the HC 250 family.

HPE StoreVirtual technology that is built into the HC 250 system matches the HA features of the hypervisor (VMware Fault Tolerance, VMware HA, or Windows Failover Cluster) from a storage perspective to provide enterprise-level availability for virtualized applications and services. The HC 250 system provides superior HA beyond a single system or rack, or even the data center. Because the storage components are vSphere Metro Storage Cluster (vMSC) certified, the HC 250 system is a great fit for stretched VMware vSphere clusters.

At a glance:

- It has four identically configured nodes with
    - Two Intel Xeon E5-2640v3 or E5-2680v3 processors
    - 128 GB, 256 GB, or 512 GB memory
    - Two 10 GbE SFP+ network connectivity
- The four-node ConvergedSystem 250-HC StoreVirtual requires a 200-240V power supply.
- The vSphere Metro Storage Cluster (vMSC) is a certified hyper-converged virtualization platform.

Two HPE Hyper Converged 250 Systems are available to support VMware or Microsoft virtualization technologies:

- Hyper Converged 250 for VMware vSphere
- Hyper Converged 250 for Microsoft Cloud Platform System (CPS) Standard

The HPE Hyper Converged 250 System is easy to install and use. Simple guided setup with HPE OneView InstantOn is available, and for daily management, the following components are preinstalled:

- HC 250 for VMware vSphere:
  - VMware vCenter Server
  - HPE OneView for vCenter Server plug-ins
- HC 250 for Microsoft Cloud Platform System Standard:
  - Microsoft System Center
  - HPE OneView for Microsoft System Center plug-ins

The preintegrated virtualization platform is powered by industry-leading virtualization software with data services from HPE StoreVirtual:

- Built with VMware vSphere 5.5 or 6.0
  - The preintegrated components include VMware vSphere and VMware vCenter.
  - Valid VMware vSphere Enterprise or higher and vCenter Standard licenses are required.
- Built with Microsoft Cloud System Platform Standard
  - The preintegrated components include Microsoft Windows Server 2012 R2 Datacenter, Hyper-V, System Center, System Center Operations Manager, and System Center VM Manager.
  - Valid Microsoft Windows Server 2012 R2 Datacenter and System Center 2012 R2 licenses are required and are available separately.

Features and benefits

The HPE Hyper Converged 250 systems offer the following features and benefits:

- **Storage clustering**—Storage Clustering enables you to consolidate multiple storage nodes into pools of storage. All available capacity and performance are aggregated and available to every volume in the cluster. As storage needs increase, the HC 250 can scale performance and capacity online.
- **Network RAID**—Network RAID stripes and protects multiple copies of data across a cluster of storage nodes, eliminating any single point of failure in the HC 250. Applications have continuous data availability in the event of a disk, controller, storage node, power, network, or site failure. You can choose the right Network RAID level for each data store containing virtualized workloads in the resource pool based on capacity and performance needs.
- **Thin Provisioning**—Thin Provisioning allocates space only as data is actually written, without requiring any preallocation of storage. This raises the overall utilization and efficiency of the HC 250, reduces costs, and ultimately increases the return on investment (ROI).
- **Application Integrated Snapshots**—Snapshots create thinly provisioned, instant point-in-time copies of data on a per-volume basis. Administrators access snapshots to recover individual files

from the volume or to roll back an entire volume. Built-in application integration provides automated quiescing for VMs on Microsoft Hyper-V or VMware vSphere. By using the Application Aware Snapshot Manager, you can extend consistency from VM-consistent to application-consistent snapshots for VMs with Microsoft Windows.

- **Remote Copy**—Remote Copy replicates snapshots between HC 250 systems and HPE StoreVirtual-based devices at primary and remote locations. The copies are thinly provisioned with no space reservation required. Remote Copy enables centralized backup and disaster recovery on a per-volume basis, and it leverages Application Integrated Snapshots for faster recovery. Remote Copy makes the HC 250 system the perfect fit for regional data centers.

  With additional HPE StoreVirtual VSA licenses, you can turn any server with a hypervisor in a central site into a replication target for the HC 250 system for enhanced business continuity. Using HPE StoreVirtual VSA as a replication target allows for custom disaster recovery sites. Failover can also be integrated with VMware Site Recovery Manager.

- **Adaptive Optimization (available with hybrid storage only)**—Auto-tiering technology for hybrid storage configurations on the HC 250 system helps to optimize the cost and performance of clusters. By using Adaptive Optimization technology to migrate data between storage tiers within individual storage systems, the HC 250 system provides a unique method for balancing performance and capacity within a storage pool and for lowering overall cost.
- **Adding storage resources on demand**—Any HPE StoreVirtual device (HPE StoreVirtual VSA or HPE StoreVirtual 4000) can be added to the same environment as an HC 250 system. This capability allows for more flexibility when additional storage capacity, or when storage with different performance characteristics, is needed. Adding another storage pool is easy with the HPE StoreVirtual Centralized Management Console and HPE OneView integration into hypervisor management—external storage resources are provisioned the same way as resources internal to the HC 250 platform. HPE StoreVirtual VSA Software or HPE StoreVirtual 4335 Storage is recommended for additional storage.
- **Integrated replication for disaster recovery**—The Hyper Converged 250 system includes integrated replication at no additional cost, which simplifies management with simple failover and failback between the primary production site and a standby disaster recovery site. In addition, the HC 250 for VMware vSphere includes the Site Recovery Adapter, which integrates Remote Copy with VMware Site Recovery Manager.
- **Changing configurations without incurring downtime**—Administrators can add capacity, increase performance, and grow and migrate volumes between HC 250 clusters on live systems with no application downtime.
- **Purchasing only what you need today**—You can avoid up-front costs and potential performance constraints by purchasing only what is needed today and growing the performance, capacity, and redundancy of your HC 250 system online as your requirements evolve. Purchasing only what you need when you need it also helps to simplify planning and budgeting processes.

- **Scaling performance and capacity simultaneously**—Each time new nodes or systems are added to an HC 250 environment, the capacity, performance, and redundancy of the entire storage solution increases.
- **Avoiding disruptive upgrades**—You can add resources to an HC 250 cluster nondisruptively as capacity and performance requirements increase. For best-in-class availability, applications remain online during maintenance events such as adding nodes or updating software or firmware.
- **Fast on-premises Azure cloud environment installation**—You can customize the HC 250 for Microsoft CPS Standard to make it fit your environment with just a few mouse clicks. Using HPE OneView InstantOn software, you can set up host names, IP addresses, networking configurations, and Microsoft Azure Services.
- **Easy vSphere environment**—You can customize the HC 250 for VMware vSphere to make it fit your environment with just a few mouse clicks. HPE OneView InstantOn software makes it easy to set up host names, IP addresses, and networking configurations.
- **Easy day-to-day management**—With HPE OneView for System Center, virtualization administrators stay within a familiar hypervisor management interface. They can provision new VMs, clone VMs, and manage storage resources in the same interface that they use every day.
- **Centralized Management Console**—Multiple data centers and sites can be managed from an all-inclusive "single pane of glass." All of the HC 250 advanced data services are managed from the CMC.
- **Business continuity**—Every HC 250 system includes integrated synchronous and asynchronous replication at no additional cost. This integration reduces costs and simplifies the management of HA, fault-tolerant, disaster recovery storage solutions.

Scalability and performance guidelines

Scalability and performance guidelines for the StoreVirtual 4130 and 4330 systems

Table 5-7 Scability and performanc guidelines for the StoreVirtual 4130 and 4330 systems

	4130 600GB SAS Storage	4330 450GB SAS Storage	4330 900GB SAS Storage	4330 1TB MDL SAS Storage	4330 FC 900GB SAS Storage
Store Virtual Scalablity	1–32 nodes	1–32 nodes	1–32 nodes	1–32 nodes	1–32 nodes
Number of drives	4	8	8	8	8
Capacity	600 GB 10k SAS	450 GB 10K SAS	900 GB 10K SAS	1TB 7.2K MDL SAS	900 GB 10K SAS
CPUs	1(6 core)	1(6 core)	1(6 core)	1(6 core)	1(6 core)
CPU Cache	15 MB	15 MB	15 MB	15 MB	15 MB
RAID Controllers	1	1	1	1	1
Memory	8 GB	32 GB	32 GB	32 GB	32 GB
Controller Cache	2 GB	2 GB	2 GB	2 GB	2 GB
Storage Ports (1 GbE)	4	4	4	4	4
Storage Ports (10 GbE)	NA	2 (optional)	2 (optional)	2 (optional)	2 (included)
Bandwidth (1 GbE)	4 Gbit/sec	4 Gbit/sec	4 Gbit/sec	4 Gbit/sec	4 Gbit/sec
Bandwidth (10 GbE)	NA	20 Gbit/sec	20 Gbit/sec	20 Gbit/sec	20 Gbit/sec

Scalability and performance guidelines for the StoreVirtual 4730, 4730FC, 4630, and 4335 systems

Table 5-8 Scability and performanc guidelines for the StoreVirtual 4730, 4730FC, 4630, and 4335 systems

	4730 600GB SAS Storage	4730 900GB SAS Storage	4730 FC 900GB SAS Storage	4630 900GB SAS Storage	4335 Hybrid Storage
Store Virtual Scalabitlity	1–32nodes	1–32nodes	1–32nodes	1–32nodes	1–32nodes
Number of drives	25	25	25	25	10
Capacity	600 GB 10K SAS	900 GB 10K SAS	900 GB 10K SAS	900 GB 10K SAS	400 GB SSD 900 GB 10K SAS
CPUs	1(6 core)	1(6 core)	1(6 core)	2(6 core ea)	1(8 core)
CPU Cache	15 MB	15 MB	15 MB	15 MB	15 MB
RAID Controllers	1	1	1	1	1
Memory	64 GB	64 GB	64 GB	64 GB	32 GB
Controller Cache	2 GB	2 GB	2 GB	2 GB	2 GB
Storage Ports (1 GbE)	4	4	4	NA	4
Storage Ports (10 GbE)	2 (included)	2 (included)	2 (included)	2 (Virtual Connect)	2 (included)
Bandwidth (1 GbE)	4 Gbit/sec	4 Gbit/sec	4 Gbit/sec	NA	4 Gbit/s
Bandwidth (10 GbE)	20 Gbit/sec	20 Gbit/sec	20 Gbit/sec	20 Gbit/sec	20 Gbit/sec

HPE 3PAR StoreServ

HPE 3PAR StoreServ provides a single product family to meet the Primary Storage needs of medium-sized enterprises experiencing rapid growth, large enterprises looking to support ITaaS, and global service providers that are building a hybrid or a private cloud.

Models:

- HPE 3PAR StoreServ 8000
- HPE 3PAR StoreServ 20000
- HPE 3PAR StoreServ File Controller
- HPE 3PAR File Persona

Differentiators:

- Form factor—Blade or rack
- Capacity
- Drive type
    - SAS or MDL SAS
    - SSD
- Performance
- Host connectivity
    - iSCSI or FC
    - CIFS/SMB, NFS, HTTP, or FTP (with StoreServ File Controller or with 3PAR File Persona)

HPE 3PAR StoreServ supports true convergence of block, file, and object access while offering the performance and flexibility that you need to accelerate new application deployments and support server virtualization, the cloud, ITaaS, and your future technology initiatives. This storage platform allows you to spend less time on management, gives you technically advanced features for less money, and eliminates the trade-offs that require you to sacrifice critical capabilities such as performance and scalability. With HPE 3PAR StoreServ, you can serve unpredictable and mixed workloads, support unstructured and structured data growth, and meet block, file, and object access storage needs from a single capacity store.

3PAR StoreServ portfolio

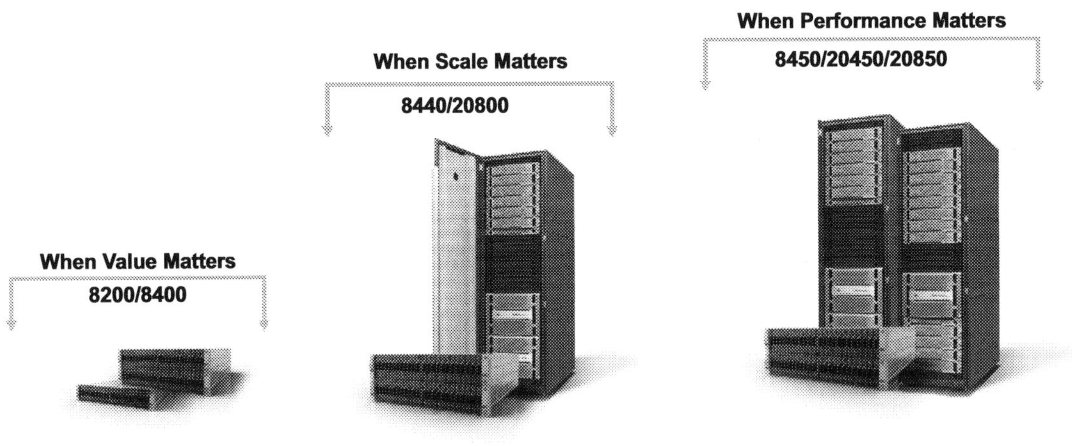

Figure 5-44 3PAR StoreServ portfolio

The all-flash HPE 3PAR StoreServ models—8450, 20450, and 20850—offer massive performance scalability, from hundreds of thousands of Input/Output Operations Per Second (IOPS) to millions of IOPS, all within a submillisecond response time (Figure 5-44). These all-flash systems uniquely offer performance without any compromises, be it the enterprise class Tier 1 data services (such as 3-data-center replication) or resiliency using their scalable (>2 nodes) architecture. These systems are well-suited for massive workload consolidation and should be the first choice for customers building their next-generation (all-flash) data centers.

The HPE 3PAR StoreServ 8200 All-Flash Starter Kit is designed to offer an incredibly affordable entry point for flash performance without limiting scalability, functionality, or data mobility.

The converged flash HPE 3PAR StoreServ models—8200, 8400, 8440, and 20800—leverage a flash-first approach that offers flash performance with the additional flexibility to add spinning media (HDDs) as a tier. In an all-flash configuration, these systems also deliver hundreds of thousands to millions of IOPS within a submillisecond response time.

- The 8440 and 20800 are hyperscalable models supporting multi-PiB raw capacity scalability.
- The 8200 and 8400 models balance performance and scalability and offer an economical starting point for customers who are looking to take advantage of the HPE 3PAR architecture.

It is important to remember that all HPE 3PAR StoreServ models, from the 8200, the 20800, to the previous generations of HPE 3PAR StoreServ, are built on a single architecture, run the exact same HPE 3PAR Operating System (HPE 3PAR OS), and offer a common set of enterprise data services. HPE 3PAR StoreServ models can natively replicate and federate among each other without the need for any external replication or a virtualization appliance.

Note

We will only cover the HPE 3PAR StoreServ 8000 series in this chapter.

Technology, business value, and customer benefits

HPE 3PAR StoreServ 8000 comes in four models:

- 8200
- 8400
- 8440—Converged Flash Array
- 8450—All Flash Array

Hardware to lead the midrange segment:

- HPE 3PAR Gen5 ASIC
- Intel Ivy Bridge Processors

- Onboard 16 Gb/s FC
- 12 Gb SAS back-end

Pricing positioning:

- Internal positioning
  - Similar pricing compared to HPE 3PAR StoreServ 7000c models
  - 15%–30% price gap between the 8440 and 20800 models, and between the 8450 and 20850 models
- Competitive positioning
  - Neutral pricing compared to the market share leader
  - All-flash penetration pricing (10%–30% less)
- Cabling
  - AOC cables for interrack connections

HPE 3PAR Gen5 ASIC

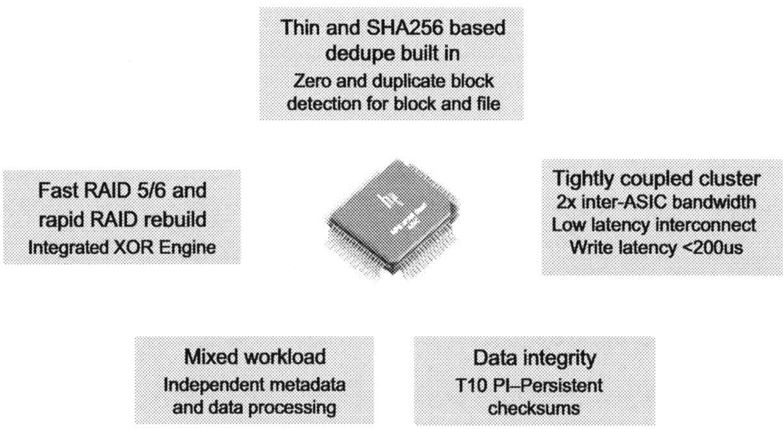

Figure 5-45 HPE 3PAR Gen5 ASIC

The HPE 3PAR StoreServ 8000 systems use the fifth and latest generation of the HPE 3PAR ASIC, the HPE 3PAR Thin Express ASIC (Figure 5-45). The HPE 3PAR Thin Express ASIC is engineered and designed for Solid-State performance.

Note
HPE 3PAR Gen5 ASIC is also used in HPE 3PAR StoreServ 20000 systems.

This ASIC enables the new HPE 3PAR 8000 and 20000 series to deliver up to 5X improvement in system bandwidth and faster exclusive or (XOR) operations. It works in parallel with the CPU, evenly processing the I/O workload across the node active-mesh, scale-out architecture to ensure lower latency and better system bandwidth.

The HPE 3PAR ASIC also features a uniquely efficient, silicon-based zero-detection and deduplication mechanism that gives HPE 3PAR StoreServ systems the power to perform inline deduplication and remove allocated but unused space with minimal impact to performance.

These ASICs also deliver mixed-workload support to alleviate performance concerns and reduce traditional array costs. Transaction- and throughput-intensive workloads run on the same storage resources without contention, thereby cutting array purchases in half. This is particularly valuable in virtual server environments, where HPE 3PAR StoreServ boosts VM density so you can reduce physical server purchases.

The Thin Express ASIC also enables Persistence Checksum that delivers T10-PI (Protection Information) for end-to-end data protection (against media and transmission errors), with no impact to applications or host operating systems.

HPE 3PAR Gen5 ASIC enhancements

Figure 5-46 HPE 3PAR Gen5 ASIC

Enhancements to the HPE 3PAR Gen5 ASIC include the following (Figure 5-46):

- Faster communication across the cluster
  - 4.0 GB/s ASIC-to-ASIC bandwidth in each direction
  - 224 GB/s total backplane bandwidth
- Higher memory bandwidth
  - Two memory channels per ASIC (two ASICs per node)
  - 42 GB/s peak data cache bandwidth per node (2.5x Gen4 ASIC)

- Modular, balanced scalability
  - Optimized scaling of node-to-node links, PCI-e links, and memory channels with a number of Gen5 ASICs on a node
  - PCIe Gen2 link at 5 Gb/s
- More parallel XOR RAID calculations
  - 14 Direct Memory Access (DMA) engines total per node (7 per ASIC)—16% more than the Gen4 ASIC (6 per ASIC)
- ASIC-enabled system features
  - Deduplication—SHA256 engine in each DMA engine
  - End-to-end Data Integrity Field (E2E DIF)—DMA engines that can generate, strip, and check T10-DIF, 8B-aligned transactions

HPE 3PAR StoreServ Software suites

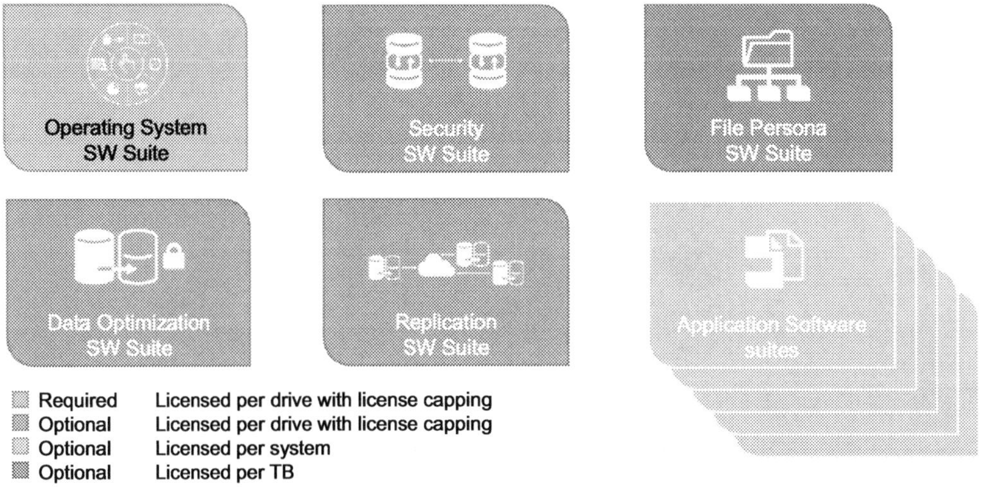

Figure 5-47 HPE 3PAR StoreServ Software suites

HPE provides an extensive selection of features for HPE 3PAR StoreServ (Figure 5-47). For convenient ordering, the 3PAR StoreServ 8000 provides the features in suites.

Two types of software licensing methods are used with the HPE 3PAR StoreServ 8000. Some software titles are licensed per system; other titles are licensed by drive.

- With system-based licensing, one license covers the whole array, independent of the configuration or capacity.
- Drive-based licensing is licensed per installed drive. A software title with drive-based licensing includes two licenses to use (LTUs): a base LTU that enables the software feature for the system, and a drive LTU that licenses the use of one drive. For each software title, you must purchase one base LTU per title per array and one drive LTU, up to a cap, for every drive that is installed in the array, independent of the drive type.
    - For the HPE 3PAR StoreServ 8200, the drive LTUs cap at 48.
    - For the HPE 3PAR StoreServ 8400, the drive LTUs cap at 168.
    - For the HPE 3PAR StoreServ 8440, the drive LTUs cap at 320.
    - For the HPE 3PAR StoreServ 8450, the drive LTUs cap at 168.

After reaching the cap, you do not need to purchase any more drive LTUs for that title.

The available software suites for HPE 3PAR StoreServ systems are as follows:

- HPE 3PAR Operating System Software Suite
- HPE 3PAR Replication Suite
- HPE 3PAR Data Optimization Software Suite
- HPE 3PAR File Persona Suite
- HPE 3PAR Security Suite
- HPE 3PAR Application Software Suite for VMware
- HPE 3PAR Application Software Suite for Hyper-V
- HPE 3PAR Application Software Suite for Exchange
- HPE 3PAR Application Software Suite for Oracle
- HPE 3PAR Application Software Suite for SQL
- HPE StoreOnce Recovery Manager Central

HPE 3PAR Operating System Software Suite

This foundational software suite is required for all new HPE 3PAR StoreServ systems. It gives you everything you need to get your HPE 3PAR StoreServ systems up and running quickly and efficiently.

This software suite is powered by the HPE 3PAR Gen5 ASIC and HPE 3PAR Thin Technologies, including HPE 3PAR Thin Provisioning, HPE 3PAR Thin Persistence, HPE 3PAR Thin Conversion, and HPE 3PAR Thin Deduplication.

Performance acceleration is assured by HPE 3PAR Adaptive Flash Cache, which reduces application response times. Network simplification and security are covered with VLAN tagging. Simplified management is enabled by the HPE 3PAR Operating System, HPE 3PAR StoreServ Management Console, HPE 3PAR Host Explorer, and HPE SmartStart software. HPE 3PAR System Reporter and HPE 3PARInfo software are designed to track performance and capacity utilization trends for multiple HPE 3PAR StoreServ Systems.

Other highlights of this suite include HPE 3PAR Full Copy, autonomic rebalancing capabilities that help you make the best use of future capacity expansions, and support for standard multipathing software for HA in clustered environments. A one-year license for HPE 3PAR Online Import is included to enable migration from HPE EVA, EMC Storage14, or HDS Storage15 systems.

Two important features for HPE 3PAR StoreServ 8000 and 20000 are Persistent Ports and Persistent Checksum.

- HPE 3PAR Persistent Ports is a Tier 1 resiliency feature that allows for nondisruptive online software upgrades on HPE 3PAR StoreServ systems without relying on multipathing software. This feature brings increased Tier 1 resiliency across the StoreServ product line, including the midrange platform, by ensuring that host paths remain online throughout the software upgrade process. The use of standard N_Port ID Virtualization (NPIV) technology, as defined by the Technical Committee T11 in the FC-Link Services (FC-LS) specification, enables a single host-facing StoreServ port to assume the identity of additional predesignated partner ports, thus enabling transparent switchover of host path connections. Persistent Port technology is also designed to work with iSCSI and FCoE connectivity.

  Starting from HPE 3PAR Operating System 3.1.3, the Persistent Ports feature also autonomically fails over a front-end controller node port that experiences a physical connection loss from a cable pull or a failed switch. This unique capability is currently available for FC connectivity only.

- HPE 3PAR Persistent Checksum is a data integrity feature built in to the HPE 3PAR Operating System that provides data integrity to ensure end-to-end data protection with a T10 DIF. Persistent Checksum addresses media and transmission errors that can be caused by any component in the I/O stack, and it provides additional protection above FC, iSCSI, and cyclic redundancy check (CRC) transmissions. With supported HBAs on hosts, T10-DIF tags are added and verified on the HPE 3PAR StoreServ system. When host HBAs are not supported, T10-DIF tags are added and verified on the array target ports and back-end HBAs before data is written to the array drives. This technology is only available on systems with the Gen5 ASIC.

HPE 3PAR Replication Suite

This suite bundles HPE 3PAR Virtual Copy with HPE 3PAR Remote Copy software, which are also sold separately, for all HPE 3PAR StoreServ models.

- HPE 3PAR Virtual Copy software protects and shares data affordably, with rapid recovery using reservationless, nonduplicative, and copy-on-write snapshots.
- HPE 3PAR Remote Copy offers simple and cost-effective data protection for efficient multitenant disaster recovery.

Also included in this bundle are Peer Persistence, which enables automatic, transparent failover over metropolitan distances using the Remote Copy Synchronous mode, and HPE 3PAR Cluster Extension Software, which enables automatic failover across data centers using the Remote Copy Asynchronous mode.

HPE 3PAR Data Optimization Software Suite

This software bundle combines HPE 3PAR Dynamic Optimization, HPE 3PAR Adaptive Optimization, HPE 3PAR Priority Optimization, and HPE 3PAR Peer Motion software.

- HPE 3PAR Dynamic Optimization delivers the required service levels for the lowest possible cost throughout the data life cycle.
- HPE 3PAR Adaptive Optimization improves storage utilization by enabling cost-optimized storage tiering.
- HPE 3PAR Priority Optimization assures service levels with Quality of Service (QoS) controls for mission-critical applications.
- HPE 3PAR Peer Motion enables load balancing at will, which allows the movement of data and workloads between arrays to be initiated without impacting applications, users, or services.

The four software titles bundled in this suite are also sold separately for all HPE 3PAR StoreServ models. Depending on purchase date, the titles included in this suite might vary. Check the HPE 3PAR Software QuickSpecs for complete details.

HPE 3PAR File Persona Suite

This software suite enables a rich set of file protocols and an Object Access API to extend the spectrum of primary storage workloads that are natively addressed by HPE 3PAR StoreServ systems with Converged Controllers. With this solution, the architectural benefits of HPE 3PAR StoreServ can be extended to the following use cases: enterprise file sync and share; home directory consolidation; group, departmental, and corporate shares; and custom cloud applications.

HPE 3PAR Security Suite

This software suite bundles HPE 3PAR Virtual Domains and HPE 3PAR Virtual Lock software. With this suite, you can segregate access and deliver robust storage services for different applications and user groups, with additional security attached to the retention of storage volumes.

HPE 3PAR Application Software Suite for VMware

This suite includes everything you need to make your VMware environment more agile and efficient:

- HPE 3PAR Recovery Manager for VMware
- Host Explorer for VMware
- vSphere APIs for Storage Awareness (VASA) support
- Three essential plug-ins:
    - VMware Site Replication Manager (SRM) Adapter
    - HPE 3PAR VAAI plug-in, which enables HPE 3PAR StoreServ Storage support for advanced storage operations that come with VMware's vSphere Storage APIs for Array Integration (VAAI)
    - HPE 3PAR Management Plug-in for VMware View

HPE 3PAR Application Software Suite for Hyper-V

You can protect your Microsoft Hyper-V environment with the HPE 3PAR Recovery Manager for Microsoft Hyper-V and the HPE 3PAR VSS Provider software included in this software bundle.

HPE 3PAR Application Software Suite for Exchange

This bundle gives you the essentials for using Microsoft Exchange, including HPE 3PAR Recovery Manager for Exchange and the HPE 3PAR VSS Provider software.

HPE 3PAR Application Software Suite for Oracle

This suite includes everything you need to protect Oracle databases, including HPE 3PAR Recovery Manager for Oracle and Oracle Space Reclamation capabilities.

HPE 3PAR Application Software Suite for SQL

You can protect Microsoft SQL databases with HPE 3PAR Recovery Manager for Microsoft SQL and the HPE 3PAR VSS Provider software.

HPE StoreOnce Recovery Manager Central

To combine the performance of snapshots with the protection of backups, this software integrates HPE 3PAR StoreServ with HPE StoreOnce Backup. It provides a converged availability and flat backup service that augments traditional backup processes. With this automated, nonintrusive software, the simplicity and performance of local and remote snapshots can be combined with the reliability and cost-effective retention of deduplicated backups.

3PAR Data at Rest Encryption

Figure 5-48 Securing data at rest

Data is perhaps the most important asset for organizations in today's digital age (Figure 5-48). Companies are looking to protect data against theft and misuse while meeting compliance requirements.

The HPE 3PAR StoreServ systems comply with the standards set forth by the US National Institute of Standards and Technology (NIST) and US Federal Information Processing Standard (FIPS) 140-2.

These systems feature Data at Rest encryption that helps protect valuable data through self-encrypting drive (SED) technology. SED drives are HDDs and SSDs with a circuit (ASIC) built into the drive's controller chipset that automatically encrypts and decrypts all data being written to and read from the media. HPE 3PAR StoreServ supports Full Disk Encryption (FDE) based on the Advanced Encryption Standard (AES) 256 industry standard. The encryption is part of a hash code that is stored internally on physical media. All encryption and decryption is handled at the drive level and needs no other external mechanism. Authentication keys are set by the user and can be changed at any time.

The Local Key Manager (LKM) included with the HPE 3PAR StoreServ encryption license is used to manage all drive encryption keys within the array and provides a simple management interface. In the event of a drive failure or the theft of a drive, a proper key sequence must be entered to gain access to the data stored within the drive.

When an SED drive is no longer powered on, the drive goes into a locked state and requires an authentication key to unlock the drive when power is restored. Without the key, access to the data on the SED is not possible.

For enhanced encryption support on the HPE 3PAR StoreServ systems, HPE offers FIPS 140-2 compliant SED drives that enable you to use an external Enterprise Secure Key Manager (ESKM). ESKM is deployed whenever customers use encrypted storage or communication methods to protect their sensitive information. Herein, they store and serve keys to unlock the data stored on FIPS 140-2 compliant drives within the HPE 3PAR StoreServ systems with strong access controls and security.

FIPS 140-2 compliance provides you with the satisfaction of knowing that your data is securely stored on the HPE 3PAR StoreServ array. Key Management on the array, with either LKM or ESKM coupled with FIPS drives, offers a safe environment in which to securely store your data.

Adaptive Flash Cache and Express Writes

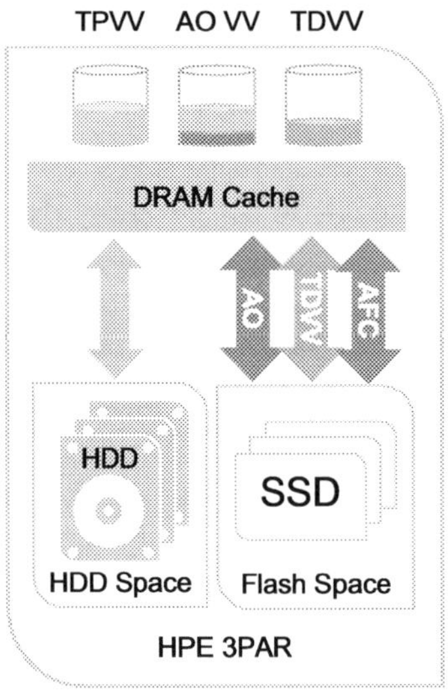

Figure 5-49 Adaptive Flash Cache diagram

Adaptive Flash Cache leverages HPE 3PAR first-in-class virtualization technologies to extend read cache (Figure 5-49). This functionality enables you to dedicate a portion of the SSD capacity as an augmentation of the HPE 3PAR array primary cache to reduce the application response time for read-intensive I/O workloads. It also improves performance when periodically reading data from a slower tier of HDDs.

Flash cache works by caching the "hot" data:

- Data is fetched into DRAM from HDDs to fulfill host requests.
- When data is removed from DRAM, it is copied (admitted) into flash cache.

- Future host reads are redirected to and served from flash cache.
- Data in flash cache is assigned different "temperatures" so that hot data remains in flash cache while cold data is removed (evicted) in the background.

HPE 3PAR Adaptive Flash Cache is included as part of the HPE 3PAR Operating System Suite 3.2.1 and later, and it is supported on all HPE 3PAR StoreServ arrays that have a blend of SSDs and HDDs.

Adaptive Flash Cache can increase the overall random read IOPS that an array can deliver by "unloading" the back end of the array, potentially resulting in increased write throughput on the array. Using SSDs as Level-2 read cache to hold random read data that has aged out of DRAM cache is a cost-effective way of keeping more random read data on fast media to improve overall system performance.

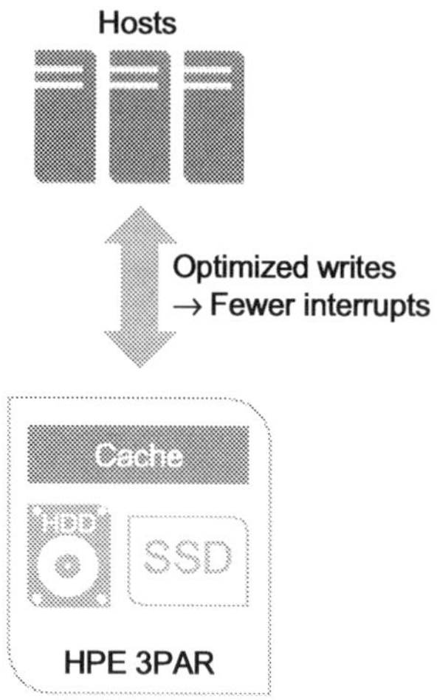

Figure 5-50 Express writes diagram

Express Writes is a 3PAR OS protocol optimization that delivers lower CPU interrupts per I/O transaction (Figure 5-50). This optimization results in higher IOPS and reduced latency for writes, and it does not require any changes on the initiator side. All supported HBAs and hosts will benefit from this optimization.

This feature is available on all HPE 3PAR StoreServ arrays at no additional cost and with no additional hardware. It requires HPE 3PAR OS 3.2.1 or later.

HPE 3PAR Express Writes represents a series of optimizations aimed at improving host write latency. This improvement is achieved through the HPE-3PAR OS by optimizing SCSI transactions and thus improving the number of interrupts per transaction. The result is improved array CPU utilization and lower host write latency.

Express Writes technology enables the HPE 3PAR StoreServ array to achieve synchronous write latency below 200 microseconds. Depending on the workload, hosts might see an overall decrease in write latency of up to 20%, which could result in an increase in IOPS and throughput.

HPE 3PAR StoreServ 8000 hardware building blocks

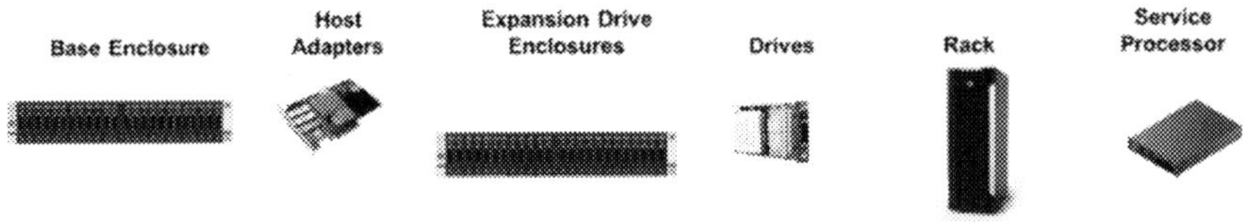

Figure 5-51 HPE 3PAR StoreServ 8000 hardware building blocks

HPE 3PAR StoreServ 8000 solutions consist of the following building blocks (Figure 5-51):

- Base enclosure
- Host adapters
- Expansion drive enclosures
- Drives
- Rack
- Service Processor

StoreServ 8400 node pair

Figure 5-52 StoreServ 8400 rear view

The following components and interfaces can be found on the rear side of the StoreServ 8400 array (Figures 5-52 and 5-53):

- Power-cooling module and battery pack
- 1 GbE MGMT+RC/FP
- 16 Gb FC ports
- 12 Gb SAS ports
- Ejector handle
- Node status LEDs
- Cluster link ports
- HBA slot
- MFG console port

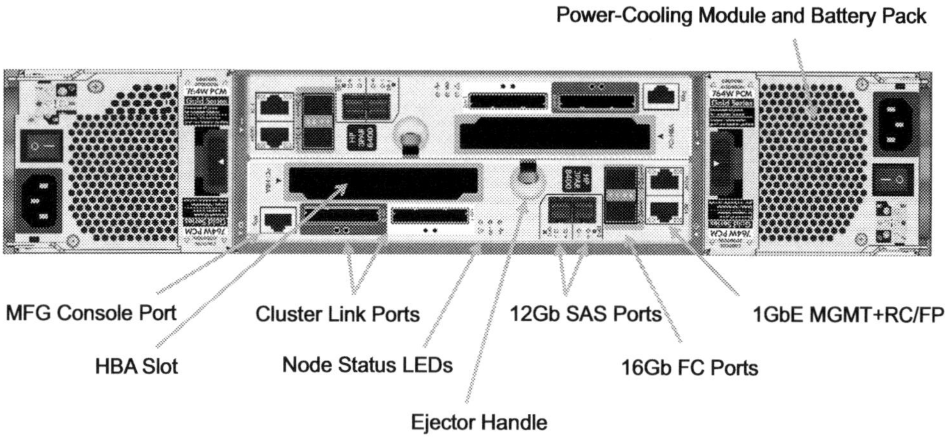

Figure 5-53 StoreServ 8400 components

HPE 3PAR StoreServ 8000 12 Gb SAS drive enclosures

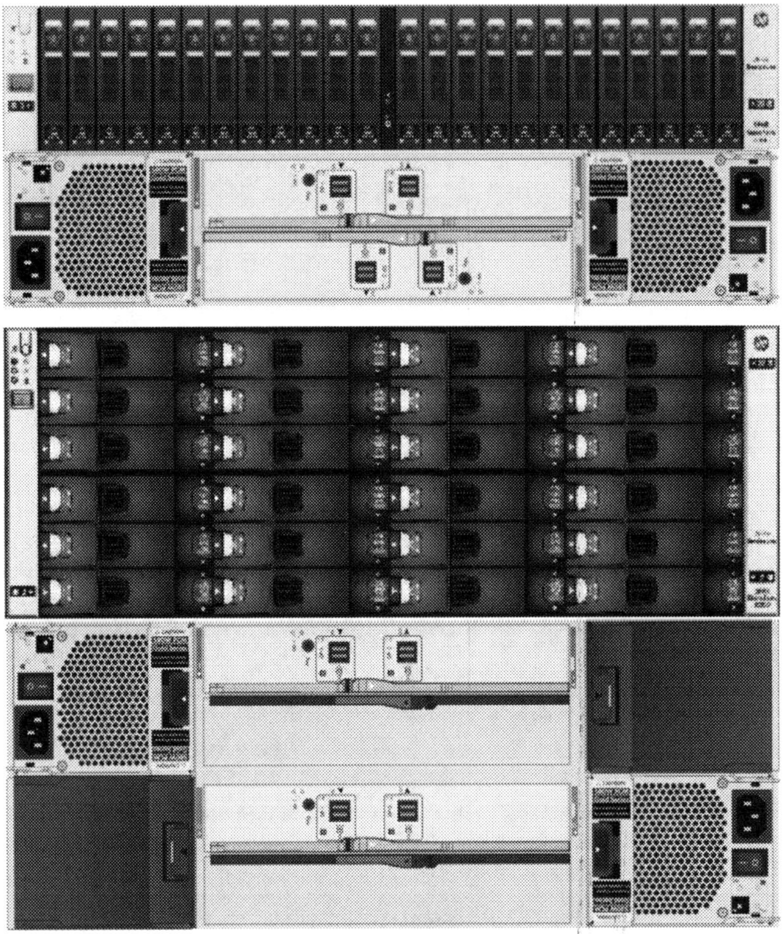

Figure 5-54 HPE 3PAR StoreServ 8000 12 Gb SAS drive enclosures

Another key element of the HPE 3PAR StoreServ system is the drive enclosure, or drive chassis, which serves as the capacity building block within the system (Figure 5-54).

Two HPE 3PAR StoreServ enclosure models are available:

- 2U 24-drive 2.5 in SFF 12 Gb
- 4U 24-drive 3.5 in LFF 12 Gb

The drive enclosures use the same drive carriers as the StoreServ 7000 systems, but they no longer use the "M67xx" model names; the name 3PAR StoreServ 8000 Drive Enclosure is used now.

No StoreServ 7000 6 Gb expansion enclosures are supported on StoreServ 8000 systems.

Full-mesh cluster interconnect

Figure 5-55 HPE 3PAR StoreServ 8000 full-mesh cluster interconnect

Backplane interconnects within servers have evolved dramatically over the years (Figure 5-55). Most, if not all, server and storage array architectures have traditionally employed simple bus-based backplanes for high-speed processor, memory, and I/O communication. Parallel to the growth of symmetric multiprocessing (SMP)-based servers, significant investments were also made to switch architectures, which have been applied to one or two enterprise storage arrays.

The move from buses to switches was intended to address latency issues across the growing number of devices on the backplane (more processors, larger memory, and I/O systems). Third-generation full-mesh interconnects first appeared in the late 1990s in enterprise servers.

The HPE 3PAR StoreServ full-mesh backplane is a passive circuit board that contains slots for up to four or eight controller nodes, depending on the model. Each controller node slot is connected to every other controller node slot by a high-speed link (4 GB/s in each direction, or 8 GB/s total), forming a full-mesh interconnect between all controller nodes in the cluster—something that HPE refers to as a "Mesh-Active" design. These interconnects deliver low-latency, high-bandwidth communication and data movement between controller nodes through dedicated point-to-point links and a low overhead protocol that features rapid internode messaging and acknowledgment. It is important to note that, while the value of these interconnects is high, the cost of providing them is relatively low. In addition, a completely separate full-mesh network of serial links provides a redundant low-speed channel of communication for exchanging control information between the nodes.

The HPE 3PAR StoreServ 20000 systems feature an eight-node-capable backplane that supports two to eight controller nodes. HPE 3PAR StoreServ 8000 systems feature either a dual-node- or a quad-node-capable system that is essentially an equivalent of what was used in earlier enterprise-class arrays that offered the same high-speed links between nodes.

Active/Active versus Mesh-Active

Most traditional array architectures fall into one of two categories: monolithic or modular. In a monolithic architecture, being able to start with smaller, more affordable configurations (scaling down) presents challenges. Active processing elements must be implemented redundantly, and they are also segmented and dedicated to distinct functions such as host management, caching, and RAID or drive management. For example, the smallest monolithic system might have a minimum of six processing elements (one for each of three functions, which are then doubled for redundancy of each function). In this design, with its emphasis on optimized internal interconnectivity, users gain the Active/Active processing advantages of a central global cache. For example, LUNs can be coherently exported from multiple ports. However, these architectures typically involve higher costs relative to modular architectures.

In traditional modular architectures, users can start with smaller and more cost-efficient configurations. The number of processing elements is reduced to just two because each element is multifunctional, handling host, cache, and drive management processes.

One trade-off for this cost-effectiveness is the cost or complexity of scalability. Because most traditional modular designs only support two nodes, scalability can only be realized by replacing the existing nodes with more powerful nodes or by purchasing and managing more arrays.

Another trade-off is that although dual-node modular architectures provide failover capabilities, they typically do not offer truly Active/Active implementations, where individual LUNs can be simultaneously and coherently processed by both the controllers.

The HPE 3PAR StoreServ architecture was designed to provide cost-effective, single-system scalability through a cache-coherent, multinode, clustered implementation. This architecture begins with a multifunction node design and, like a modular array, requires just two initial controller nodes for redundancy. However, unlike traditional modular arrays, enhanced direct interconnects are provided between the controllers to facilitate Mesh-Active processing. Unlike legacy Active/Active controller architectures, here each LUN (or volume) is active on only a single controller, the Mesh-Active design enables each LUN to be active on every controller in the system, thus forming a mesh. This design delivers robust, load-balanced performance and greater headroom for cost-effective scalability, overcoming the trade-offs typically associated with modular and monolithic storage arrays.

System-wide striping

Through a Mesh-Active design and system-wide striping, the HPE 3PAR StoreServ architecture can provide the best of traditional modular and monolithic architectures in addition to massive load balancing. The HPE 3PAR StoreServ Mesh-Active design allows all volumes to be active on all controllers, and it promotes system-wide striping that autonomically provisions and seamlessly stripes volumes across all system resources to deliver high, predictable levels of performance. System-wide striping of data provides high and predictable levels of service for all workload types through the massively parallel and fine-grained striping of data across all internal resources such as disks, ports, loops, cache, and processors. As a result, as the use of the system grows—or in the event of a component failure—service conditions remain high and predictable.

Unlike application-centric approaches to storage, HPE 3PAR StoreServ provides autonomic rebalancing that enables the system to evenly balance and use all available physical resources. This is particularly important with hardware upgrades because the existing data should be rebalanced and striped across new resources. On HPE 3PAR StoreServ systems, this rebalancing is done without service disruption or preplanning.

For flash-based media, fine-grained virtualization combined with system-wide striping drives uniform I/O patterns by spreading wear evenly across the entire system. Should there be a media failure, system-wide sparing also helps to guard against performance degradation by enabling a many-to-many rebuild, resulting in faster rebuilds. Because HPE 3PAR StoreServ autonomically manages this system-wide load balancing, no extra time or complexity is required to create or maintain a more efficiently configured system.

Controller node architecture

An important element of the HPE 3PAR StoreServ architecture is the controller node, and it is a powerful data movement engine that is designed for mixed workloads. Depending on the model, a single system can be modularly configured as a cluster of two to eight controller nodes. This modular approach provides flexibility, a cost-effective entry footprint, and affordable upgrade paths for increasing performance, capacity, connectivity, and availability as needs change.

In addition, the minimum dual-controller configuration means that the system can withstand an entire controller node failure without impacting data availability. Controller nodes can be added to the cluster in pairs nondisruptively, and each node is completely hot-pluggable to enable online serviceability.

HPE 3PAR StoreServ 8000 disk drives

Always consult the HPE Product Bulletin for the current list of supported disk drives.

Table 5-9 HPE 3PAR StoreServ 8000 disk drives

HDD SFF	HDD LFF	SSD
HPE 3PAR 8000 2TB SAS 7.2K SFF Drive	HPE 3PAR 8000 2TB SAS 7.2K LFF Drive	SFF
	HPE 3PAR 8000 4TB SAS 7.2K LFF Drive	HPE 3PAR 8000 480GB SAS MLC SFF SSD
HPE 3PAR 8000 600GB SAS 10K SFF Drive	HPE 3PAR 8000 6TB SAS 7.2K LFF Drive	HPE 3PAR 8000 920GB SAS MLC SFF FE SSD
HPE 3PAR 8000 1.2TB SAS 10K SFF Drive	HPE 3PAR 8000 6TB SAS 7.2K LFF FE HDD	
HPE 3PAR 8000 1.8TB SAS 10K SFF Drive		HPE 3PAR 8000 480GB SAS cMLC SFF SSD
HPE 3PAR 8000 1.2TB SAS 10K SFF FE Drive		HPE 3PAR 8000 1.92TB SAS cMLC SFF SSD
		HPE 3PAR 8000 3.84TB SAS cMLC SFF SSD
HPE 3PAR 8000 300GB SAS 15K SFF Drive		HPE 3PAR 8000 1.92TB SAS cMLC SFF FE SSD
HPE 3PAR 8000 600GB SAS 15K SFF Drive		
HPE 3PAR 8000 600GB SAS 15K SFF FE Drive		LFF
		HPE 3PAR 8000 480GB SAS cMLC LFF SSD

StoreServ 8000 models and system features

StoreServ 8000 models:

- 8200
- 8400
- 8440
- 8450 All Flash Array (AFA)

Table 5-10 StoreServ 8000 models and system features

Model	8200	8400	8440	8450 AFA
CPU	6C Ivy Bridge	6C Ivy Bridge	10C Ivy Bridge	10C Ivy Bridge
Total system memory Control/data	2N: 32 GB/32 GB 1333 MHz	2N: 32 GB/32 GB 4N: 64 GB/64 GB 1333 MHz	2N: 128 GB/64 GB 4N: 256 GB/128 GB 1333 MHz	2N: 128 GB/64 GB 4N: 256 GB/128 GB 1333 MHz
Drives supported	Mixed HDD and SSD	Mixed HDD and SSD	Mixed HDD and SSD	SSD only
Maximum drives 2N/4N	HDD: 240 SSD: 120	HDD: 288/576 SSD: 120/240	HDD: 480/960 SSD: 240/480	SSD: 240/480
Raw capacity	750 TiB	1200/2400 TiB	1500/3000 TiB	922/1843 TiB
Host I/O	16 Gb FC 10 Gb iSCSI/FCoE 1 GbE/10 Gb Ethernet File Persona			

HPE 3PAR StoreServ 8000 key performance metrics

The HPE 3PAR StoreServ 8000 systems can maintain high and predictable performance in multitenant environments because of architectural innovations that use all available array hardware resources at all times. This design helps to eliminate resource contention, support mixed workloads, and enhance caching algorithms to accelerate performance and reduce latency.

Load balancing

Purpose-built for the enterprise and for virtual and cloud data centers, the HPE 3PAR StoreServ architecture is unlike legacy controller architectures in that the Mesh-Active system design allows each volume to be active on any controller in the system. High-speed, full-mesh interconnections join multiple controller nodes to form a cache-coherent Active/Active cluster. As a result, the system delivers symmetrical load balancing and utilization of all controllers, with seamless performance scalability by adding more controllers and disk drives to the system.

Mixed-workload support

The HPE 3PAR StoreServ controller node design separates the processing of SCSI control commands from data movement. This allows transaction-intensive and throughput-intensive workloads to run on the same storage resources without contention, thereby supporting massive consolidation and multitenancy. This means that, for example, the system can concurrently handle an OLTP application and an extremely bandwidth-consuming data warehousing application.

This capability is made possible by the HPE 3PAR ASIC, which offloads data processing from the control processor, where metadata is processed. By pathing and processing data and metadata separately, transaction-intensive workloads are not held back behind throughput-intensive workloads. Compared to the architectures of traditional storage vendors—including many of today's all-flash arrays—the HPE 3PAR StoreServ platform delivers excellent performance consistently, even in mixed-workload scenarios.

Example performance metrics are shown in Table 5-11. Always verify performance metrics in NinjaSTARS.

Table 5-11 **Example performance metrics**

Platform	Workload	Performance metric
8440/8450	8K 100% Random Read, RAID 1, CPVV, Node Local	>1M IOPS at <1ms
8440/8450	8K 100% Random Write, RAID 1, CPVV, Distributed	>400K IOPS at <1ms
8440/8450	256K 100% Seq. Read, RAID 1, CPVV, Node Local	24 GB/s
8400	8K 100% Random Read, RAID 1, CPVV, Node Local	650K IOPS at <1ms
8400	256K 100% Seq. Read, RAID 1, CPVV, Node Local	20 GB/s

Performance metrics were measured on a configuration based on

- 4 nodes per array
- 48 MLC SSDs per node pair
- Built-in SAS ports
- 1 FC HBA per node
- All drive enclosures and hosts balanced across all ports

HPE 3PAR StoreServ Software object scalability

The two kinds of virtual volumes (VVs) are base volumes and snapshot volumes.

- Base volume—This type can be considered to be the "original" VV. It is either a fully provisioned VV, a thinly provisioned VV, or a thinly provisioned deduplicated VV. In other words, it directly maps all the user-visible data.
- Snapshot volume—This type is created using HPE 3PAR Virtual Copy software. When a snapshot is first created, all of its data is mapped indirectly to the data in the parent volume. When a block is written to the parent, the original block is copied from the parent to the snapshot data space, and the snapshot points to this data space instead. Similarly, when a block is written in the snapshot, the data is written in the snapshot data space, and the snapshot points to this data space. These snapshots are copy-on-write (COW) snapshots.

VVs have three types of space:

- **User space**—Represents the user-visible size of the VV (the size of the SCSI LUN seen by a host) and contains the data of the base VV.
- **Snapshot data space**—Is used to store modified data associated with snapshots. The granularity of snapshot data mapping is 16 KB pages.
- **Snapshot admin space**—Is used to save the metadata (including the exception table) for snapshots.

Each of the three space types is mapped to logical disks (LDs), and all of the LDs are striped across all controller nodes; thus, VVs can be striped across multiple nodes for additional load balancing and performance.

The size limit for an individual virtual volume is 16 TB. A VV is classified by using one of the following provisioning types:

- **Fully provisioned VV (FPVV)**—Has either no snapshot space or deprecated, statically allocated snapshot space.
- **Thinly provisioned VV (TPVV)**—Has space for the base volume allocated from the associated common provisioning group (CPG) and snapshot space allocated from the associated snapshot CPG (if any).

  On creation, 256 MB per node is allocated to a TPVV. Storage is allocated on demand in the snapshot data area as required by the host operation being performed. The snapshot admin area contains the metadata indexes that point to the user data in the shared data (SD) area. Because the shared administration (SA) metadata must be accessed to locate the user data, the indexes are cached in policy memory to reduce the performance impact of the lookups.

  TPVVs associated with a common CPG share the same LDs and draw space from that pool as needed, allocating space on demand in small increments for each controller node. Because the volumes that draw space from the CPG require additional storage, the HPE 3PAR Operating System automatically extends existing LDs or creates new LDs until the CPG reaches the user-defined growth limit, which restricts the maximum size of the CPG.
- **Thinly deduped VV (TDVV)**—Behave similarly to TPVV volumes except that TDVVs within the same CPG share common pages of data. The data shared is determined by the inline deduplication mechanism. TDVVs are supported only on CPGs that use SSDs as a tier of storage.
- **Commonly provisioned VV (CPVV)**—Has space that is fully provisioned from the associated CPG, and the snapshot space is allocated from the associated snapshot CPG.

Scalability limits for HPE 3PAR StoreServ 8000 models are listed in Table 5-12.

Table 5-12 Scalability limits for HPE 3PAR StoreServ 8000 models

Object	8200	8400	8440	8450	20450	20800	20850
Base VV	32k	32k	64k	64k	64k	64k	64k
Total VV	64k	64k	128k	128k	128k	128k	128k
Maximum VLUNs	128k	128k	256k	256k	256k	256k	256k
16 GB/s FC port initiators	256	256	256	256	256	256	256
iSCSI, FCoE port initiators	256	256	256	256	256	256	256
FC system initiators	1024	4096	4096	4096	8192	8192	8192
iSCSI, FCoE, system initiators	1024	2048	2048	2048	8192	8192	8192
Remote Copy maximum VVs—Sync	800	800—2n 2400—4n	800—2n 2400—4+n	800—2n 2400—4+n	800—2n 2400—4+n	800—2n 2400—4+n	800–2n 2400—4+n
Remote Copy maximum VVs—Async periodic	2400	2400—2n 6000—4n	2400—2n 6000–4 + n	2400—2n 6000—4n	2400—2n 6000—4n	2400—2n 6000—4n	2400—2n 6000—4n
Peer Persistence maximum VVs	600	600	600	600	600	600	600

Maximum raw capacity requirements

Table 5-13 shows the maximum raw capacity for hybrid systems.

Table 5-13 Maximum raw capacity for hybrid systems

Maximum raw capacity	7200c	7400c	7440c	8200	8400	8440	20,800
Node pair (TB)	500	800	1000	750	1200	1500	1500
System (TB)	500	1600	2000	750	2400	3000	6000

Table 5-14 shows maximum raw capacity for all-flash systems.

Table 5-14 Maximum raw capacity for all-flash systems

Maximum raw capacity	7450c	8450 (480 maximum SSDs)	20,450	20,850
Node pair (TB)	260.8	921.6	983	983
System (TB)	460.8	1843	1966	3932

All-flash array system limits are based on the assumption that there will be 3.84 TB cMLC SSD.

Physical scale

A maximum capacity 3PAR StoreServ 8440 SFF system can consist of 40 2U enclosures, which would nearly fill two 42U racks. An 8440 with a large number of LFF enclosures could fill up to three racks (assuming 4 TB HDDs are used).

File and object offerings for HPE 3PAR StoreServ

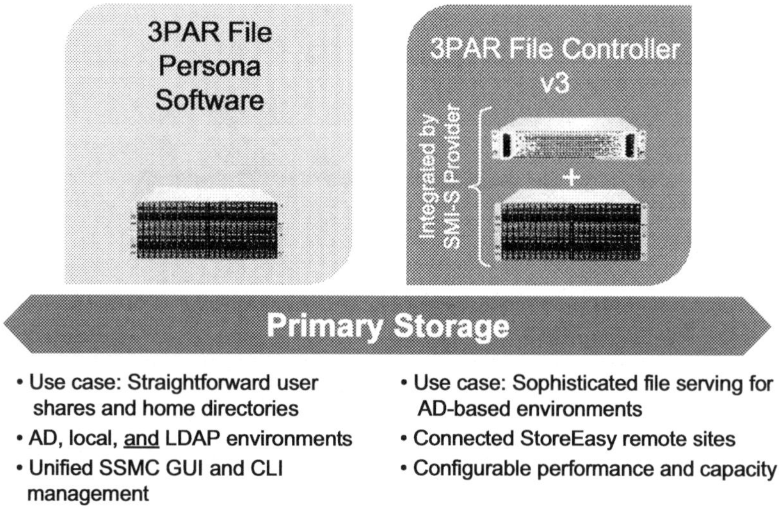

Figure 5-56 File and object access solutions for 3PAR StoreServ systems

Different file and object access solutions are available in the HPE Storage portfolio (Figure 5-56). Converged file services are extending the already rich set of file and object storage offerings for 3PAR StoreServ in the Primary File Storage portfolio.

3PAR File Persona Software provides converged file services on all 8000 systems, bringing in truly converged block, file, and object access. The first release of converged file services is targeted for

- Straightforward user shares and home directories with up to 3000 concurrent home directory users
- Active Directory (AD), local, and Lightweight Directory Access Protocol (LDAP) authentication (where LDAP is an addition in converged file services compared to File Controller)
- Unified GUI and CLI management using the StoreServ Management Console (SSMC), a new management console with the look and feel of HPE OneView

HPE 3PAR File Controller is a gateway model based on Windows Storage Server 2012 R2. It is in front of the 3PAR block array and it is designed to serve. Features of HPE 3PAR File Controller include the following:

- Sophisticated file serving for AD-based environments, including the home directories, groups shares, and application workloads
- Connection to StoreEasy remote sites with BranchCache
- Configurable performance and capacity to support up to 20,000 home directory users concurrently

HPE 3PAR StoreServ File Controller v3

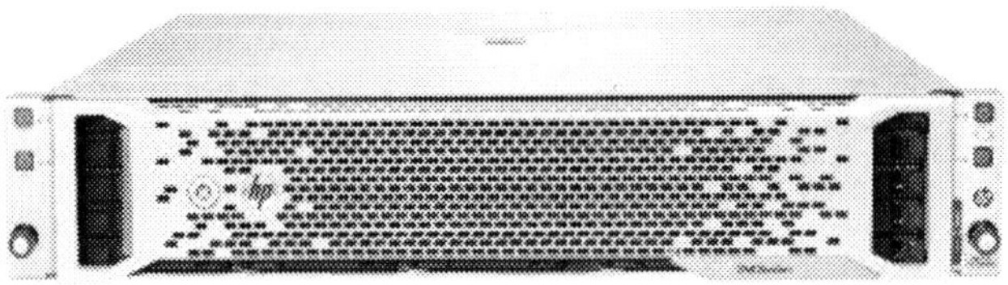

Figure 5-57 HPE 3PAR StoreServ File Controller v3 front view

The HPE 3PAR StoreServ File Controller v3 system is optimized, efficient, secure, and highly available storage to address the file storage challenges of medium-to-large business and branch office SAN environments (Figure 5-57). This system is built on industry-leading HPE ProLiant DNA and Microsoft Windows Storage Server, so it integrates easily into new and existing SAN environments with a straightforward, consistent management experience for IT generalists or storage administrators.

The HPE 3PAR StoreServ File Controller v3 system sets the standard for performance and expandability in gateway services for 3PAR arrays. It consists of the 3PAR StoreServ File Controller v3 system 2U chassis and one or two 3PAR StoreServ File Controller v3 single nodes. A 3PAR StoreServ File Controller v3 system with two nodes provides HA by clustering the nodes together. Up to four 3PAR StoreServ File Controller v3 systems can be configured together as an eight-node cluster.

Figure 5-58 HPE 3PAR StoreServ File Controller v3 rear view (second node optional)

The HPE 3PAR StoreServ File Controller v3 is a 2U rack-mount system with one or two nodes, based on the HPE Apollo r2600 server (version 2 was a 2U, single-server system; Figure 5-58). It supports up to 20,000 users per node, and the capacity provided depends on the attached 3PAR StoreServ array.

Primary protocols and interconnects include Server Message Block (SMB) (2.0, 2.1, 3.0, and 3.02), NFS (v2, v3, and v4.1), WebDAV, HTTP/HTTPS, and FTP/FTPS, iSCSI.

- **SMB 3.0 protocol**—Using new advanced functionality, leverage high-performance, continually available file shares to deliver feature-rich storage for business-critical applications and virtual servers.
- **SMB 3 Multichannel**—Enable the aggregation of network bandwidth and network fault tolerance when multiple paths are available between SMB clients and file servers.
- **NFS v4.1**—For Windows Storage Server 2012 R2, Server for NFS has been updated to support continuous availability. The improvements make it possible to reliably store and run VMware ESX on a VM that is running Windows Server 2012 by using the NFS protocol to share files. Additional improvements in failover clustering take advantage of better resilience to hardware and software outages that might afflict individual cluster nodes. In addition, failover clustering isolates VMware ESX hypervisor hosts from intermittent glitches during failovers.
- **iSCSI**—Some applications require block storage. iSCSI Target Server can provide these applications with continuously available block storage. Because the storage is remotely accessible, iSCSI Target Server can also consolidate block storage for central or branch office locations. By using boot-capable NICs or a software loader, you can deploy hundreds of diskless servers. This capability is ideal for large deployments of identical operating system images, such as a Hyper-V server farm or High Performance Computing (HPC) clusters.

HPE 3PAR StoreServ File Controller v3 comes with Windows Storage Server 2012 R2 Standard Edition, MS Failover Cluster, File Controller Dashboard, Sub-file Deduplication, Snapshots, Replication, File Classification, File Screening, Quotas, Reporting, Server Manager, and an iLO Advanced License.

Optional software includes HPE CLX for hands-free failover and failback; HPE LiveVault for backup to cloud; Third-party Double-Take Availability for scalable replication and disaster recovery, and iTernity Compliant Archive Solution (iCAS) for legally compliant data archiving.

HPE 3PAR File Persona Software Suite

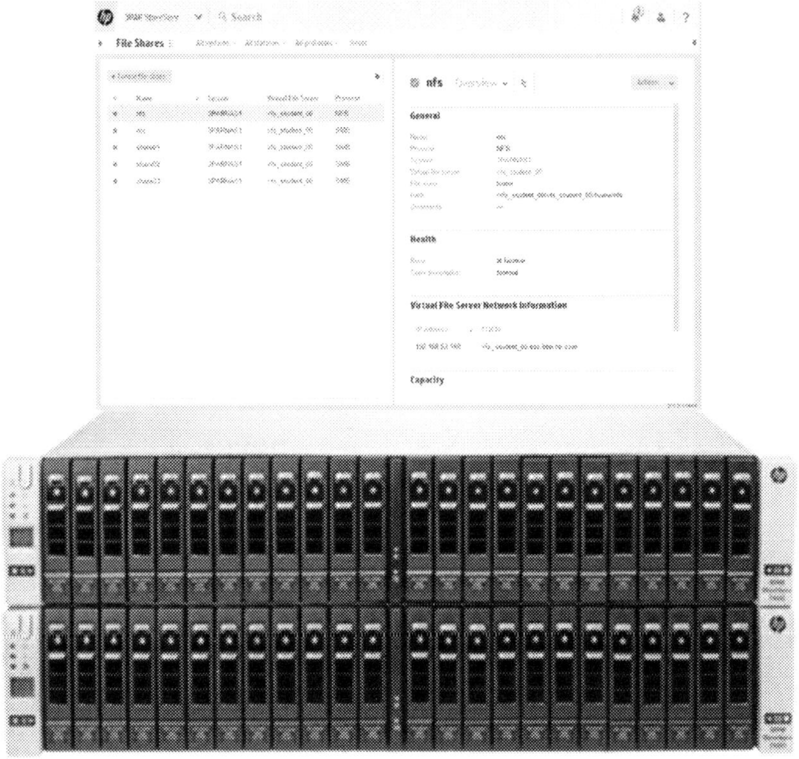

Figure 5-59 HPE 3PAR StoreServ system and user interface for sharing folders

Benefits and features (Figure 5-59)

- Up to 256 TB aggregate file capacity with user and capacity quota policies
- AD, local, and LDAP authentication plus Windows infrastructure integration with DFS Namespace and Microsoft Management Console support
- Rich protocol support with SMB 3.0, 2.1, 2.0, and 1.0; NFSv4.0, v3; and Object Access API
- Data optimization with thin provisioning, Adaptive Optimization, Dynamic Optimization, and Adaptive FlashCache
- Data protection with antivirus support, snapshots, Remote Copy replication, network share and NDMP-based backup, and Data at Rest Encryption
- Productized as Software Suite LTU containing all file and object capabilities

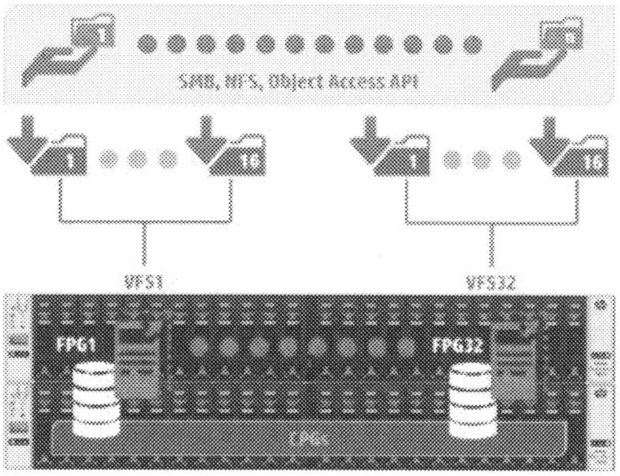

Figure 5-60 Logical view of HPE 3PAR File Persona managed objects

The HPE 3PAR File Persona software can be enabled on a 3PAR storage system node pair with an optional license (Figure 5-60). It requires either a two-port 10 GbE NIC or a four-port 1 GbE NIC to be installed in the system or the onboard 1 GbE RCIP port to be enabled for File Persona.

The HPE 3PAR File Persona software is designed for client workloads such as home directory consolidation, group and department shares, corporate shares, and custom cloud applications.

File data services include User Authentication Services; capacity and user/group Quota Management; File Store Snapshots with user-driven file restore; and Antivirus Scan Services for integration with third-party antivirus software.

Example—Using the HPE 3PAR Management Console

The following example shows you how to create a new volume on an HPE 3PAR StoreServ system by using the HPE 3PAR Management Console.

- You can start the 3PAR Management Console by using the desktop icon (Figure 5-61).

Figure 5-61 3 PAR Management Console desktop icon

- Login credentials are required (Figure 5-62).

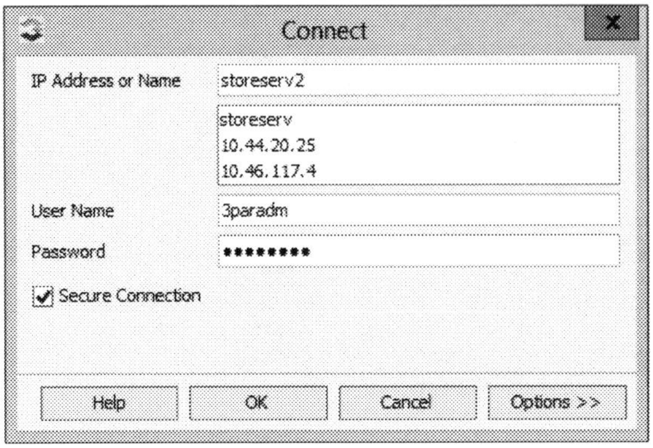

Figure 5-62 Login credentials

- The Summary screen shows the model, serial number, OS version, and health statistics of systems, including components in a Degraded or Failed state (Figure 5-63).

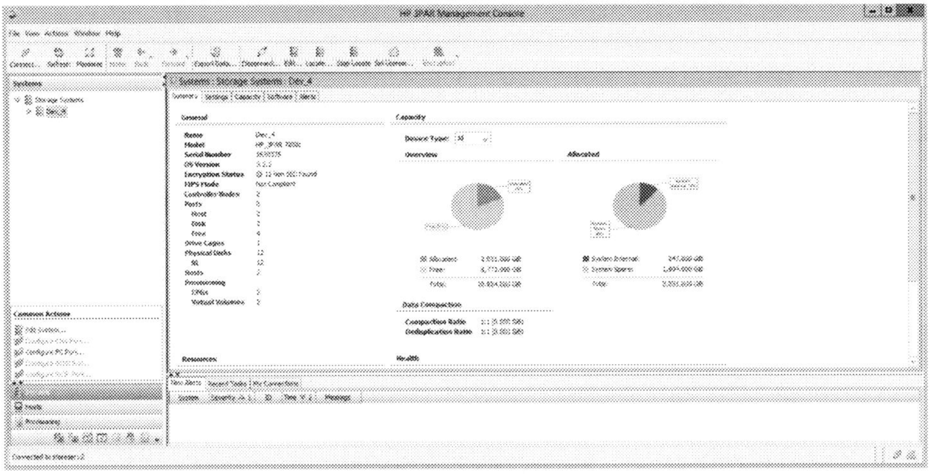

Figure 5-63 The Summary screen

- 3PAR Storage system section, which is opened from the left-hand pane, shows various details, such as how many controller nodes the solution has or what its health status is (Figure 5-64).

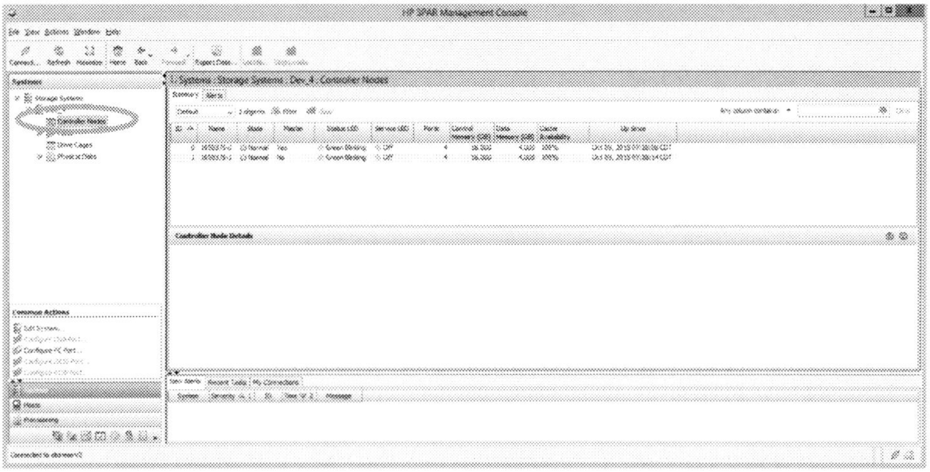

Figure 5-64 3PAR Storage system section—Controller Nodes

- The remaining subtrees in the **Systems** section show how many physical drives are installed and how many drive cages are used (Figure 5-65).

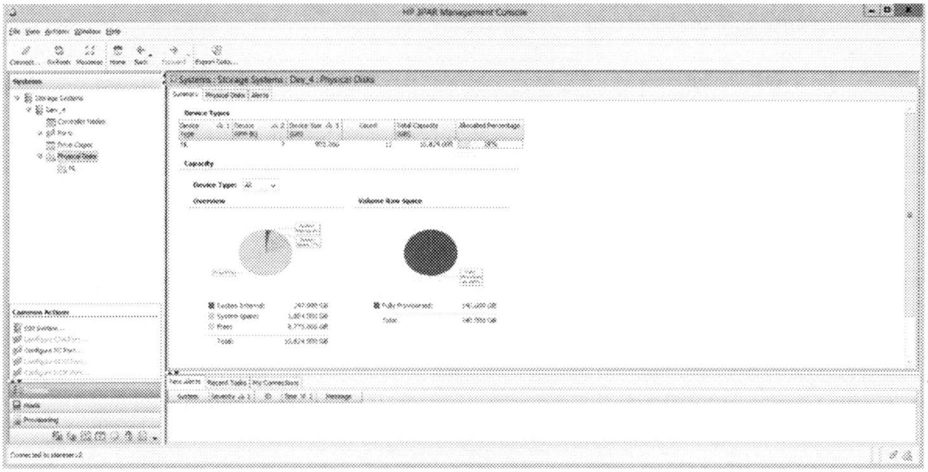

Figure 5-65 **Systems** section—Physical Drives

- The **Provisioning** link in the left-hand pane and the Summary screen show how many 3PAR Common Provisioning Groups (CPGs) and virtual volumes are present and how much free capacity is available (Figure 5-66).

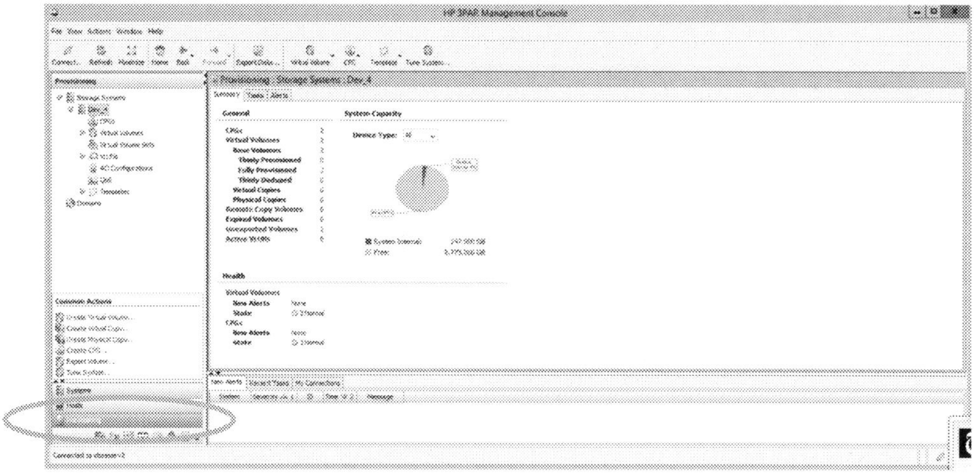

Figure 5-66 The **Provisioning** link and the Summary screen

- To create new volume, you need to use a CPG. Available CPGs can be found by expanding the **CPGs** subtree in the 3PAR Storage system in the left pane (Figure 5-67).

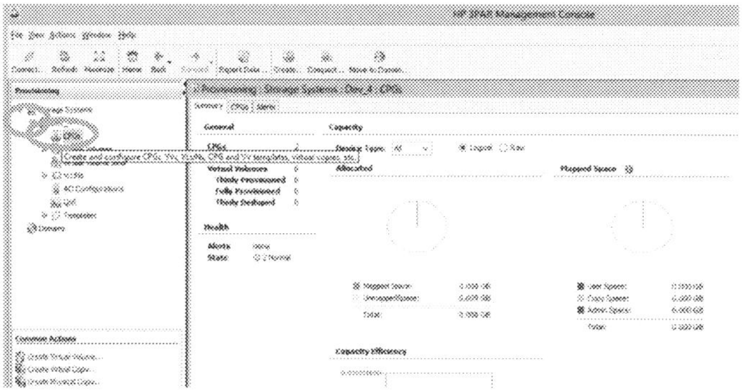

Figure 5-67 3PAR Storage Systems—CPGs

- You can select CPGs from the CPGs tab, where volumes are also visible.

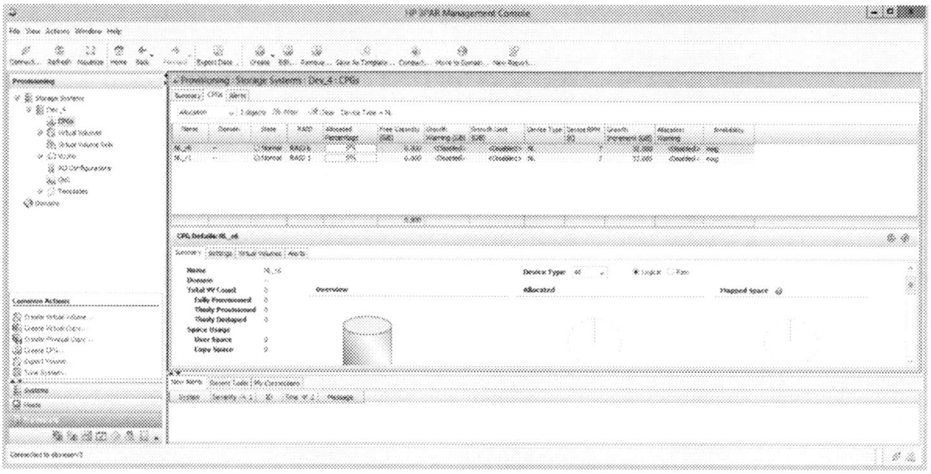

Figure 5-68 Provisioning—CPGs

- To create a new volume, right-click **Virtual Volumes** in the left pane and then select **Create Virtual Volume**.

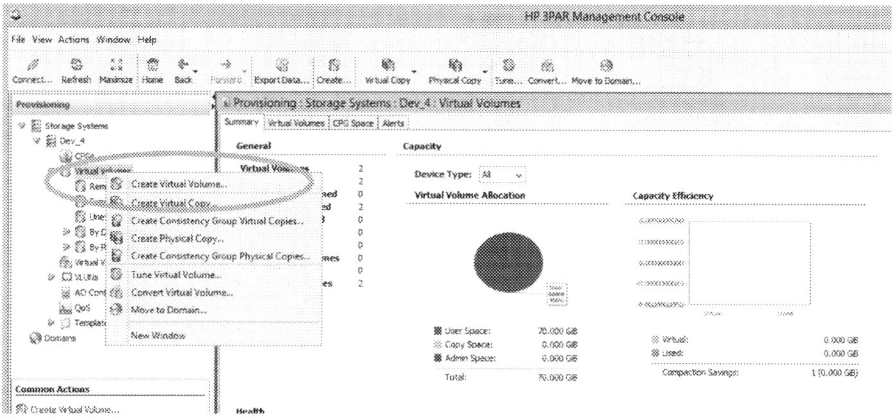

Figure 5-69 Storage Systems—**Create Virtual Volume**

The Create Virtual Volume wizard starts (Figure 5-70).

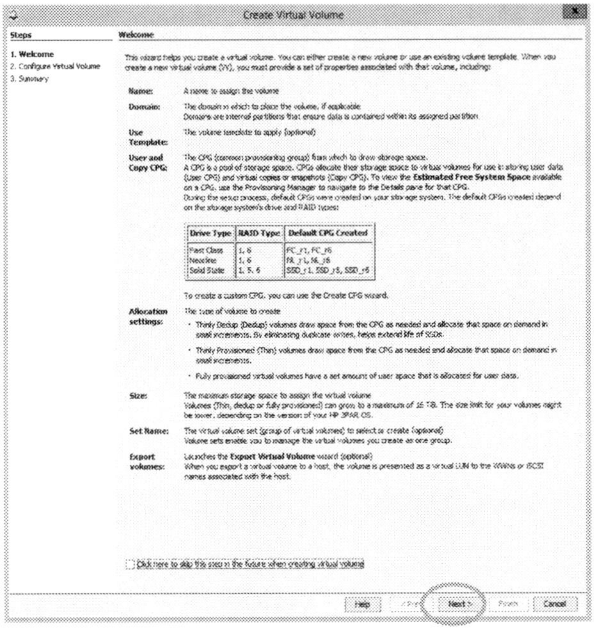

Figure 5-70 Create Virtual Volume wizard

- Then the name, allocation settings, size, and additional settings can be changed for the newly created volume (Figure 5-71).

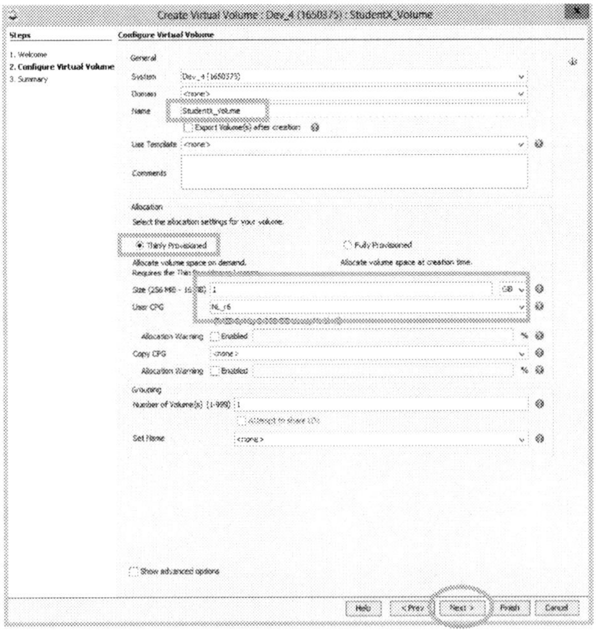

Figure 5-71 Configure Virtual Volume

- Review the Summary screen before clicking the **Finish** button (Figure 5-72).

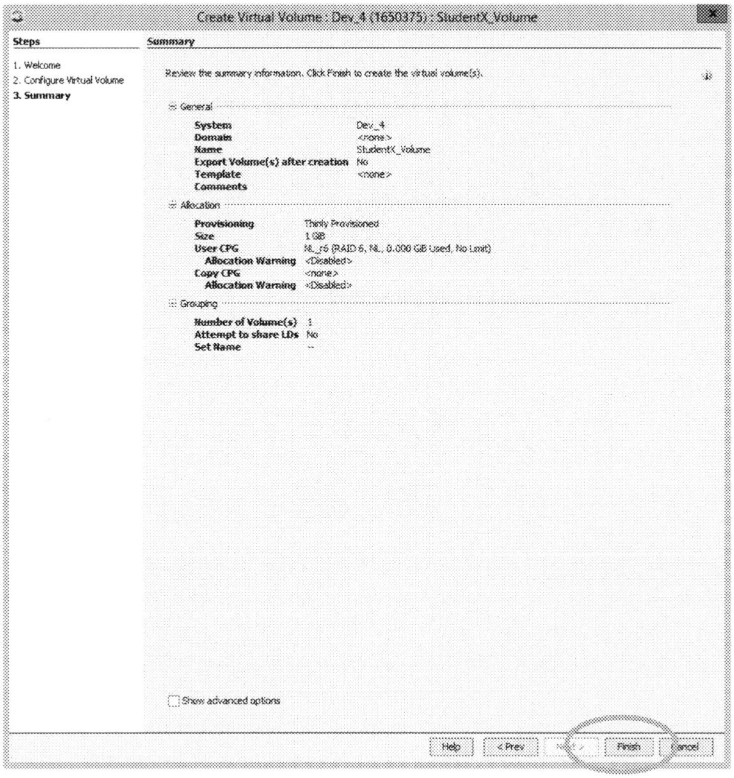

Figure 5-72 The Summary screen

- The new volume is listed in the Virtual Volumes tab, including the Worldwide Name (WWN) for the volume (Figure 5-73).

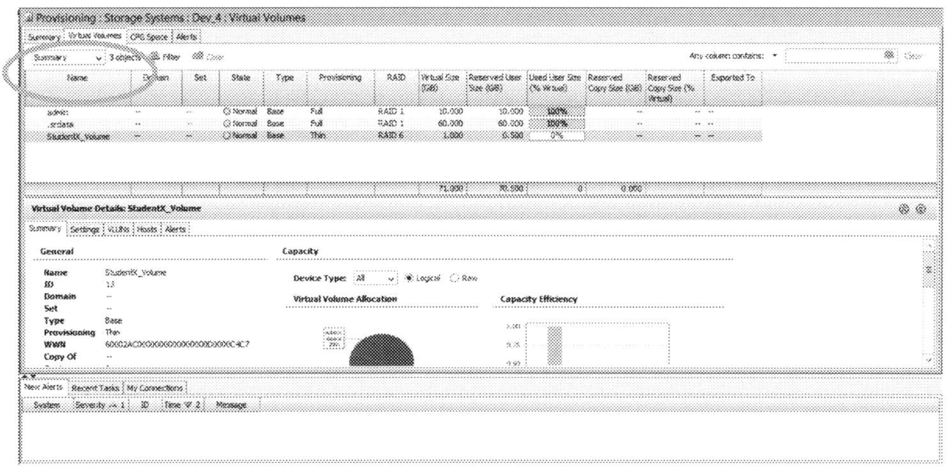

Figure 5-73 New volume is listed in the Virtual Volumes tab

Additional tasks that are necessary to provision the volume to the server are the host creation process, exporting the volume to the host, and partitioning and formatting in the operating system.

Example—Using the HPE 3PAR StoreServ Management Console (web-based)

Another management interface for HPE 3PAR solutions is the web-based HPE 3PAR StoreServ Management Console. You can access the console from this icon on the desktop (Figure 5-74).

Figure 5-74 HPE 3PAR StoreServ Management Console

- Login credentials are required to use the console (Figure 5-74).

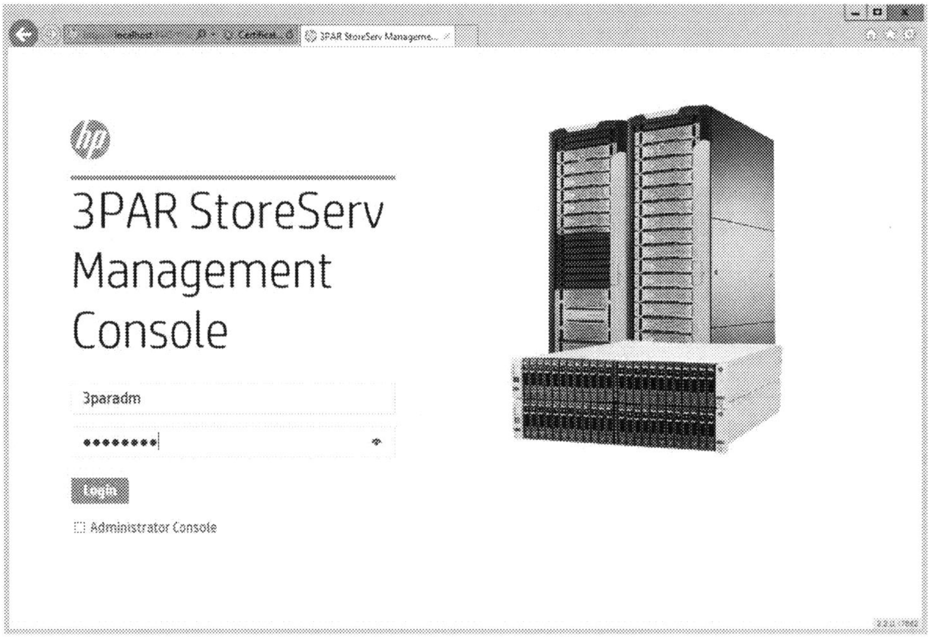

Figure 5-74 Login credentials

- The dashboard (Figure 5-75) shows you which storage systems the console is monitoring, the total and free capacity, and additional details.

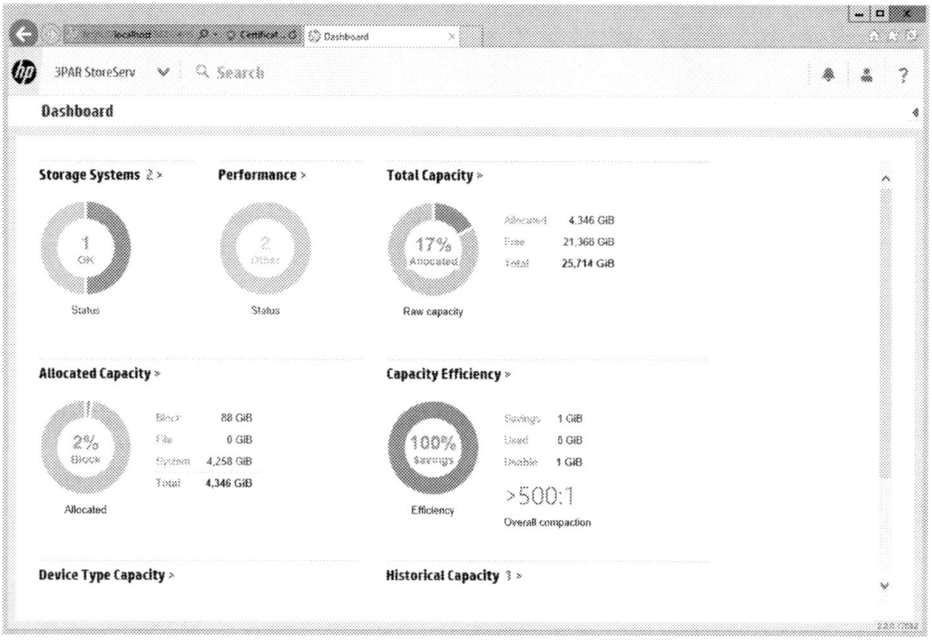

Figure 5-75 Dashboard

- Virtual Volumes (Figure 5-76) can be accessed from the menu in the Block Persona section.

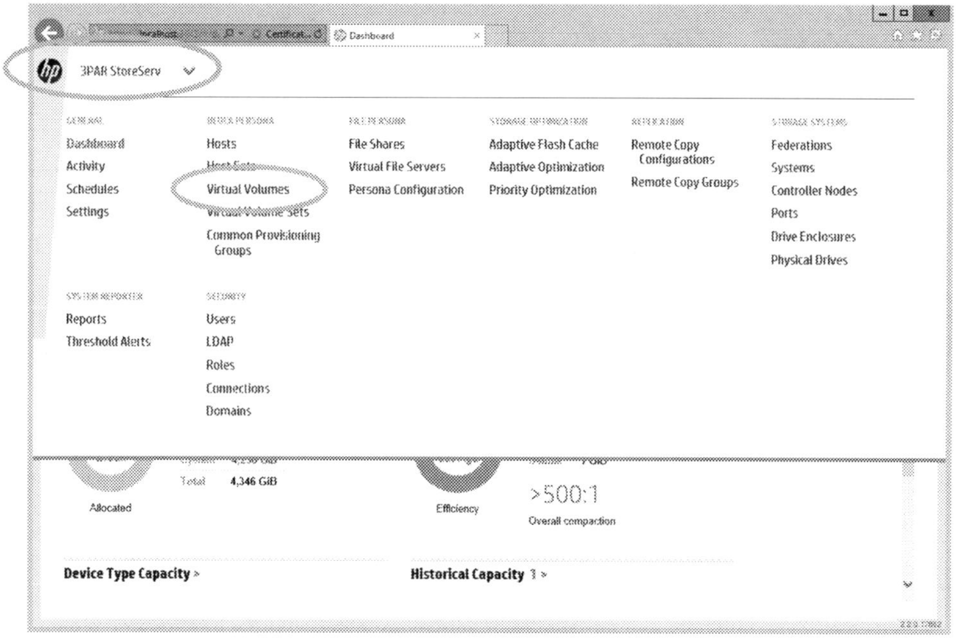

Figure 5-76 Virtual Volumes

- Details about volume are then available from the left-hand pane (Figure 5-77).

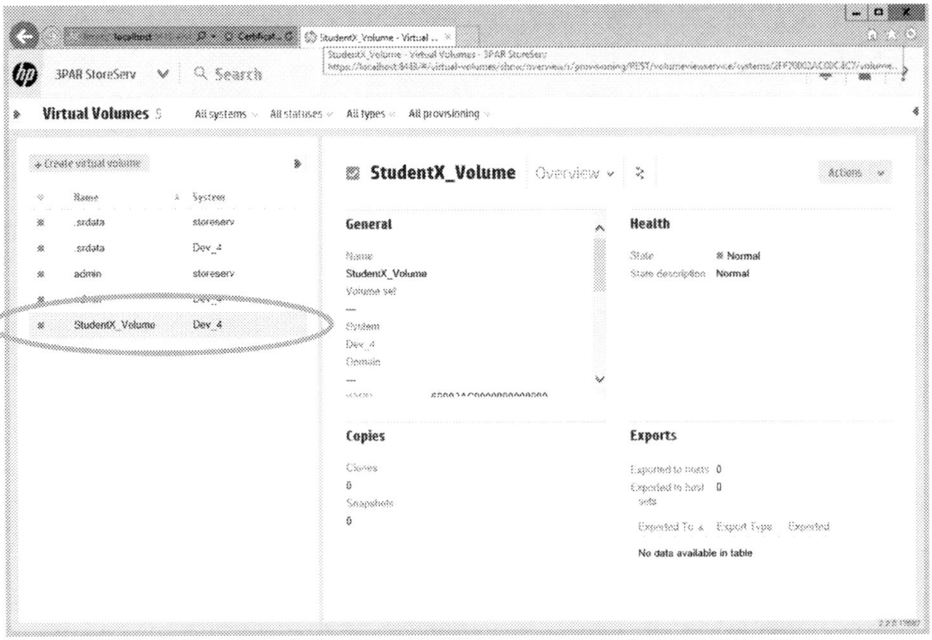

Figure 5-77 StudentX_Volume—details about volume

- The Capacity section (Figure 5-78) shows various details about a volume, such as its capacity efficiency.

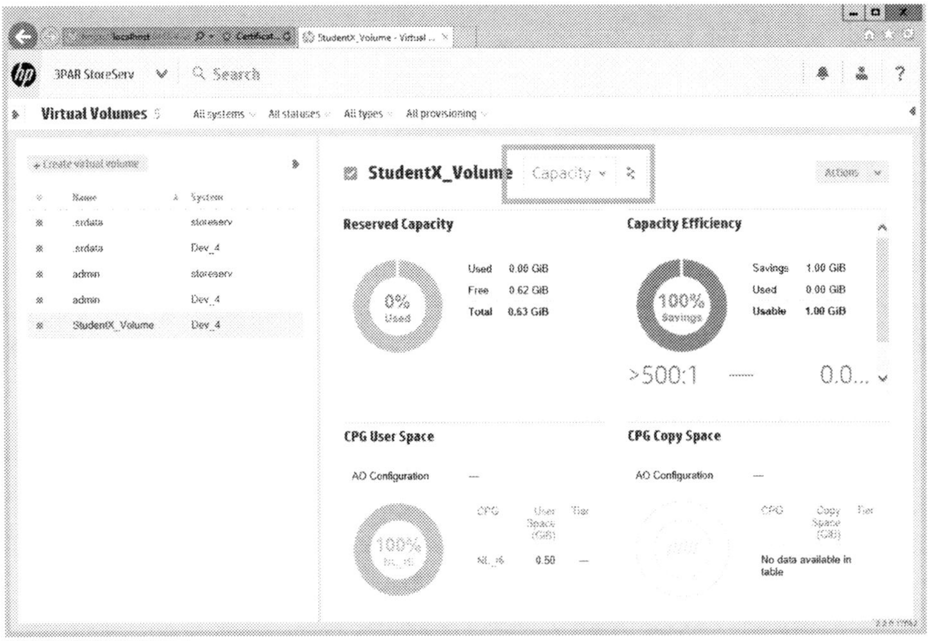

Figure 5-78 The Capacity section

- To create snapshots from the volume, right-click the **Actions** menu and select Create Snapshot (Figure 5-79).

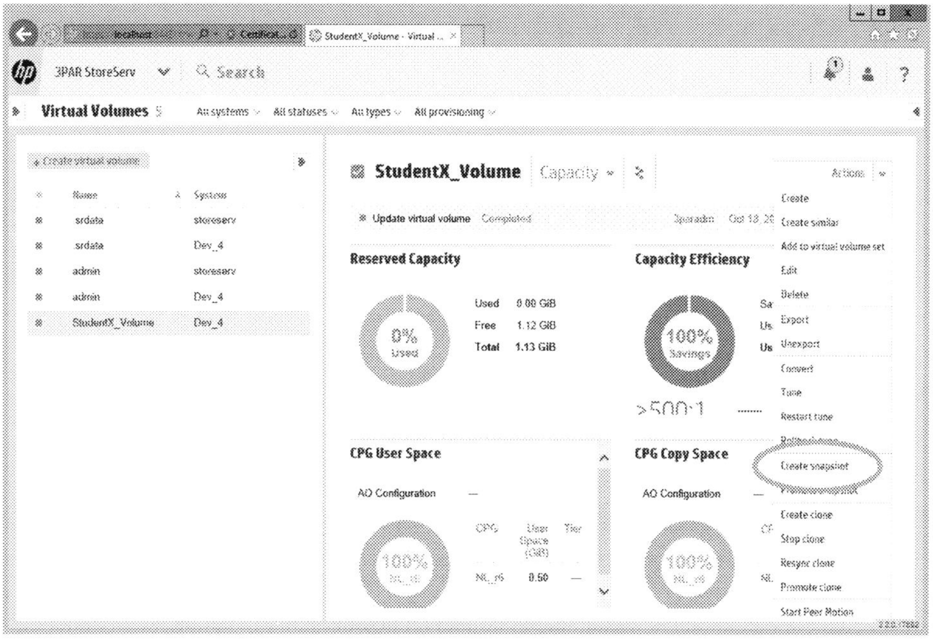

Figure 5-79 To create snapshots

- Visible VVs can be filtered. To see snapshots only, change the Type to **Snapshot** (Figure 5-80).

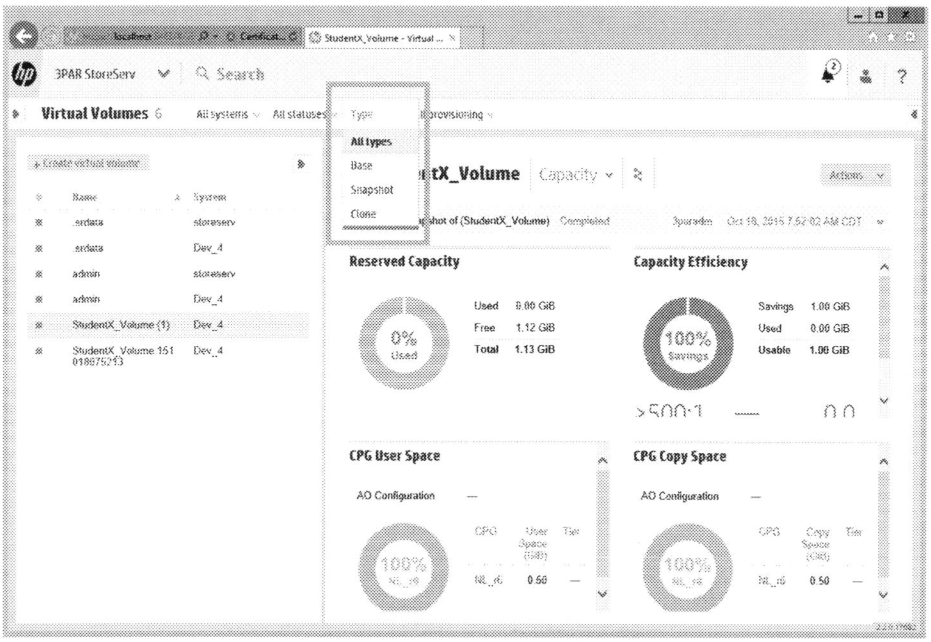

Figure 5-80 To see snapshots only

- You can select **Map** to review snapshot properties such as capacity information or its relationship to other objects (Figure 5-81).

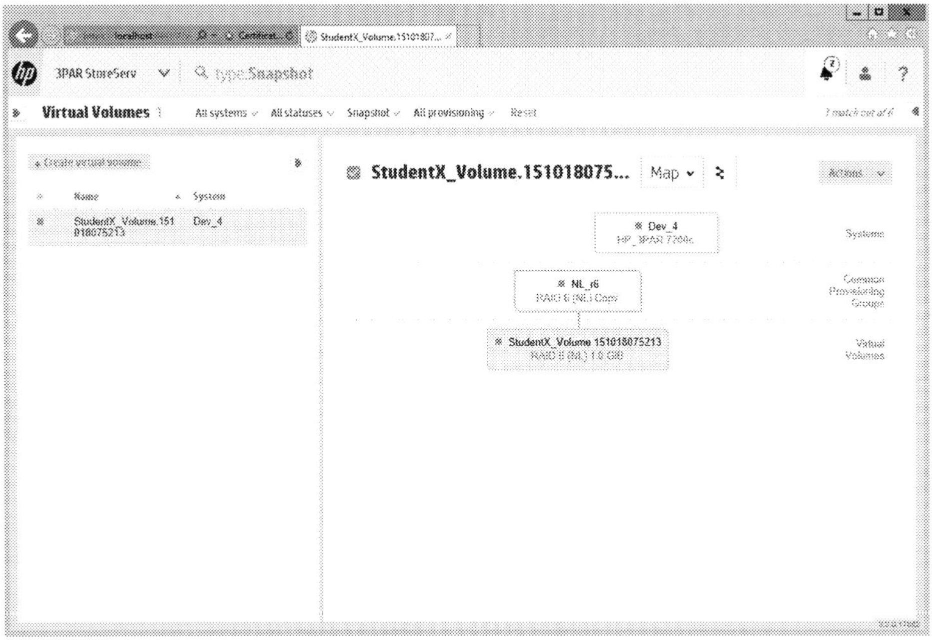

Figure 5-81 To review snapshot properties

- You can revert the content of original volume back to Snapshot by selecting **Promote Snapshot** in the **Actions** menu (Figure 5-82).

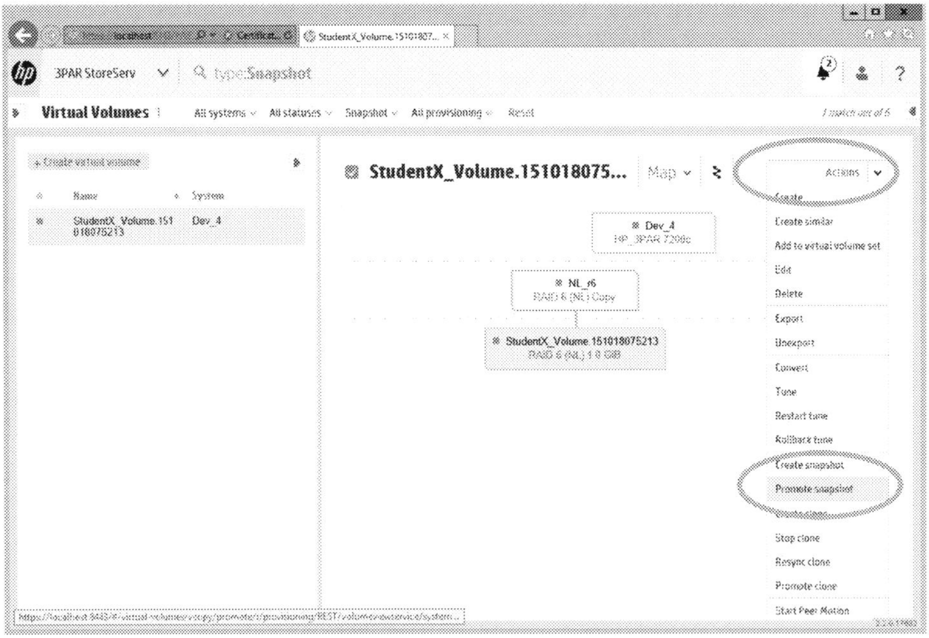

Figure 5-82 Revert the content of original volume back to Snapshot

Example—Using HPE OneView to configure and monitor storage

HPE OneView can be used to manage and monitor 3PAR Storage. The following example shows how to use the HPE OneView user interface to perform several typical tasks related to storage.

- You can access HPE OneView using a web browser (Figure 5-83).

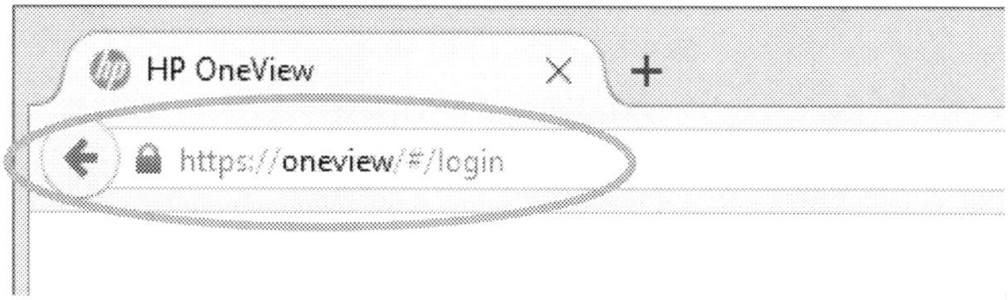

Figure 5-83 Access HPE OneView

- Login credentials are required (Figure 5-84).

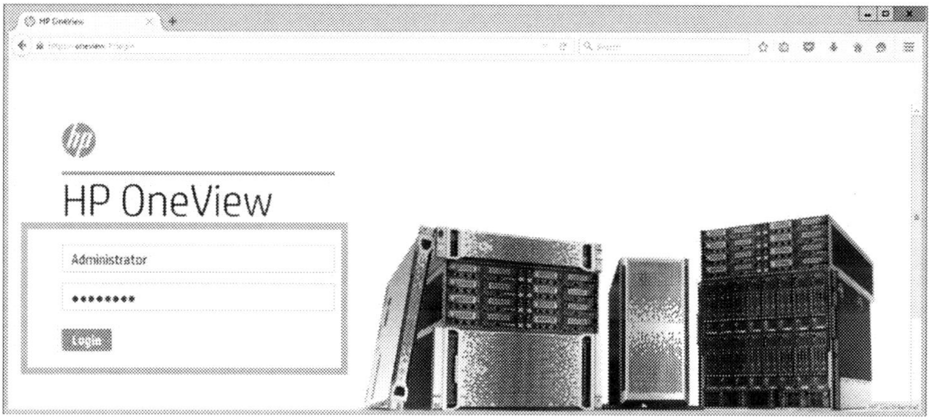

Figure 5-84 Login credentials

- Storage Systems can be configured from main menu (Figure 5-85).

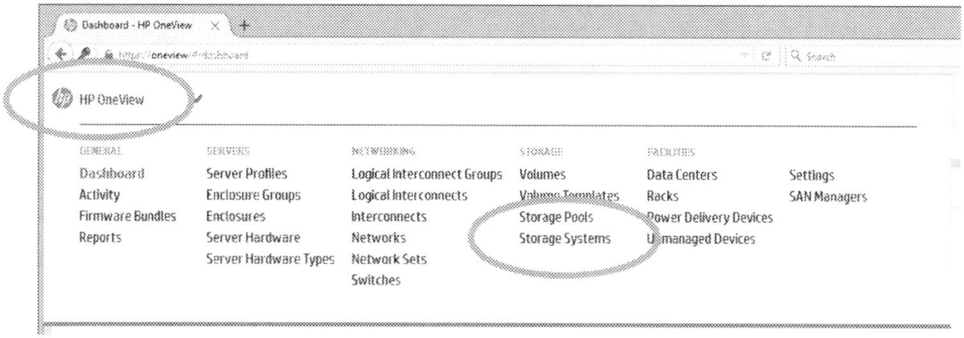

Figure 5-85 Storage Systems configured from main menu

- You can connect HPE OneView to the 3PAR system with the **Add storage system** menu option (Figure 5-86).

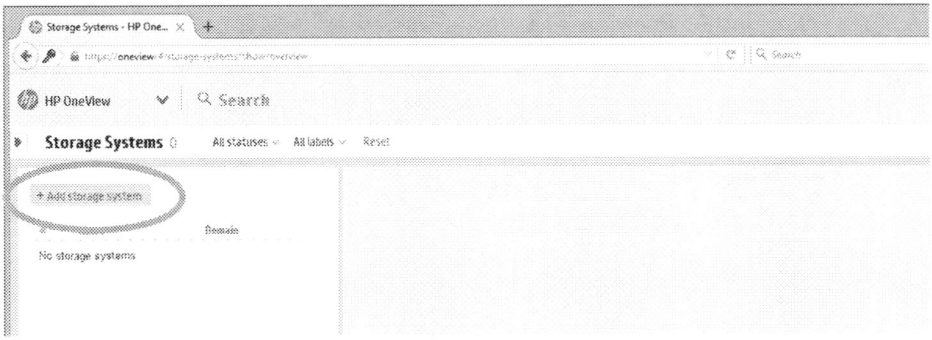

Figure 5-86 **Add storage system** menu option

- You must enter the IP address of the 3PAR storage system and your login credentials (Figure 5-87).

Figure 5-87 IP address and login credentials

- Another group of configuration objects in HPE OneView is Storage Pools (Figure 5-88).

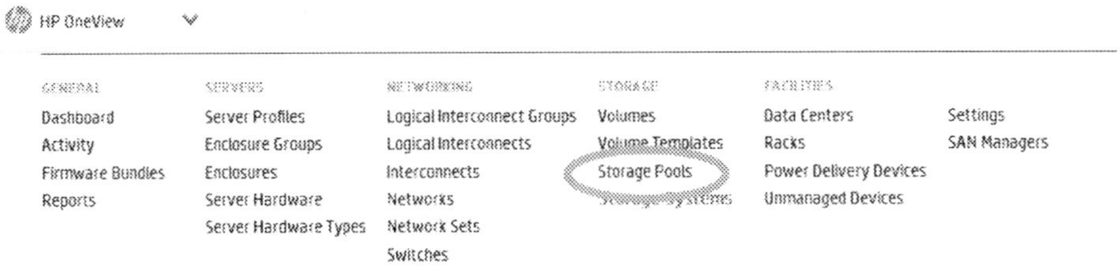

Figure 5-88 Storage Pools in HPE OneView

- Clicking **Add storage pool** enables you to add a CPG to HPE OneView management (Figure 5-89 and 5-90).

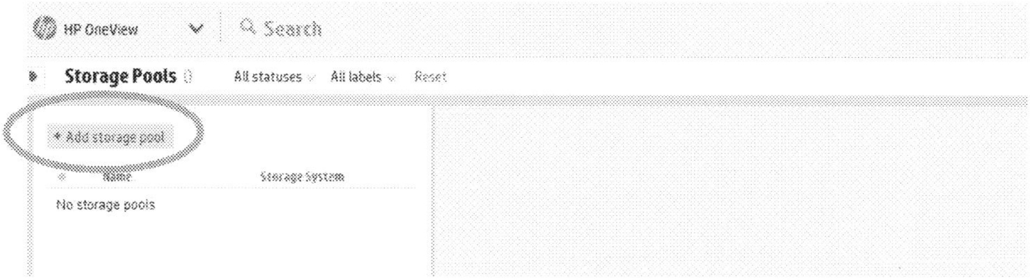

Figure 5-89 **Add storage pool**

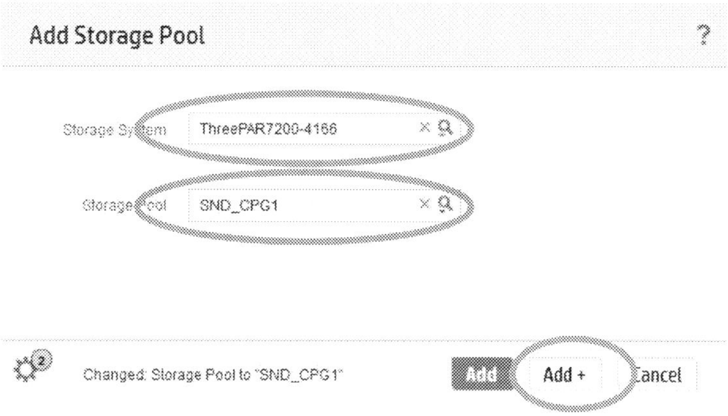

Figure 5-90 **Add storage pool** to add a CPG

- Details related to CPGs are then listed in HPE OneView interface (Figure 5-91).

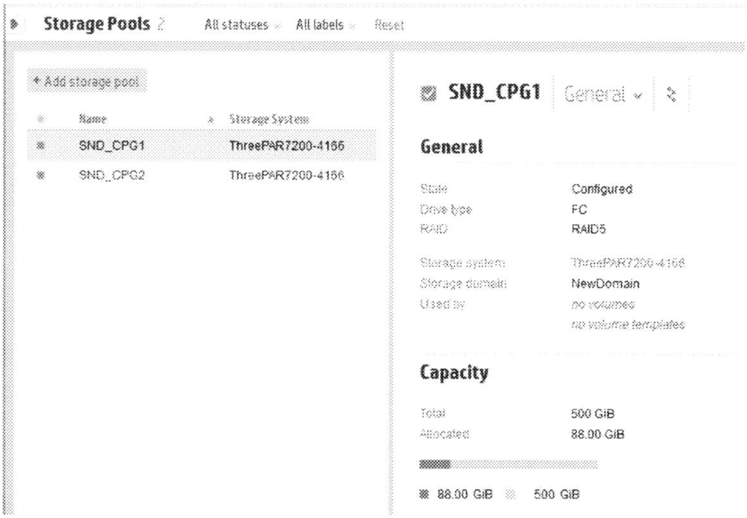

Figure 5-91 Details related to CPGs

- A Volume Template is a logical resource that enables storage administrators to create a standard configuration from which they can create multiple volumes without understanding the underlying storage infrastructure (Figures 5-92 to 5-94).

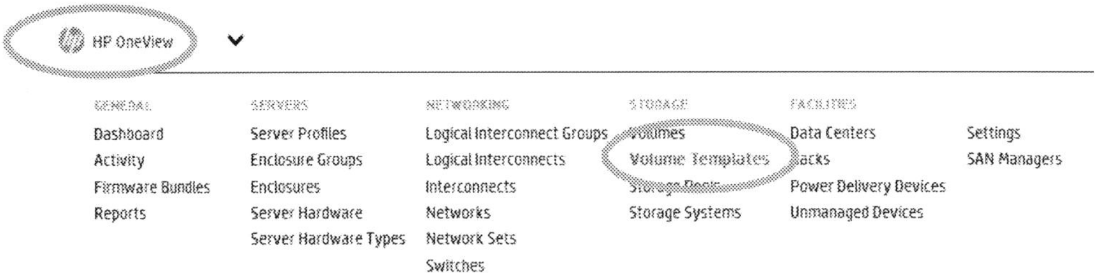

Figure 5-92 A Volume Template

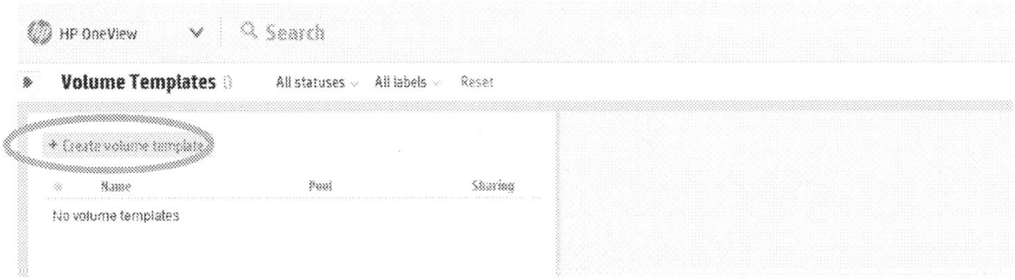

Figure 5-93 Creat volume template

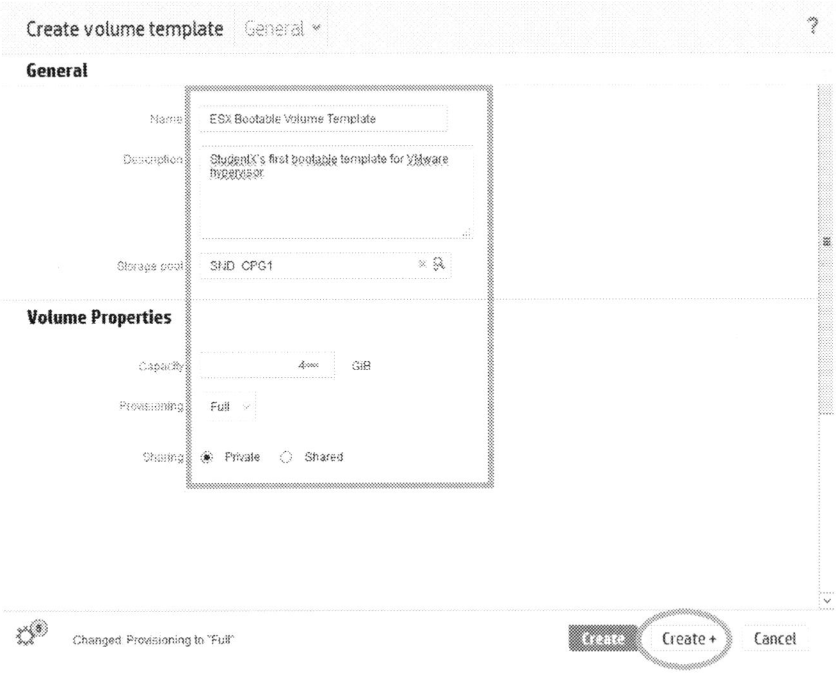

Figure 5-94 Create volume template and volume properties

- Server administrators can be forced to use templates only (Figure 5-95).

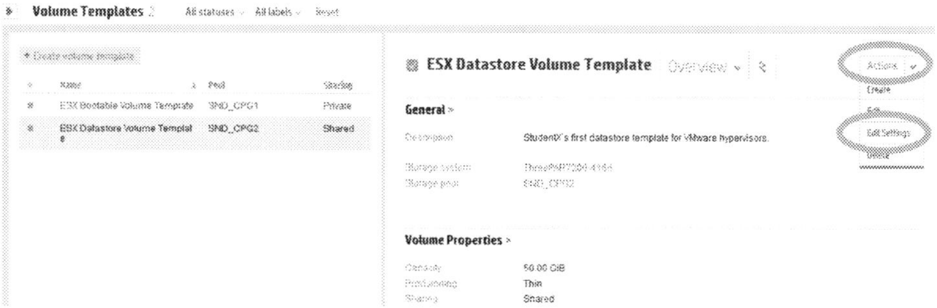

Figure 5-95 Use templates only

- New volumes are created from volumes section (Figure 5-96 to 5-98).

Figure 5-96 New volumes created

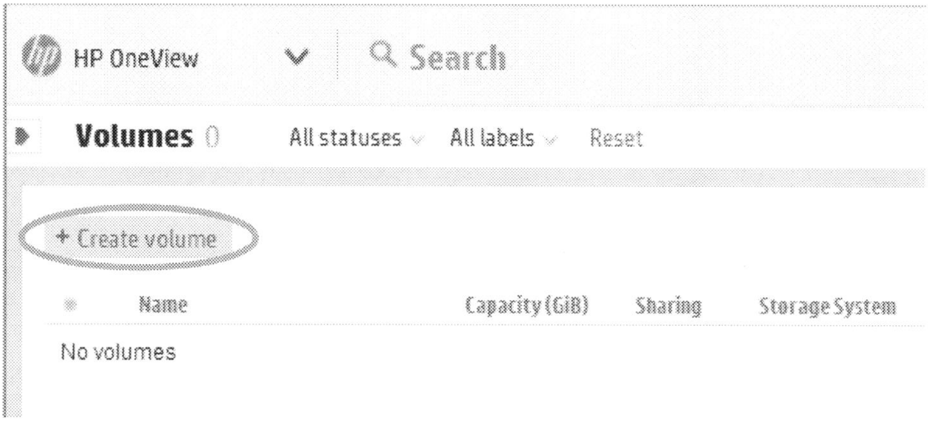

Figure 5-97 Volumes section

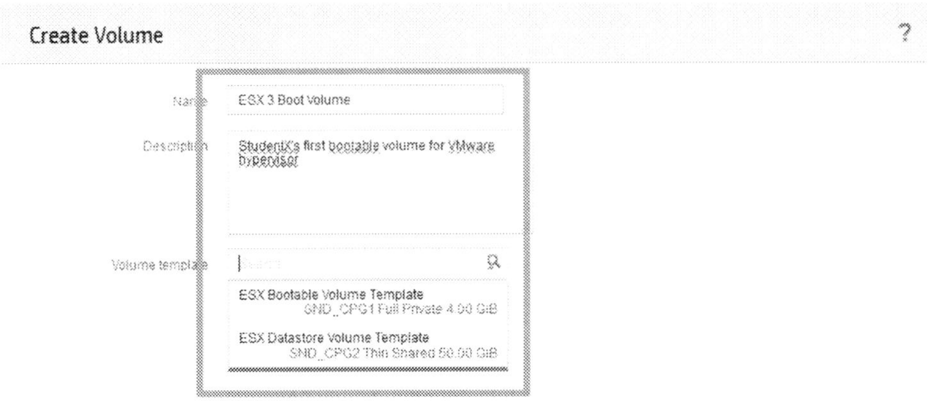

Figure 5-98 Create volumes

- To check visualizations in HPE OneView, select the **Map** view from the General section (Figure 5-99).

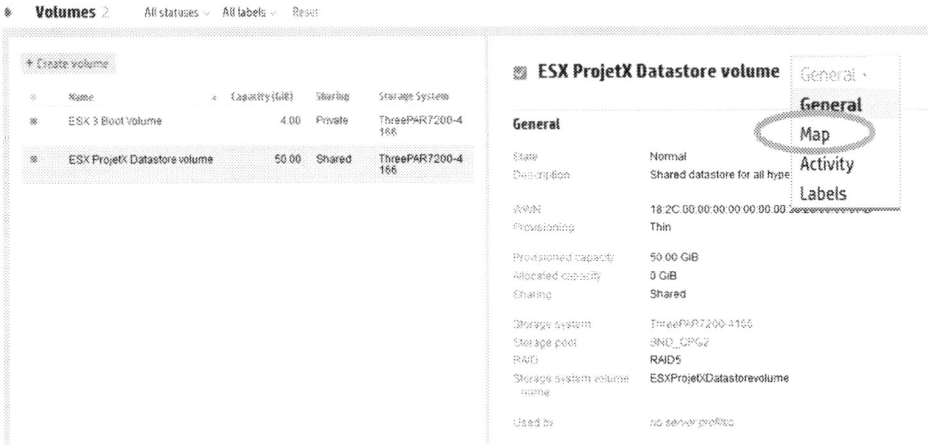

Figure 5-99 To check visualizations in HPE OneView

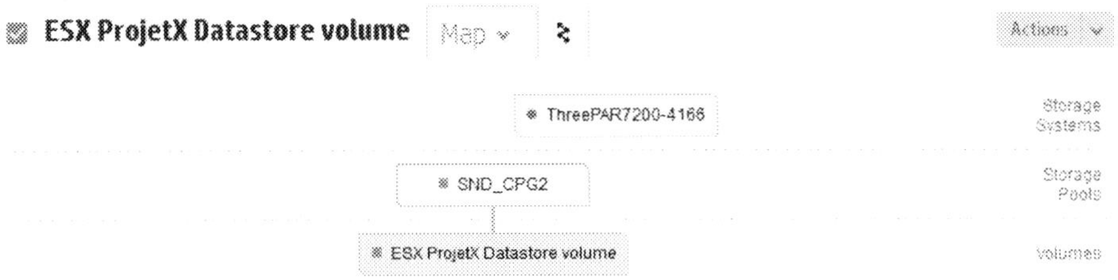

Figure 5-100 The **Map** view

- The HPE OneView Dashboard shows your changes to the storage objects (Figures 5-101 and 5-102).

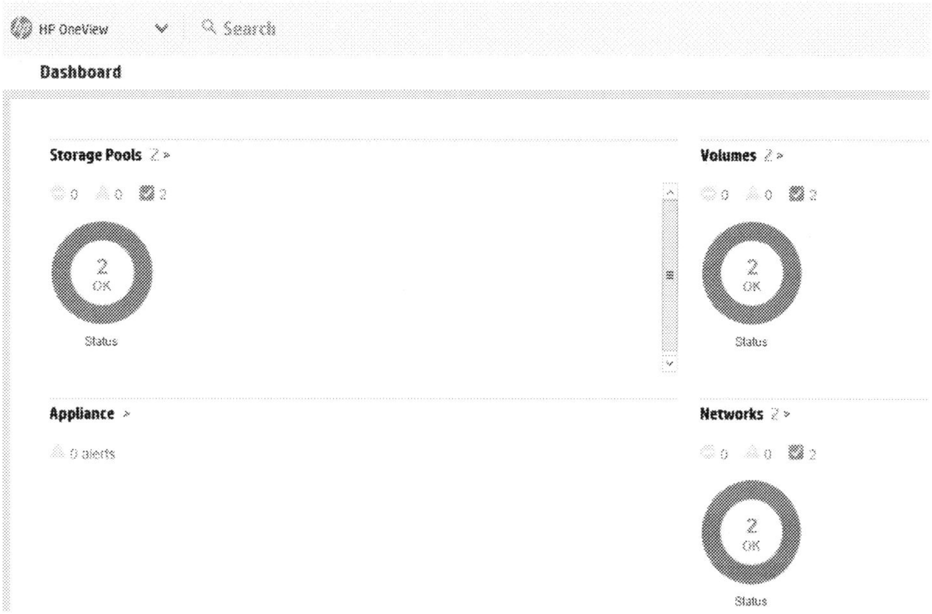

Figure 5-101 The HPE OneView Dashboard

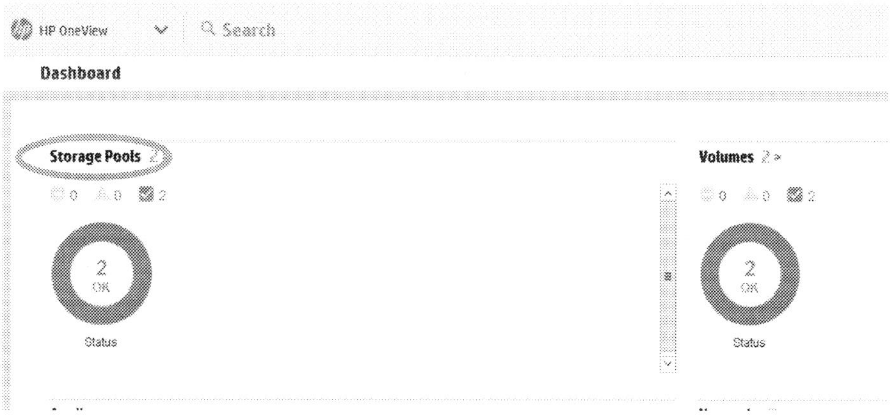

Figure 5-102 Changes to the storage objects

SIZING SAN PRODUCTS

SAN products can be ordered in various configurations so that the solution will fit customer needs.

HPE provides a variety of tools and resources to help size and configure a customer solution. Two of the best places to find online information about HPE products and solutions are the HPE website and HPE Product Bulletin.

Sizing considerations

To correctly select a disk enclosure and its connected components, evaluate the following considerations:

- Total capacity and room for growth
- Performance requirements
    - Types of disks (rotational speed and SSD)
    - Type of stored data, access method, and read/write ratio
    - IOPS, megabytes per second (MB/s)
- Required availability (node count, controller count, and RAID level)
- Features
    - File-based access, replication, and backup
- Host interface
    - SAS, iSCSI, and FC
- Requirements for installation support and enhancements to warranty services

Reference documents and sizing tools

Figure 5-103 HPE website

Sizing tools are described in more detail in Chapter 8 (Figure 5-103). You can use the following reference documents and sizing tools to select and configure the correct disk enclosure:

- HPE QuickSpecs
- HPE Product Bulletin
- HPE website:
    - http://www.hp.com/go/msa.
    - https://www.hpe.com/us/en/storage/storevirtual.html.
    - https://www.hpe.com/us/en/storage/3par.html.
    - https://www.hpe.com/us/en/integrated-systems/hyper-converged.html.
- HPE Storage Sizing Tool
- HPE NinjaSTARS for 3PAR
- HPE StoreVirtual Storage Optimizer
- MSA Storage Optimizers

HPE Storage Sizing Tool

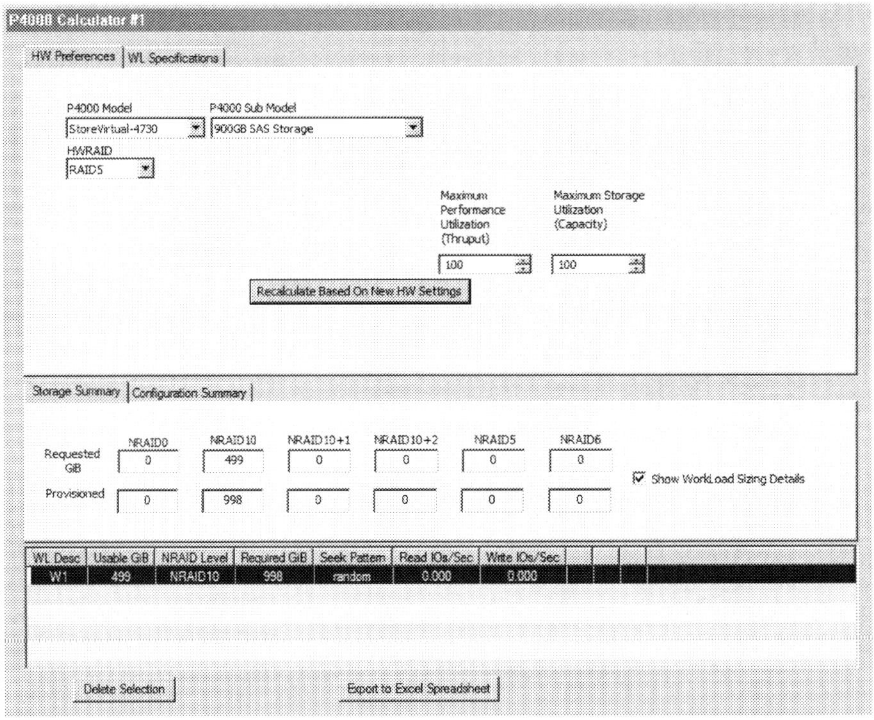

Figure 5-104 HPE Storage Sizing Tool

The HPE Storage Sizing Tool can be used to size all current supported arrays (Figure 5-104). Its results can be exported to Excel with part numbers and prices. The sizer provides several features to make sizing easier, such as the ability to create disk groups based on required capacity. The HPE Storage Sizing Tool also verifies scalability limits.

Example—Using the HPE Storage Sizing Tool to configure SAN storage

The following example shows you how to configure an all-flash HPE 3PAR StoreServ solution based on the following customer requirements:

- Two HPE BladeSystem c7000 enclosures equipped with HPE Virtual Connect FlexFabric modules
- Server blades to host VMware hypervisors and an array to provide storage for datastores holding VMs
- HPE b-series FC switches
- An empty HPE642 Rack
- Hundreds of thousands of IOPS

- 50 TB for datastores with some free room to grow
- 10 TB for file sharing from user desktops
- RAID 5

Following are the key points for using the sizing tool:

- You can start the HPE Storage Sizing Tool from the icon on the desktop.

Figure 5-105 HPE Storage Sizing Tool icon

- To see the calculators offered in the Storage Sizing Tool, select **Storage Calculators** in the left-hand menu (Figure 5-106).

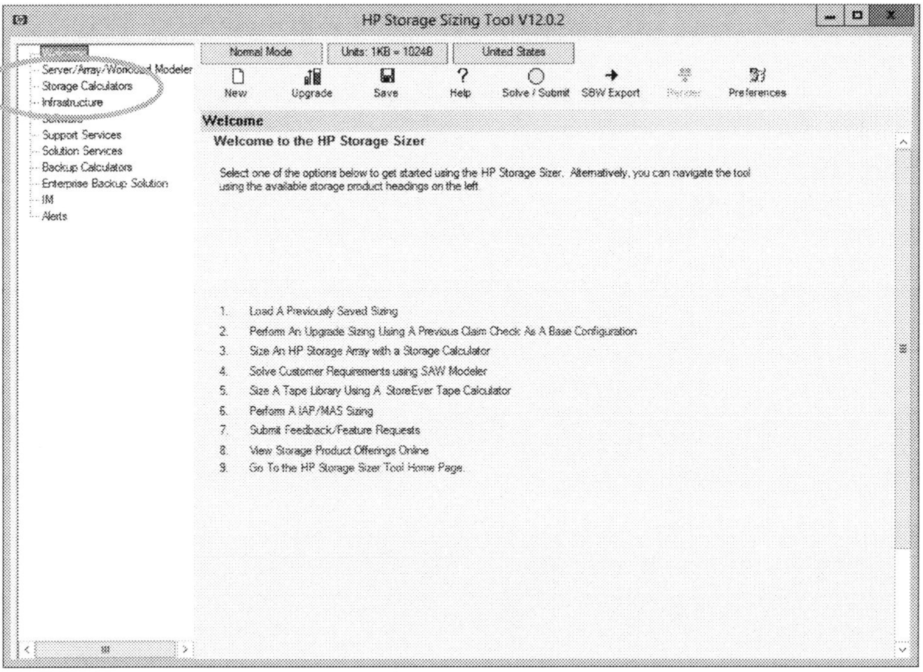

Figure 5-106 The Storage Sizing Tool—select **Storage Calculators**

- The **Launch HPE 3PAR StoreServ 7000/8000** link will start the calculator for this scenario (Figure 5-107).

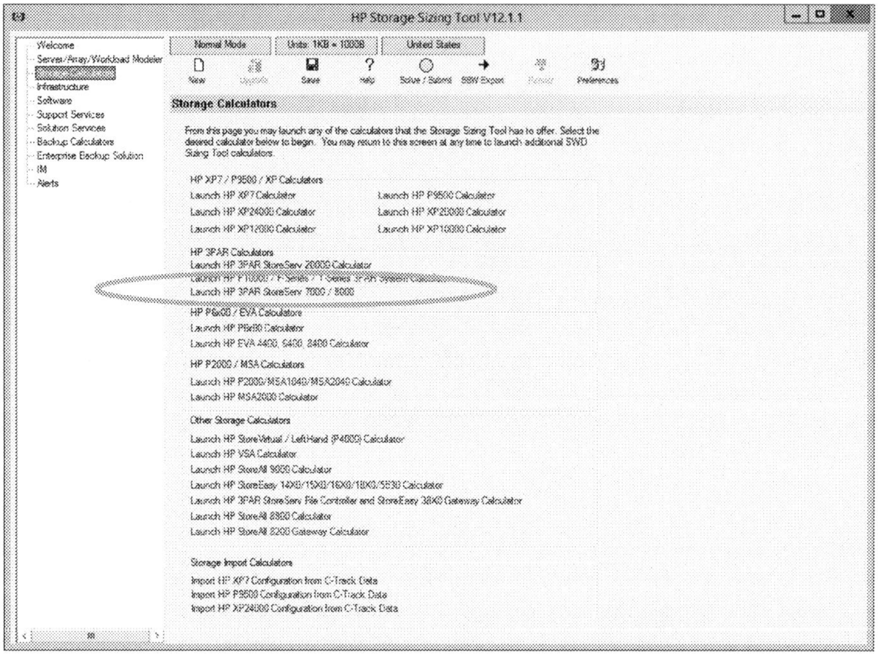

Figure 5-107 The **Launch HPE 3PAR StoreServ 7000/8000** link

- You can select an array type such as all-flash **8450** from the Array Type combo box (Figure 5-108).

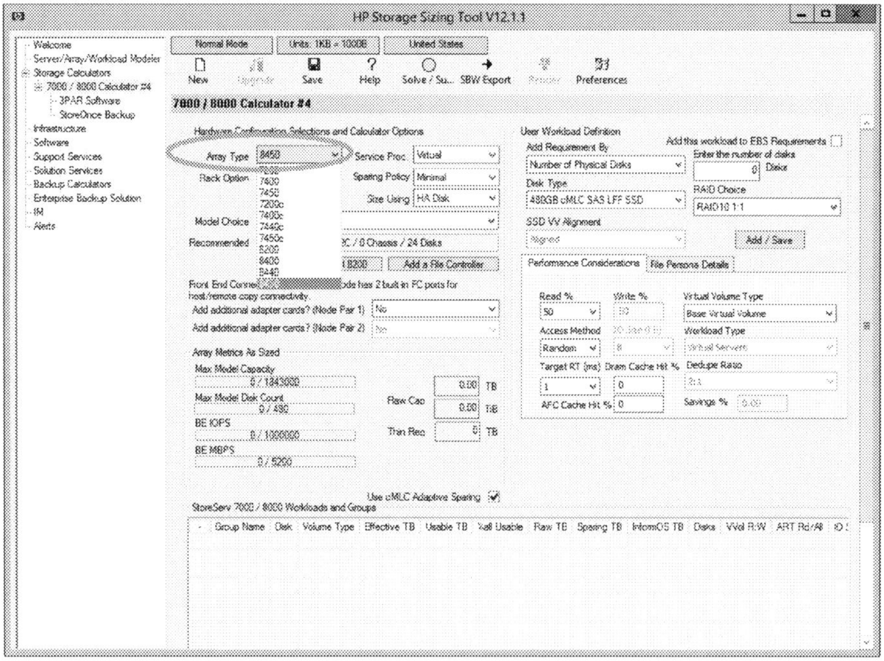

Figure 5-108 An Array Type—all-flash **8450**

- Various options such as racks can be added to the configuration (Figure 5-109).

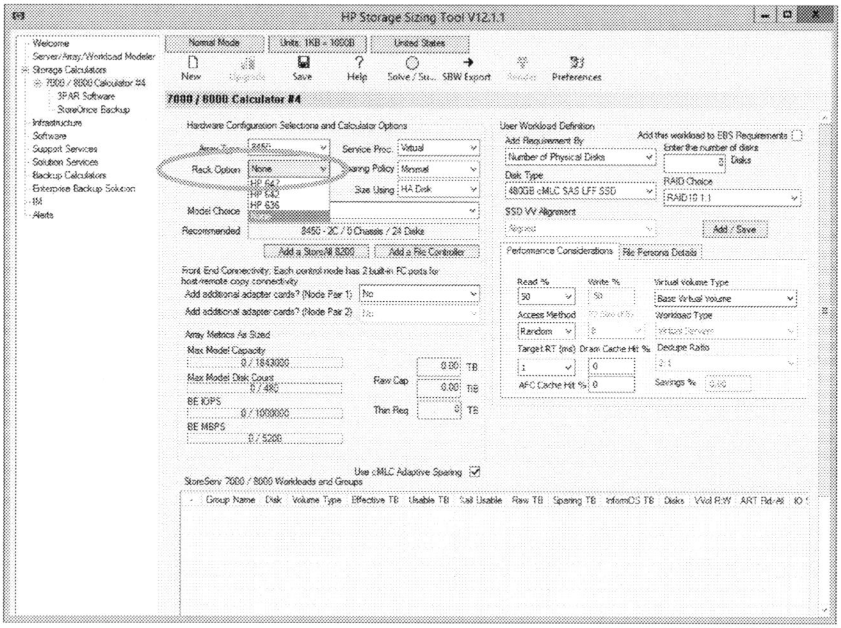

Figure 5-109 Racks options

- Different types of workloads can be configured for HPE 3PAR Array, based on number of disks, usable capacity, or IOPS (Figure 5-110).

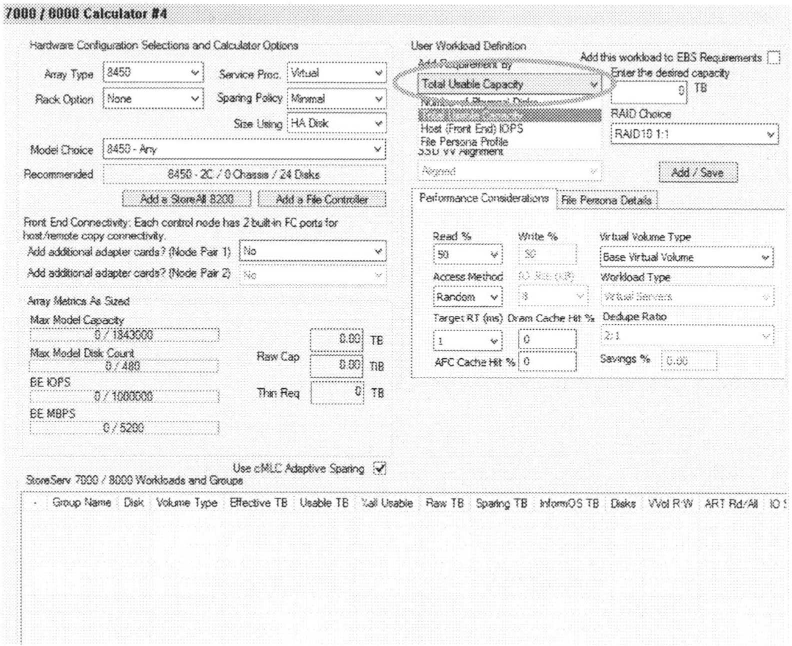

Figure 5-110 Types of workloads

- If the workload is being added based on total usable capacity, you must enter a desired capacity (Figure 5-111).

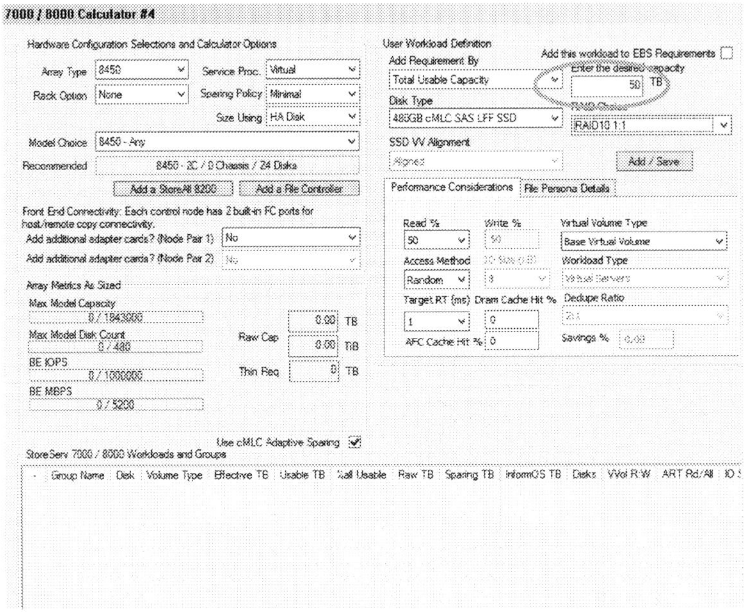

Figure 5-111 Workload added—enter a desired capacity

- The HPE Storage Sizing Tool also enables you to specify RAID levels and a data-to-parity ratio for workloads, which helps to select the proper amount of drives (Figure 5-112).

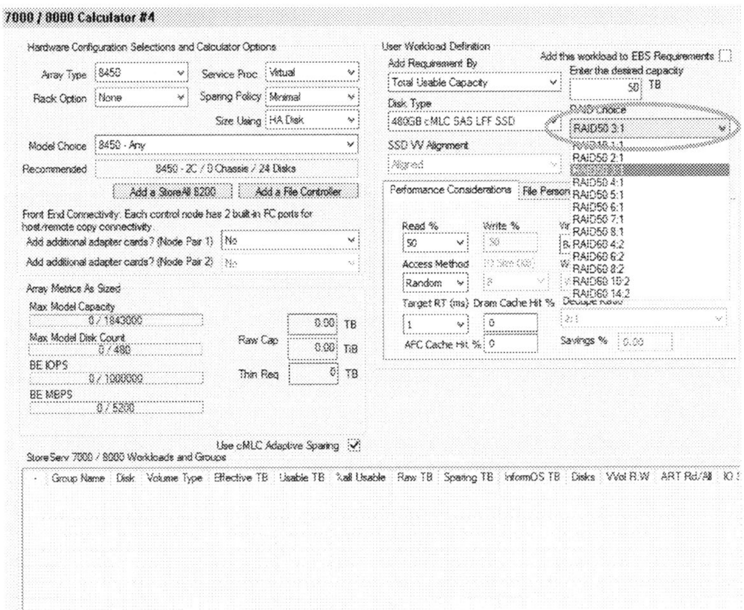

Figure 5-112 RAID choice—proper amount of drives

- Various types of drives and SSDs can be selected in the Disk Type field (Figure 5-113).

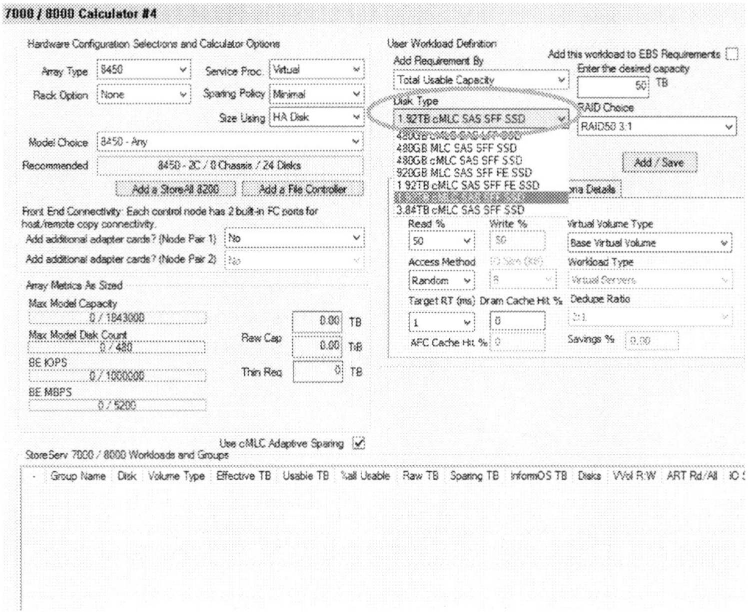

Figure 5-113 the Disk Type field—Various types of drives and SSDs

- When all of the parameters for workloads have been added, the disk group can be added by clicking the **Add/Save** button.

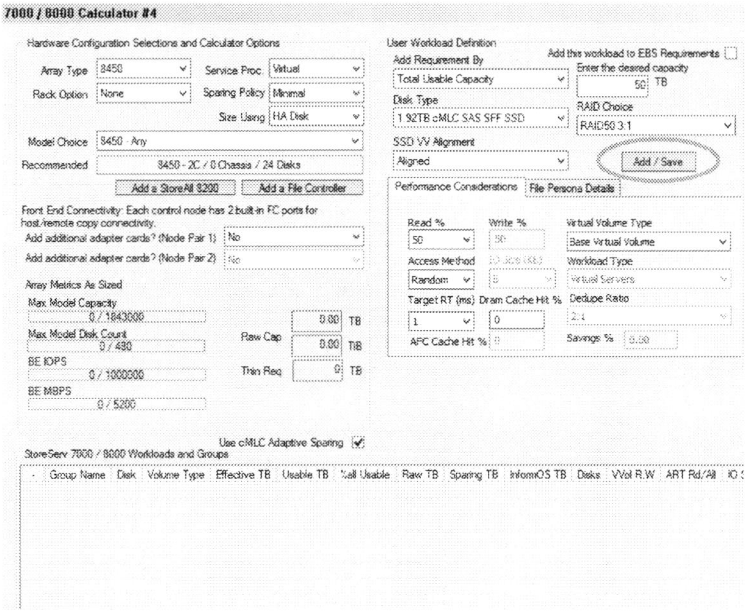

Figure 5-114 The disk group added by clicking the **Add/Save** button

You can repeat the steps to add multiple workloads to single array, using different workload definitions (Figure 5-115).

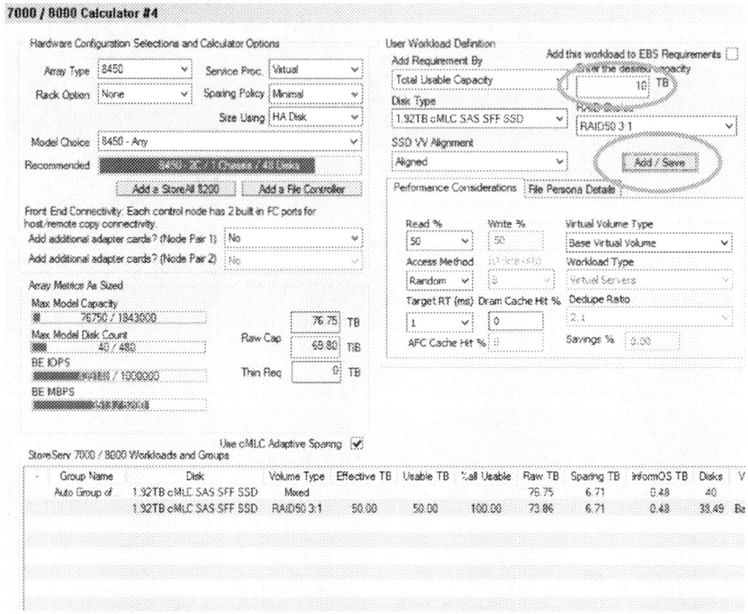

Figure 5-115 To add multiple workloads to single array—use different workload definitions

- The HPE Storage Sizing Tool also allows you to add software components (Figure 5-116).

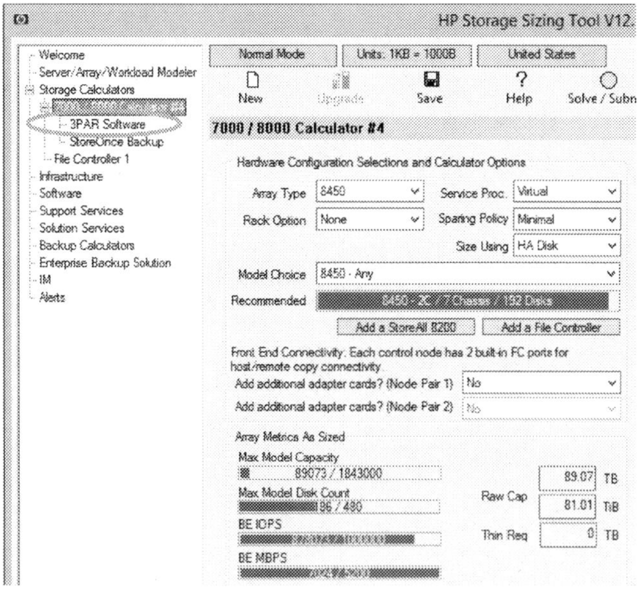

Figure 5-116 The HPE Storage Sizing Tool—add software components

- For example, products such as the **File Persona Suite** can be added to the configuration from the Product selection combo box (Figure 5-117).

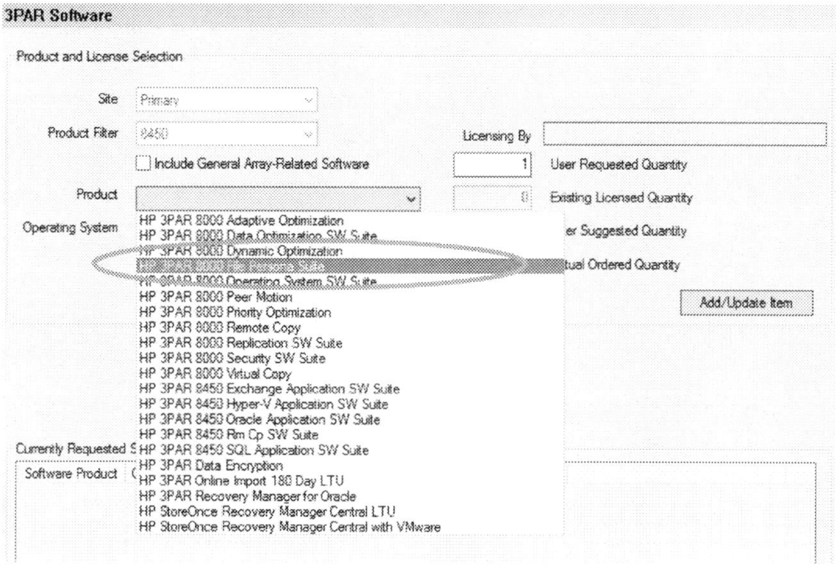

Figure 5-117 The File Persona Suite

- Software products usually require additional configuration. The **Add/Update Item** link will add the product to the configuration (Figure 5-118).

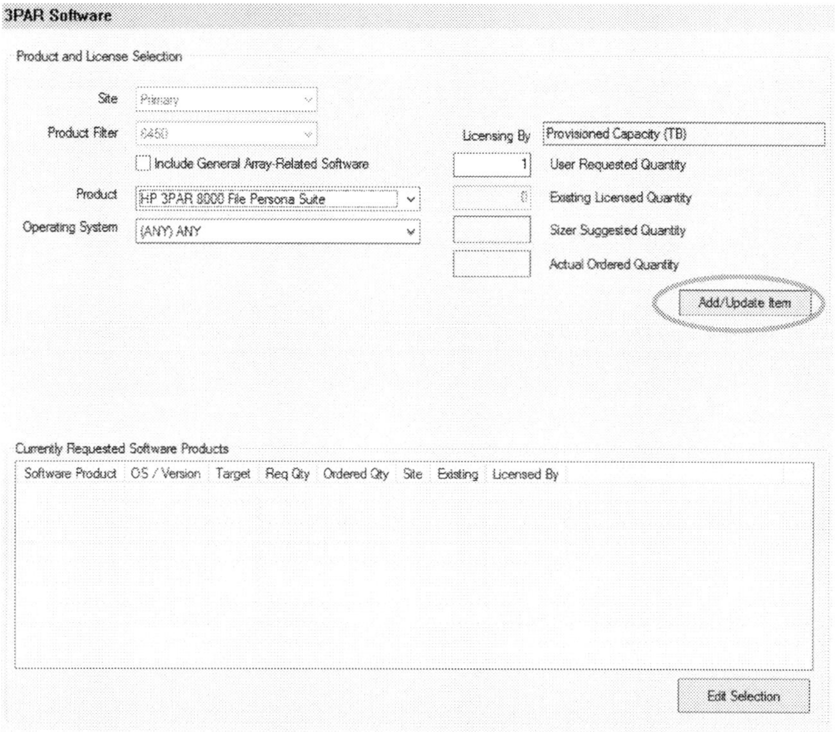

Figure 5-118 The **Add/Update Item** link

- Click the **Solve/Submit** button to create a configuration based on the user input (Figure 5-119).

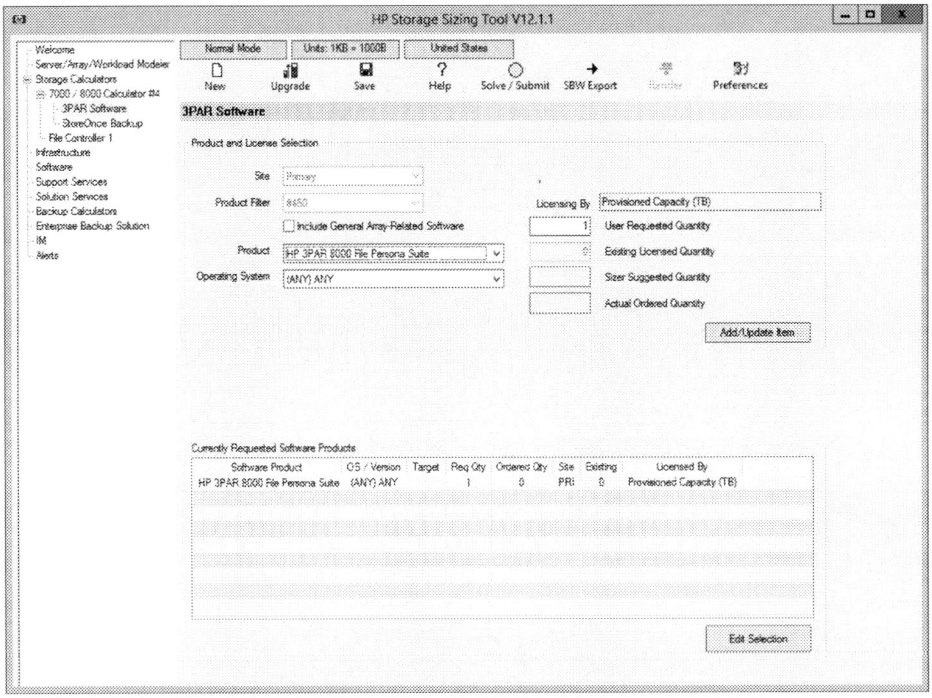

Figure 5-119 To create a configuration based on the user input

- The exported file can be opened in Microsoft Excel (Figures 5-120 and 5-121).

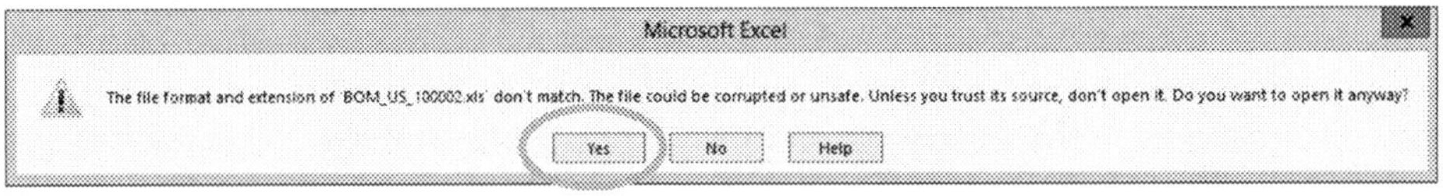

Figure 5-120 The exported file to be opened in Microsoft Excel

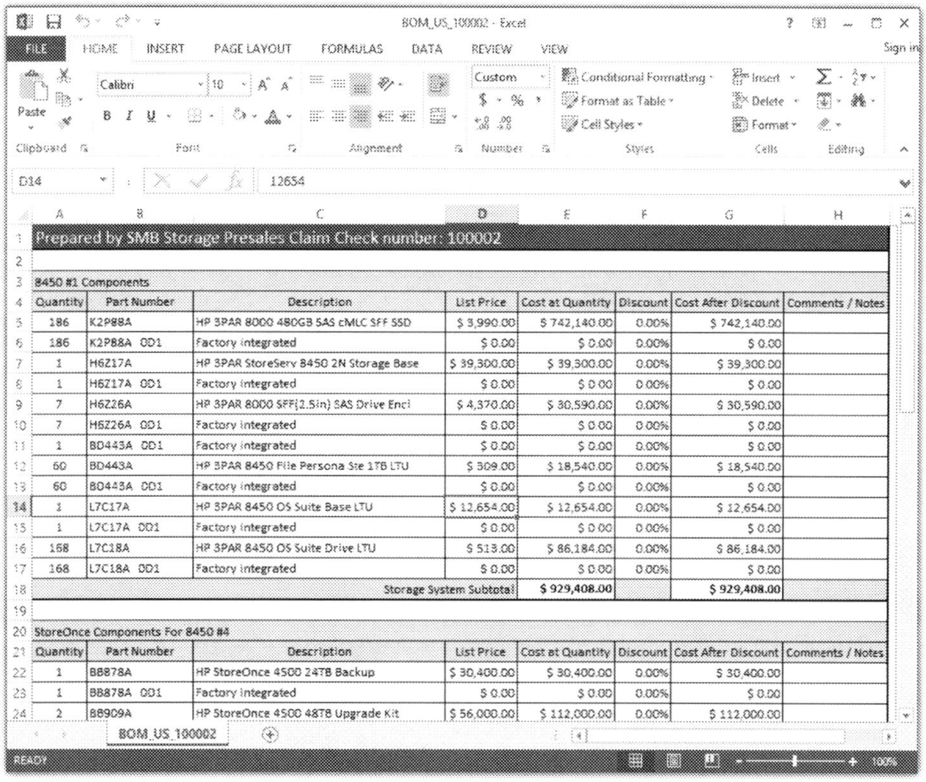

Prepared by SMB Storage Presales Claim Check number: 100002

8450 #1 Components

Quantity	Part Number	Description	List Price	Cost at Quantity	Discount	Cost After Discount	Comments / Notes
186	K2P88A	HP 3PAR 8000 480GB SAS cMLC SFF SSD	$ 3,990.00	$ 742,140.00	0.00%	$ 742,140.00	
186	K2P88A 0D1	Factory integrated	$ 0.00	$ 0.00	0.00%	$ 0.00	
1	H6Z17A	HP 3PAR StoreServ 8450 2N Storage Base	$ 39,300.00	$ 39,300.00	0.00%	$ 39,300.00	
1	H6Z17A 0D1	Factory integrated	$ 0.00	$ 0.00	0.00%	$ 0.00	
7	H6Z26A	HP 3PAR 8000 SFF(2.5in) SAS Drive Encl	$ 4,370.00	$ 30,590.00	0.00%	$ 30,590.00	
7	H6Z26A 0D1	Factory integrated	$ 0.00	$ 0.00	0.00%	$ 0.00	
1	BD443A 0D1	Factory integrated	$ 0.00	$ 0.00	0.00%	$ 0.00	
60	BD443A	HP 3PAR 8450 File Persona Ste 1TB LTU	$ 309.00	$ 18,540.00	0.00%	$ 18,540.00	
60	BD443A 0D1	Factory integrated	$ 0.00	$ 0.00	0.00%	$ 0.00	
1	L7C17A	HP 3PAR 8450 OS Suite Base LTU	$ 12,654.00	$ 12,654.00	0.00%	$ 12,654.00	
1	L7C17A 0D1	Factory integrated	$ 0.00	$ 0.00	0.00%	$ 0.00	
168	L7C18A	HP 3PAR 8450 OS Suite Drive LTU	$ 513.00	$ 86,184.00	0.00%	$ 86,184.00	
168	L7C18A 0D1	Factory integrated	$ 0.00	$ 0.00	0.00%	$ 0.00	
			Storage System Subtotal	$ 929,408.00		$ 929,408.00	

StoreOnce Components For 8450 #4

Quantity	Part Number	Description	List Price	Cost at Quantity	Discount	Cost After Discount	Comments / Notes
1	BB878A	HP StoreOnce 4500 24TB Backup	$ 30,400.00	$ 30,400.00	0.00%	$ 30,400.00	
1	BB878A 0D1	Factory integrated	$ 0.00	$ 0.00	0.00%	$ 0.00	
2	BB909A	HP StoreOnce 4500 48TB Upgrade Kit	$ 56,000.00	$ 112,000.00	0.00%	$ 112,000.00	

Figure 5-121 The exported file output

HPE NinjaSTARS for 3PAR

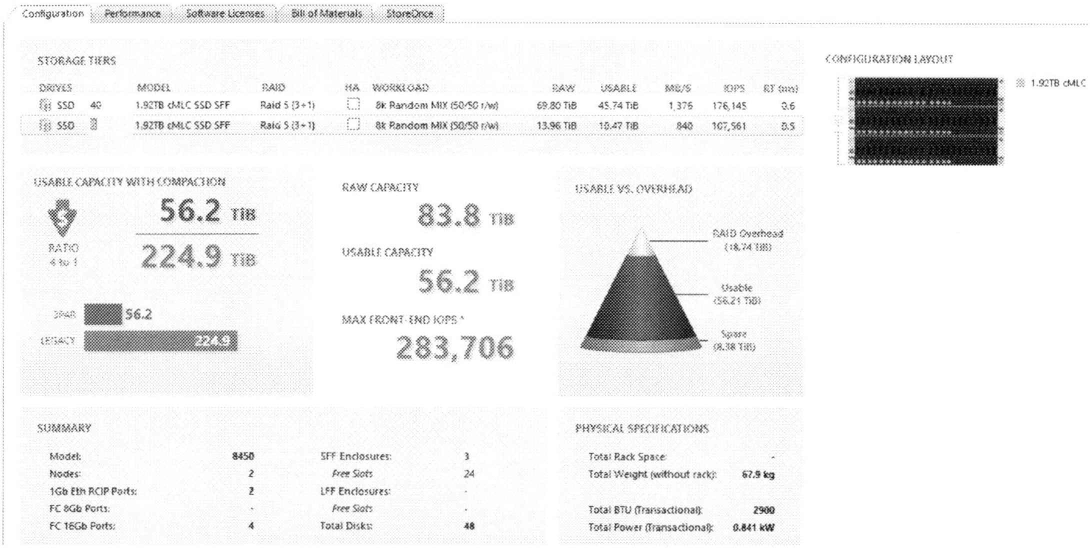

Figure 5-122 HPE NinjaSTARS for 3PAR

The next-generation HPE NinjaThin Assessment Tool is HPE NinjaSTARS for 3PAR (Figure 5-122). (STARS stands for Storage Assessment, Recommendation, and Sizing.)

NinjaSTARS for 3PAR replaces NinjaThin 7000. It includes measured data from HPE Storage Optimizer as well as other new features for increased accuracy and performance modeling. NinjaSTARS was developed for worldwide channel partners to assess legacy storage environments for customers interested in HPE 3PAR StoreServ arrays.

Note

To download the tool, navigate to the Get Thin Guarantee page on the regional Partner Portal. From the landing page, go to **Business Units→ HPE Enterprise Group → HPE Storage, Programs and Promotions**, and then **HPE 3PAR Get Thin Guarantee Program → Ninja Thin Tool**. Access NinjaSTARS for 3PAR technical training, if needed. For tool support and feedback, contact ninjastars@hpe.com.

HPE NinjaSTARS for 3PAR has built-in templates and wizards that help sales to

- Assess a customer's current storage solution
- Provide StoreServ configuration and sizing recommendations
- Meet customers' performance and growth needs
- Drive better customer conversations based on their unique IT environment

HPE NinjaSTARS for 3PAR includes the following:

- Powerful editing tools and direct import of EVA configuration data
- Highly accurate performance sizing based on measured or lab-tested data
- Written data utilization and TCO projections similar to the NinjaThin 7000 tool
- Storage Optimizer measured data for increased accuracy and performance modeling

Example—Using HPE NinjaSTARS for 3PAR

The following example shows how to use HPE NinjaStars for 3PAR to configure a solution based on this customer scenario:

The customer is planning to upgrade an EVA 8400 system to HPE 3PAR StoreServ 8440 and is currently using 600 GB 15K SAS and 2 TB NL FATA drives. The administrator can monitor the performance of the array and provide an export EVAPerf during a 4-hour window. The customer wants to use cost-optimized storage tiering.

Following are some key points for using the sizing tool:

- **NinjaSTARS for 3PAR** can be started by using the icon on the desktop (Figure 5-123).

Figure 5-123 **NinjaSTARS for 3PAR** icon

- You can create a new configuration, click **New** in the NinjaSTARS configuration ribbon (Figure 5-124).

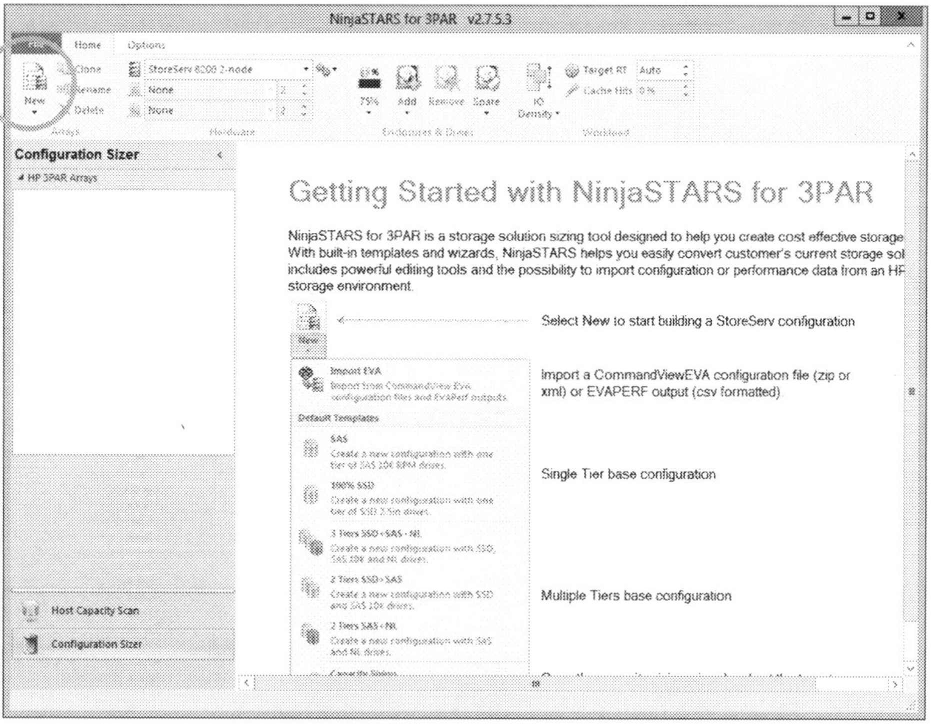

Figure 5-124 Getting started with NinjaSTARS configuration ribbon

- A key benefit of NinjaSTARS is the ability to import performance date from EVAPerf (Figures 5-125 and 5-126).

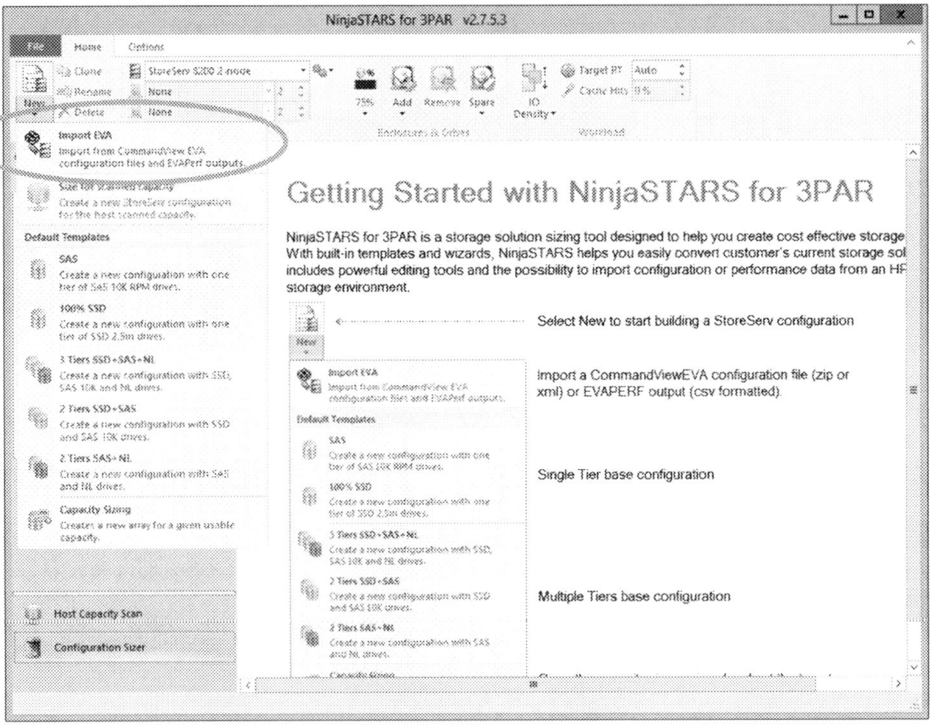

Figure 5-125 A key benefit—the ability to import performance date from EVAPerf

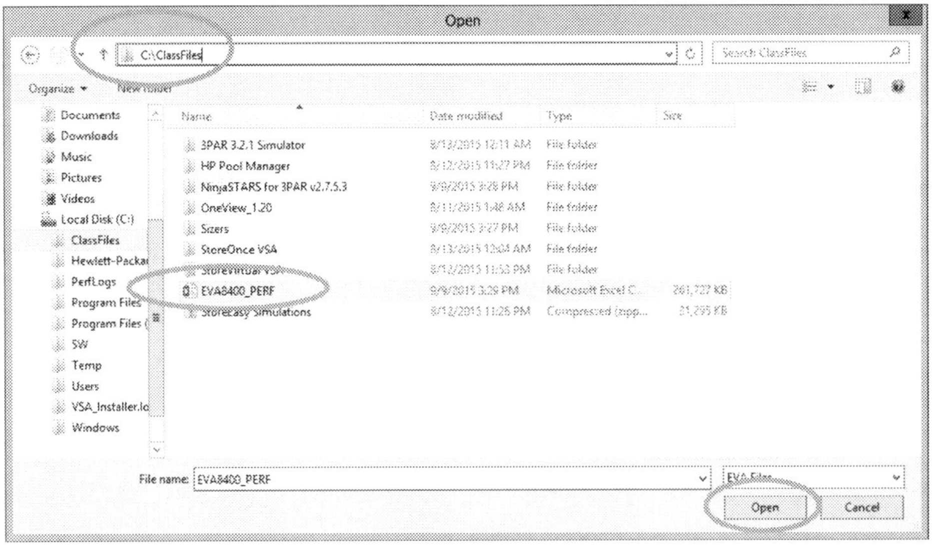

Figure 5-126 EVAPerf

- The import procedure takes some time to complete. During the import, the length of the capture window and the original model are displayed (Figure 5-127).

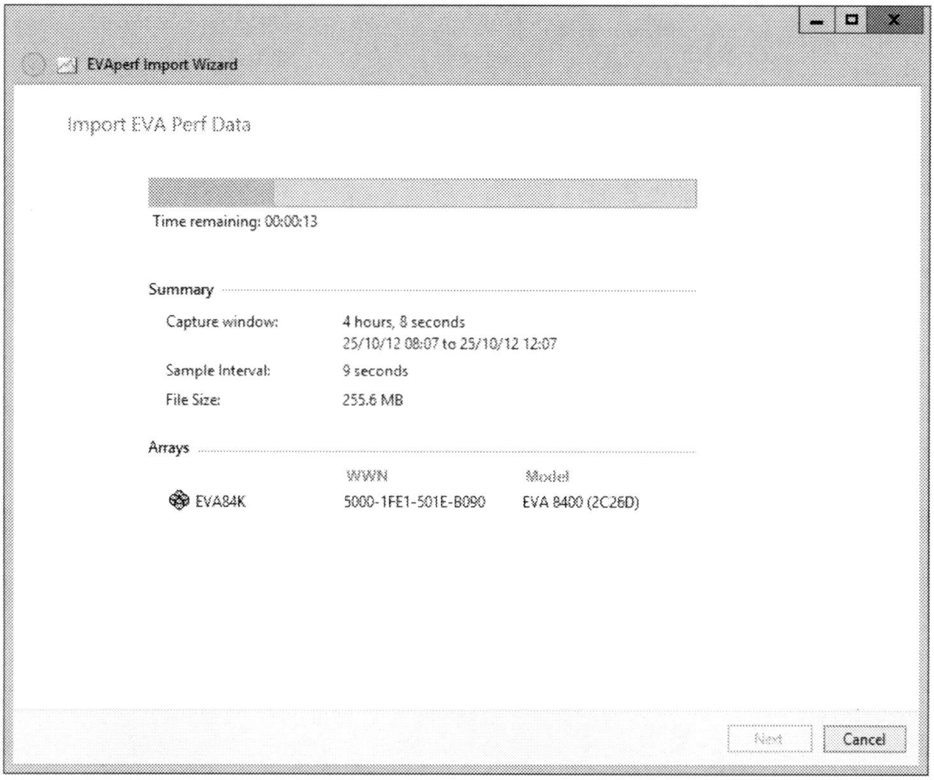

Figure 5-127 The import procedure

- After the import finishes, click the **Next** button to go to the next step (Figure 5-128).

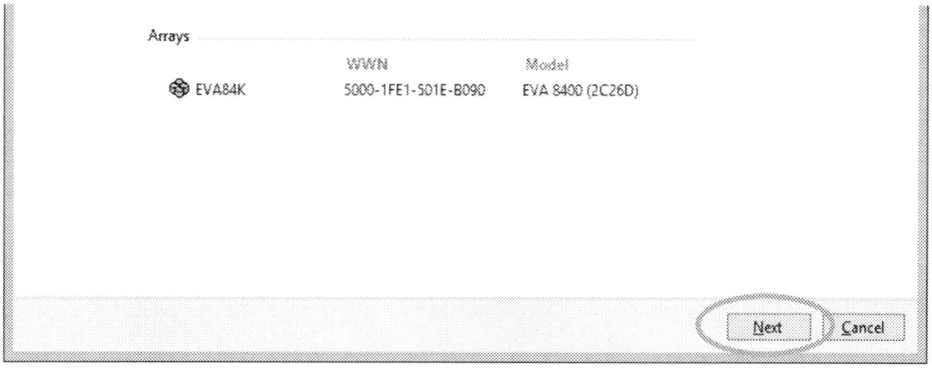

Figure 5-128 To go to the next step—click the **Next** button

- You can add missing information by clicking the empty area in the Disk Type column and then enter data as described by the customer (Figure 5-129 and 5-130).

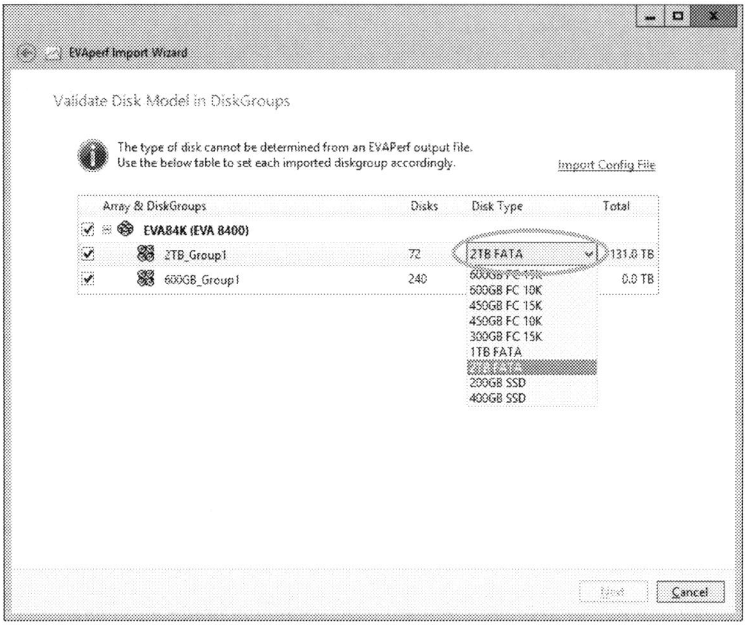

Figure 5-129 Add missing information

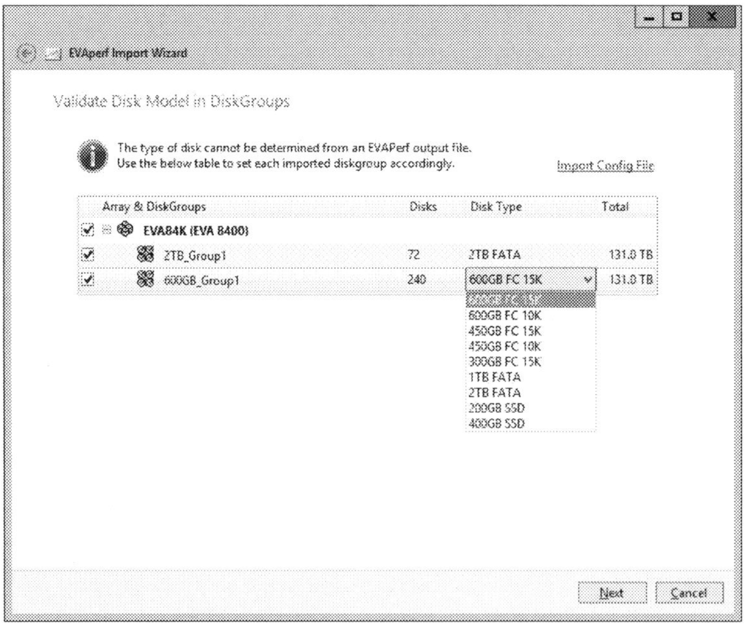

Figure 5-130 An EVAPerf output file

- The errors will disappear (Figure 5-131).

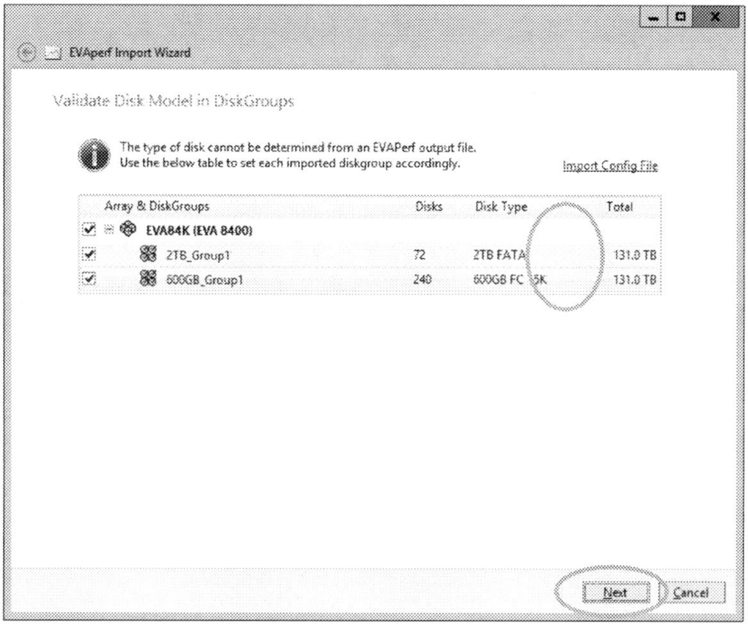

Figure 5-131 Table to set each imported diskgroup accordingly

- You can review the performance sizing goals before closing the wizard with the **Finish** button (Figure 5-132).

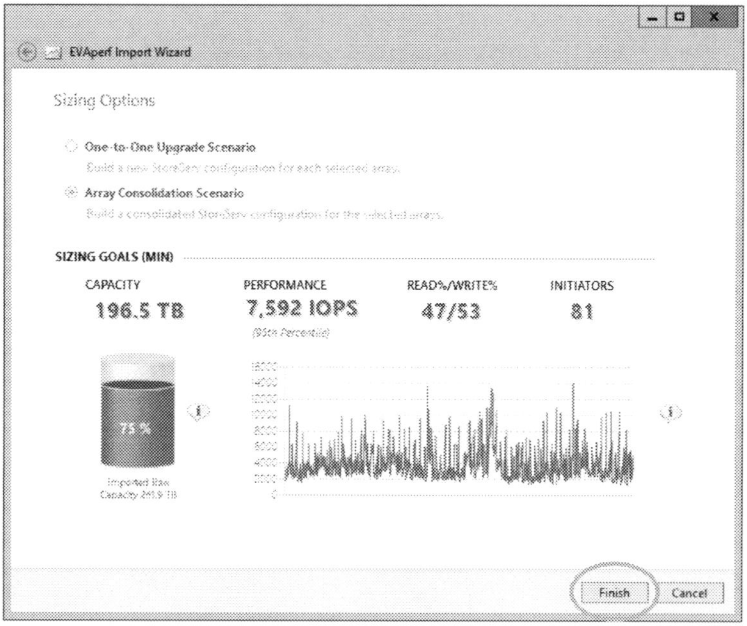

Figure 5-132 Sizing options

- You can change the array type based on customer preferences (Figure 5-133).

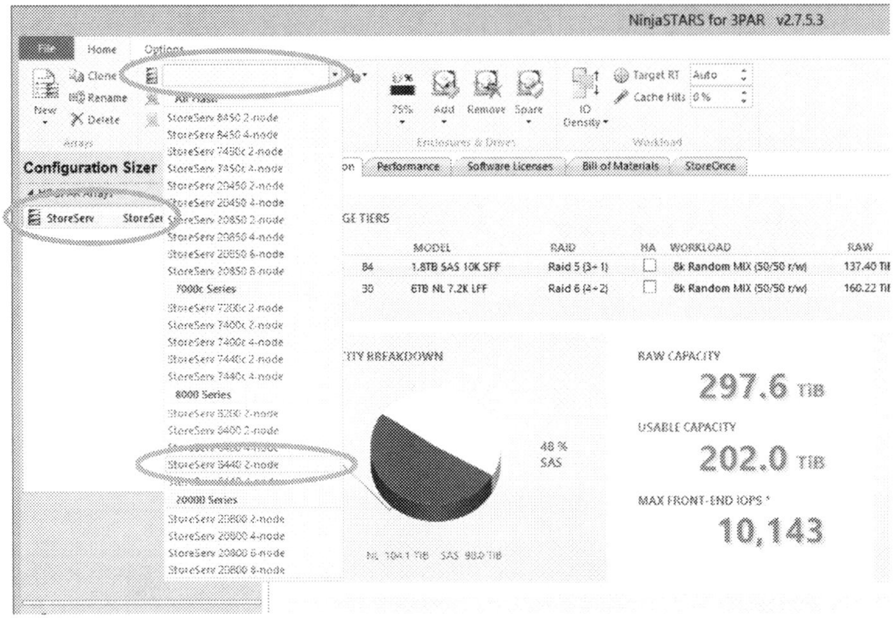

Figure 5-133 The array type based on customer preferences

- The sizer will display your proposed configuration (Figure 5-134).

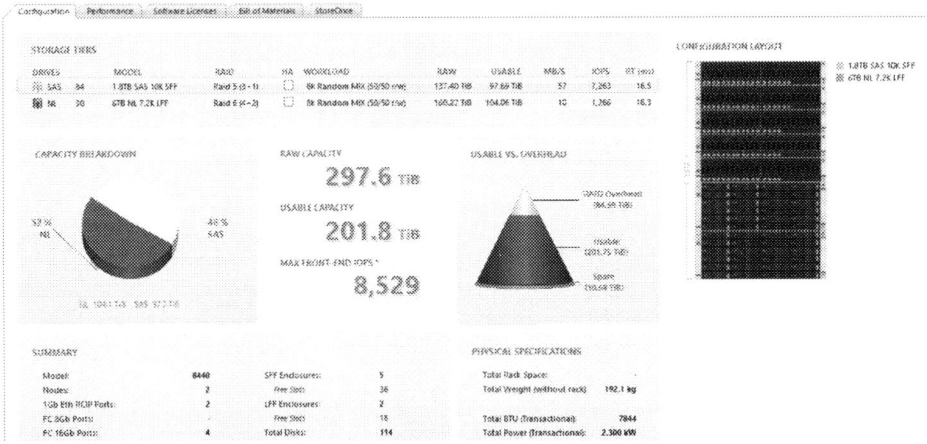

Figure 5-134 Proposed configuration

- You can click the **Performance** tab to see additional information such as IOPS (Figure 5-135).

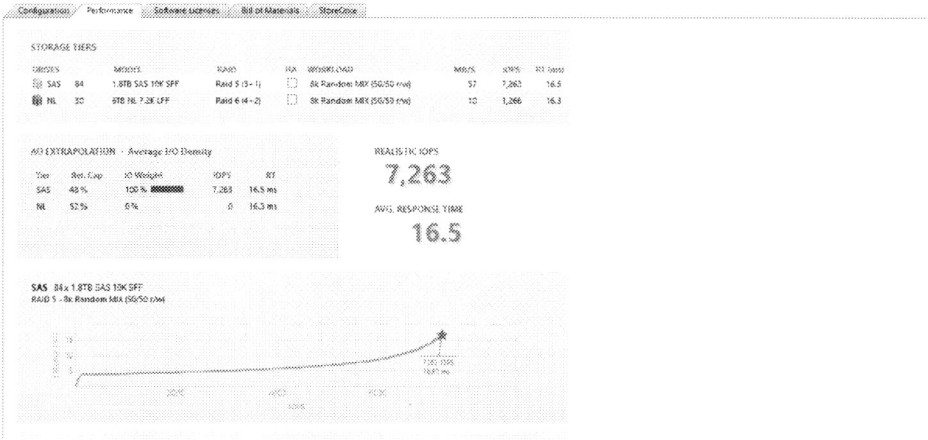

Figure 5-135 To see additional information

- The **Software Licenses** tab provides additional software component selections (Figure 5-136).

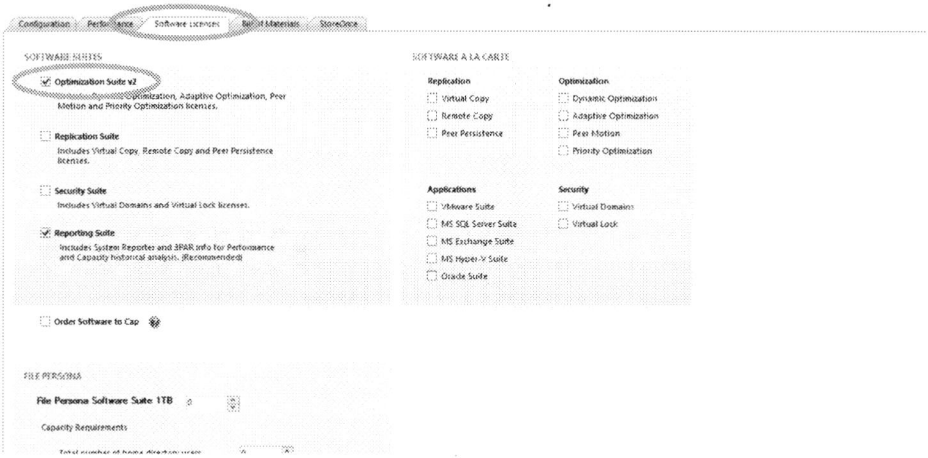

Figure 5-136 The **Software Licenses** tab

- The **Bill of Materials** tab shows recommended part numbers and quantities (Figure 5-137).

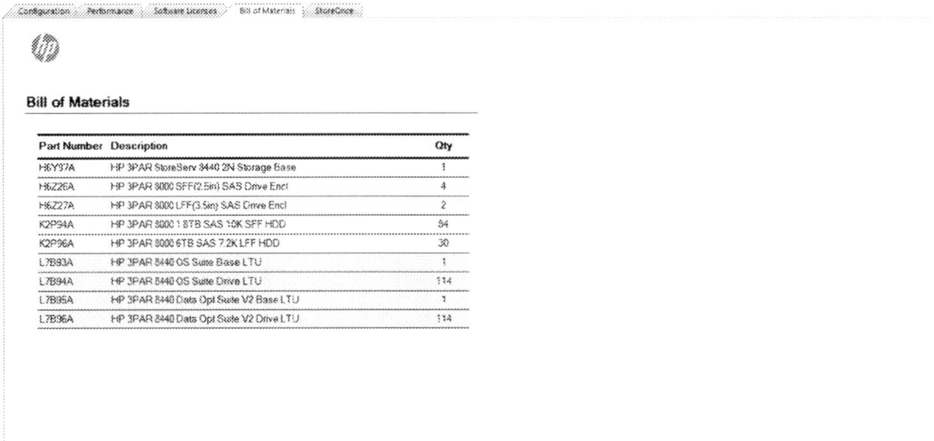

Part Number	Description	Qty
H6Y97A	HP 3PAR StoreServ 8440 2N Storage Base	1
H6Z26A	HP 3PAR 8000 SFF(2.5in) SAS Drive Encl	4
H6Z27A	HP 3PAR 8000 LFF(3.5in) SAS Drive Encl	2
K2P94A	HP 3PAR 8000 1.8TB SAS 10K SFF HDD	84
K2P96A	HP 3PAR 8000 6TB SAS 7.2K LFF HDD	30
L7B93A	HP 3PAR 8440 OS Suite Base LTU	1
L7B94A	HP 3PAR 8440 OS Suite Drive LTU	114
L7B95A	HP 3PAR 8440 Data Opt Suite V2 Base LTU	1
L7B96A	HP 3PAR 8440 Data Opt Suite V2 Drive LTU	114

Figure 5-137 **Bill of Materials** tab

- You can save the final configuration in various formats (Figure 5-138).

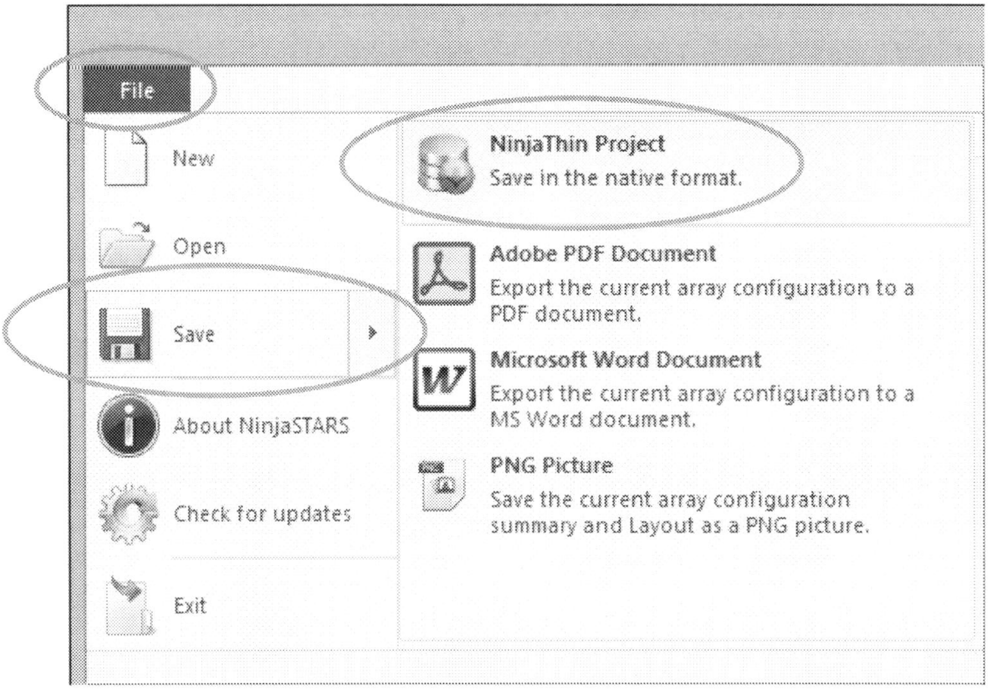

Figure 5-138 Various formats for saving

Table 5-15 HPE 3PAR StoreServ 7000 and 8000 feature comparison

	HPE 3PAR StoreServ 7000	HPE 3PAR StoreServ 8000
ASIC	Gen4	Gen5
Processors	Sandy Bridge	Ivy Bridge
	2x 6-core 1.8 GHz or	2x 6-core 2.2 GHz or
	2x 8-core 2.3 GHz	2x 10-core 2.4 GHz
Control memory	1 DIMM	2 DIMMs
PCI Express	Gen 2–4 GB/s	Gen 3–8 GB/s
Built-in FC ports	8 Gb	16 Gb
SAS backend	6 Gb	12 Gb (copper and optical)
IOPS random read	Up to 900k IOPS 4Kb	Up to 1000k IOPS 8Kb
Sequential read bandwidth	Up to 5.8 GB/s	Up to 24 GB/s
Maximum initiators supported	Up to 1024 or 2048	Up to 2048 or 4096
Maximum raw capacity	Up to 2000 TiB	Up to 3000 TiB

Note

Not all differences are listed, such as bigger caches on the 8000 series. Most characteristics depend on the particular model.

Learning check

The following questions will help you to measure your understanding of the material presented in this chapter. Read all the choices carefully because there might be more than one correct answer. Select or write the correct answers for each question.

1. A SAN requires all components to come from a single vendor.

   ☐ True

   ☐ False

2. Name at least three components of a SAN solution.

   ....

   ....

   ....

   ....

   ....

3. What are the differences between HPE MSA 1040 Storage and HPE MSA 2040 Storage?

   ....

   ....

   ....

   ....

   ....

4. The maximum capacity of the MSA 1040 device is 16 TB.

   ☐ True

   ☐ False

5. The HPE MSA 2040 device can be configured to achieve more than 100k IOPS for random reads.

   ☐ True

   ☐ False

6. Name at least three components of the HPE StoreVirtual solution.

 ....

 ....

 ....

 ....

 ....

7. HPE StoreVirtual Systems are built on the HPE ProLiant platform.

 ☐ True

 ☐ False

8. HPE 3PAR StoreServ nodes are connected in a ring topology.

 ☐ True

 ☐ False

9. Recommend two products capable of providing file and object access to HPE 3PAR StoreServ.

 ....

 ....

 ....

 ....

 ....

10. HPE NinjaSTARS for 3PAR supports imported performance data from HPE EVA systems.

 ☐ True

 ☐ False

For answers, see Chapter 5 Answers on page 440.

6 SAN Infrastructure

OBJECTIVES

In this chapter, you learn to:

✓ Identify the products in the HPE storage area network (SAN) infrastructure portfolio

✓ Explain the unique features and target customers for each product family:

- ✓ HPE StoreFabric Fibre Channel (FC) switches
- ✓ HPE StoreFabric FC host bus adapters (HBAs) and converged network adapters (CNAs)

✓ Describe specifications and performance limits of cables, transceivers, and port licenses

✓ Explain how to size SAN infrastructure products

✓ Locate reference information for the HPE SAN infrastructure portfolio

INTRODUCTION

This chapter provides an overview of SAN infrastructure products. It begins with basics of HPE products in this category and then describes unique features, capacity limits, performance figures, and target customers for HPE StoreFabric products. It also describes HPE StoreFabric FC switches, server HBAs, and available accessories. Lastly, it describes reference documents and tools you can use for sizing and configuration.

SAN Infrastructure

The HPE StoreFabric portfolio offers expansive connectivity between servers and storage in a data center. You can choose from a broad selection of products that are focused on performance and resiliency and help transition and consolidate data and networks concurrently. You can deploy new data center technologies, converge unlike protocols, and move to new designs while protecting your previous investments. HPE StoreFabric products are tested, certified, and serviced for end-to-end integration into the HPE storage, networking, and server ecosystem for a superior SAN experience.

HPE Storage Networking

Figure 6-1 HPE StoreFabric products

From small and medium-sized operations to data centers and cloud, HPE StoreFabric has dynamic end-to-end solutions that solve even most frustrating storage networking challenges (Figure 6-1). The comprehensive portfolio of HPE storage networking products and accessories includes switches and directors as well as routers, adapters, transceivers, and cables for entry-level, midrange, and high-end environments. The HPE SAN infrastructure products include the following:

- SAN switches
- FC directors
- HBAs
- Cables, transceivers, and port licenses
- Software

Important

Not all HPE StoreFabric products are suitable for small and medium-sized business environments.

HPE StoreFabric product portfolio

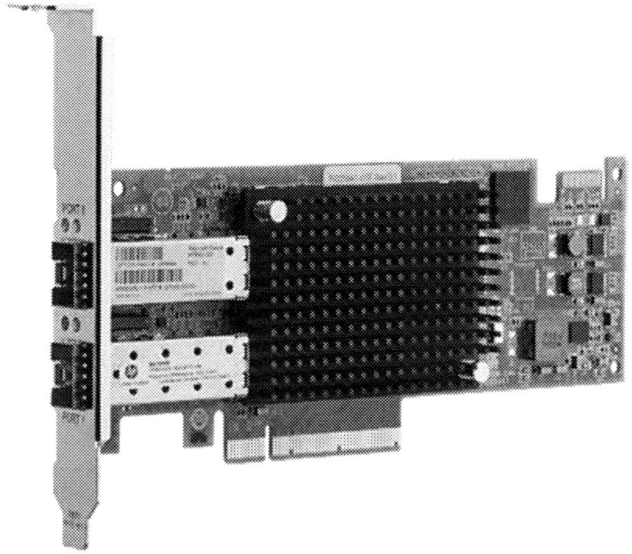

Figure 6-2 HPE StoreFabric Host FC HBA

The HPE StoreFabric product portfolio supports a wide range of data center protocols such as FC, Ethernet (1 GbE/10 GbE), FC over IP (FCIP), Internet Small Computer System Interface (iSCSI), and FC over Ethernet (FCoE) (Figure 6-2).

The extensive portfolio includes FC and multipurpose or converged protocol switches, transceivers, and cables. These products provide the scalability, flexibility, performance, security, and interoperability that mission-critical data and applications demand from a storage networking environment.

HPE StoreFabric FC switches

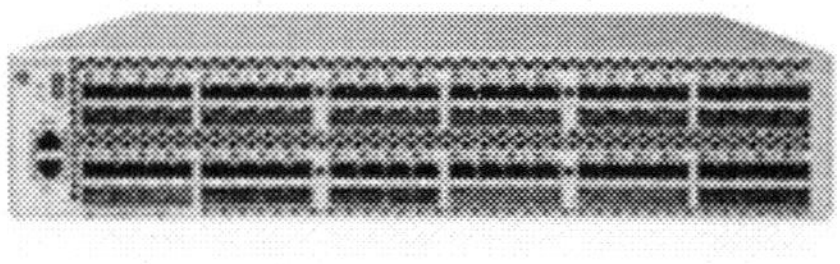

Figure 6-3 HPE StoreFabric FC switches

These switches combine ease of use, extensive interoperability testing, flexibility, and improved security with cost-effective pricing (Figure 6-3).

The HPE StoreFabric FC switches support demanding virtualized environments with 16 Gb/s (Gen5) FC performance.

HPE StoreFabric FC directors

Figure 6-4 HPE StoreFabric FC directors

You can build highly scalable networks that address data growth and application demands and that transform current SAN fabrics into cloud-optimized SANs (Figure 6-4).

HPE advanced director-class switches, coupled with network management software, provide breakthrough performance, high availability, and energy efficiency, with long-term investment protection.

HPE StoreFabric Host FC adapters and CNAs

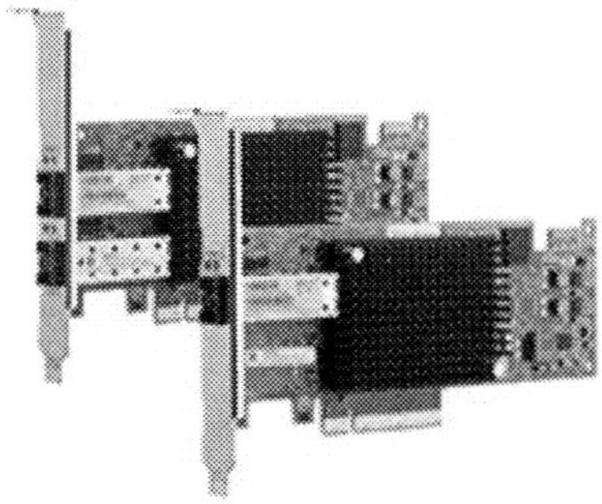

Figure 6-5 HPE StoreFabric Host FC adapters and CNAs

You can eliminate I/O bottlenecks in environments with greater virtual machine (VM) density and bandwidth requirements (Figure 6-5).

HPE HBA deliver next-generation performance that can help your business gain a higher return on IT investment.

These devices are also part of an integrated solution for HPE Storage and servers. HBA firmware updates are easy when using the HPE Smart Update utility.

Cables, transceivers, and port licenses

Figure 6-6 HPE FC transceivers

A variety of pay-as-you-go accessories for storage networking save time and money (Figure 6-6).

For example, HPE optical cables and hardware reduce space and costs. Port license upgrades enable you to scale the SAN easily.

Why choose HPE StoreFabric Storage Networking?

HPE StoreFabric Storage Networking offers the following advantages:

- **Integrated**—The products are tested, certified, and serviced for the HPE storage, server, and networking ecosystem.
- **Optimized**—The products bring lower costs, better performance, and efficiency for virtual server and cloud environments.
- **Comprehensive**—the portfolio includes directors, switches, multiprotocol routers, data migration appliances, adapters, cables, and SAN software.
- **Converged**—You can protect legacy investments while combining unlike protocols (FC, 1 GbE/10 GbE, FCIP, iSCSI, and FCoE).

HPE StoreFabric FC switches

Product lines:

- H-Series—Entry-level and midrange switches
- B-Series—Entry-level and midrange switches
- C-Series—Midrange switches
- HPE BladeSystem—Embedded switches

Differentiators:

- Form Factors
    - BladeSystem
    - Rack
- Protocol support
    - FC
    - FCIP
    - FCoE
    - Ethernet
    - Converged Enhanced Ethernet (CEE)
    - Inter-Switch Link Trunking (ISL)
    - Serial Attached SCSI (SAS)
- Maximum numbers of ports
    - 16, 20, 22, 24, 48, more than 80
- Extra features
    - SAN extension
    - Stackable
    - Encryption

The HPE StoreFabric FC switch portfolio offers benefits to businesses of all sizes. These switches can provide improved performance and functionality, faster I/O, data convergence at the server level, larger network pipes for aggregation, better longevity of solutions, fewer disruptive upgrades, and more room to scale.

HPE offers an ideal range of FC and multiprotocol or converged protocol switches for SAN and LAN that deliver the scalability, performance, and broad interoperability that mission-critical data and applications demand. Combining ease of use, flexibility, and improved security with cost-effective pricing, these switches are key enablers of a dynamic and adaptive storage networking environment. Also, they provide an ideal platform for adaptation as storage networking needs grow and change—in terms of both basic connectivity and advanced SAN or converged fabric functionality.

HPE partners with industry-leading vendors to make sure our customers have a robust, end-to-end storage networking solution. You can choose from broad portfolio and implement the FC switch that best meets your storage requirements.

Whether you want to put a single-switch solution into action in a new SAN installation, adopt a FC SAN for the first time, add switches to a core/edge network as an aggregation point for remote or branch offices, or meet the needs of a hyperscale, private cloud storage environment, the broad HPE portfolio can help in making the best choice. From entry-level to midrange and from embedded to enterprise, our range of switches caters to the diverse and unique requirements of storage networking environments.

Important
Not all HPE StoreFabric FC switches are covered in this chapter.

H-Series: Entry-level switches

HPE 8/20q FC Switch and 8 Gb Simple SAN Connection Kit

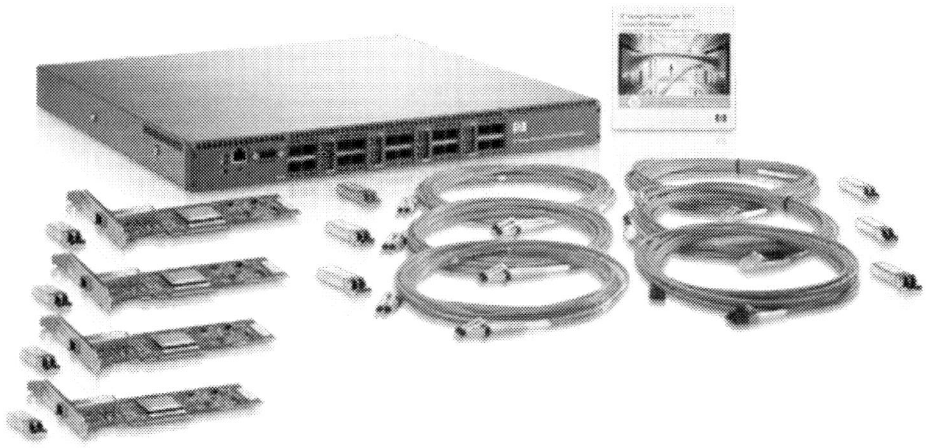

Figure 6-7 HPE 8/20q FC Switch and 8 Gb Simple SAN Connection Kit

Features of the HPE 8/20q FC Switch (Figure 6-7) include the following:

- Topology: 8 Gb/s FC
- Management: A complete SAN management system
- Bandwidth: 340 Gb/s
- Number of ports: 8–20 ports (depending on the model)
    - Upgradable in 4-port increments
- Aggregate switch bandwidth: 340 Gb/s

- Included management features:
    - SAN Connection Manager (SCM)
    - QuickTools
    - Command line interface
- Warranty: 1/1/1

The HPE 8/20q SAN Switch offers the simplest and most affordable way to share the capacity of FC storage devices among multiple servers. You can connect as few as two devices or scale up to 20 FC devices with a single switch. You can mix any combination of servers, disk storage, or backup storage for maximum flexibility and investment protection.

The HPE 8/20q also supports legacy 4 Gb and 2 Gb FC SAN devices and the ability to interconnect multiple FC switches to expand a SAN. The HPE 8/20q SAN Switch interoperates with existing Brocade and Cisco switches, making the H-Series 8/20q a valuable contributor to any SAN.

The HPE 8/20q SAN Switch supports all of the HPE FC storage arrays and FC backup solutions, including the following:

- P2000 Series Disk Arrays (MSA family)
- P6000 Series Disk Arrays (EVA family)
- 3PAR Series Disk Arrays
- StoreVirtual 4000 Series FC (LeftHand family with FC connectivity)
- Enterprise Backup Solutions

With the dependency on fewer storage devices in a consolidated SAN and with the increasing number of virtualized servers on fewer server hardware platforms, the need to build redundancy into SANs is paramount. The HPE 8/20q SAN Switch switches and H-Series HBAs can be configured as redundant FC paths to reduce the downtime associated with a single failed FC component.

B-Series: Entry-level switches

HPE SN3000B 16 Gb FC Switch

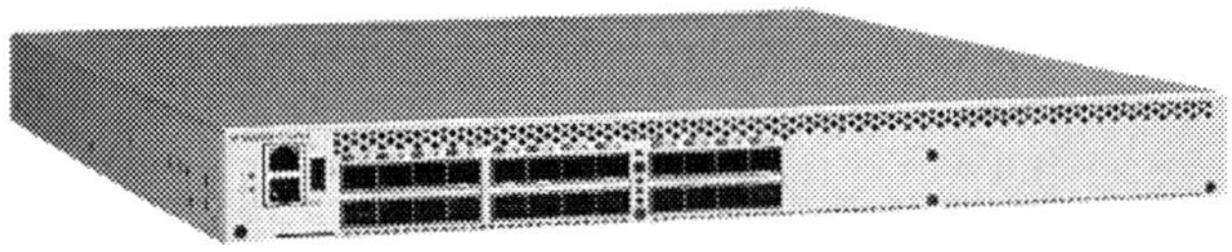

Figure 6-8 HPE SN3000B 16 Gb FC Switch

Features of the HPE SN3000B 16 Gb FC Switch (Figure 6-8) include

- 16 Gb/s
- 12 or 24 ports
    - Ports on Demand
- 768 Gb/s maximum bandwidth
- Redundant and hot swappable power supplies
- Adaptive Networking
    - Ingress Rate Limiting
    - Quality of Service (QoS)
    - Traffic Isolation Zones
- One-year on-site warranty, next-business-day (NBD) response

To remain competitive, IT organizations must keep pace with ever-increasing workloads without a similar increase in their budgets or their resources. Although virtualization has provided some relief by enabling the benefits of faster deployment and consolidation, it also tends to put additional stress on data center networks. In addition, the move toward cloud computing, which promises greater efficiency and a more service-oriented business model, means that these networks will face even greater demands.

The HPE SN3000B FC Switch meets the demands of hyperscale, private cloud storage environments by delivering market-leading 16 Gb FC technology and capabilities that support highly virtualized environments. Designed to enable maximum flexibility and investment protection, the SN3000B Switch is configurable in 12 or 24 ports and supports 4, 8, or 16 Gb/s speeds in an efficiently designed 1U package. It also provides a simplified deployment process and a point-and-click user interface, making it both powerful and easy to use. The SN3000B Switch offers low-cost access to industry-leading SAN technology while providing "pay-as-you-grow" scalability to meet the needs of an evolving storage environment.

The SN3000B FC Switch is available in two models:

- HPE SN3000B 16 Gb 24-port/24-port Active FC Switch
- HPE SN3000B 16 Gb 24-port/12-port Active FC Switch

Other characteristics include

- Flexible fabrics
    - 16 Gb/s optimized interswitch links (ISLs)
    - 128 Gb/s high-performance, resilient frame-based trunk
    - Port preservation for server and storage connectivity

- Operational simplicity
  - Dynamic Fabric Provisioning (DFP)
  - End-to-end optics and link validation
  - Simple deployment
  - HPE SAN Network Advisor
- Cloud-optimized performance
  - Energy-efficient advanced switching ASIC
  - Better performance on 16 Gb FC compared to earlier 8 Gb generation models (higher IOPS and twice the bandwidth)

Adaptive Networking

Adaptive Networking is a family of technologies that allows for flexible control of traffic movement within a fabric, which delivers application-aware management of fabric resources. Applications can be used with multiple protocols and multiple classes of service. Adaptive Networking includes the following features:

- **Ingress Rate Limiting**—Allows the ingress bandwidth of a port to be throttled to a rate lower than what was negotiated with the SAN node. This feature could be useful for enterprises that are offering stepped levels of service and enforcing SLAs.
- **QoS**—Enables zones with high, medium, and low priorities within a fabric, on a zone-by-zone basis. This feature can be useful for prioritizing array replication over MANs and WANs over less critical traffic.
- **Traffic Isolation Zones**—Defines paths through a fabric for some or all nodes. Failover allows a non-preferred path to be used if the preferred path fails. Traffic Isolation Zones use failover by default, but it can be disabled if traffic should stop when a preferred path fails. A Traffic Isolation Zone can be used to manually map out traffic flows within a fabric based on application, priority, and topology.

HPE 8/24 SAN Switch

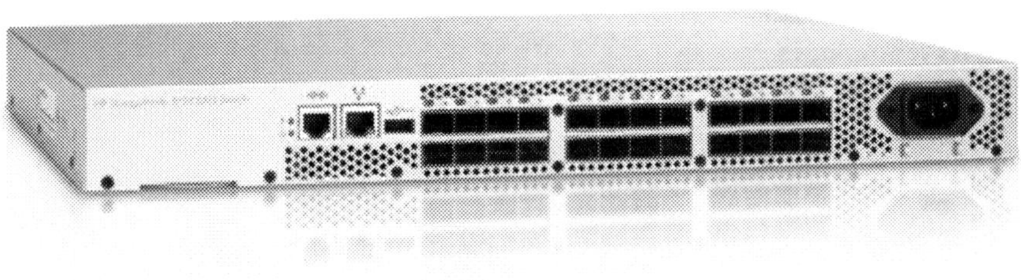

Figure 6-9 HPE 8/24 SAN Switch

Features of HPE 8/24 SAN Switch (Figure 6-9) include

- 8 Gb/s connectivity
- 16 or 24 ports
- 384 Gb/s switching capacity
- Optional Power Pack+ software bundle
  - Fabric Vision, Fabric Watch, ISL Trunking, and Extended Fabric
- One-year on-site warranty, NBD response

The HPE 8/24 SAN Switch consolidates data center connectivity for small and medium-sized enterprises. The switch ports are scalable from 16 to 24 FC ports. This switch offers high availability and lossless networking between applications and data as well as between servers and storage networks.

This switch delivers affordable and flexible 8 Gb/s connectivity and can be incorporated into an extensive core fabric or placed as an edge switch for SAN environments.

The 8/24 SAN Switch is also ideal for entry-level and small and medium-sized business customers that require a simple, yet flexible network storage connectivity solution that will meet evolving requirements. The HPE 8/24 SAN Switch provides the needed connectivity and flexibility at a price point that is affordable, while delivering industry-leading performance for long-term investment protection. This solution is easy to deploy and requires little management time, reducing operational costs.

A new configuration of this switch, bundled with small form-factor pluggable (SFP) transceivers, is now available.

The HPE 8/24 Base SAN Switch delivers 16 high-performance, autosensing 1, 2, 4, and 8 Gb/s FC ports. With powerful yet flexible capabilities, such as Ports on Demand scalability to 24 ports (in an 8-port increment), the 8/24 Base SAN Switch enables organizations to start small and grow their storage networks in a nondisruptive manner.

A fully populated 8/24 SAN Switch with 24 enabled ports provides 384 Gb switching capacity for uncongested, sustained, 8 Gb full-duplex throughput. In addition, organizations can choose to initially deploy 4 Gb/s SFPs and upgrade to 8 Gb/s SFP+ at a later date.

The 8/24 SAN Switch with 24 enabled ports can be deployed as a full-fabric switch or in Access Gateway mode, which provides connectivity to any SAN (the default mode setting is a switch). The Access Gateway mode uses N_Port ID Virtualization (NPIV) switch standards to present FC connections as logical devices to SAN fabrics. Attaching through NPIV-enabled switches and directors, the 8/24 SAN Switch with 24 enabled ports in Access Gateway mode can connect to Brocade, McDATA, Cisco, or other SAN fabrics.

Key benefits of the Access Gateway mode include

- Improved scalability for large or rapidly growing server and virtual server environments
- Simplified management through the reduction of domains and management tasks
- Fabric interoperability for mixed-vendor SAN configurations that require full functionality

An optional enhanced Power Pack+ software bundle can be purchased separately. It includes Fabric Vision, Fabric Watch, ISL Trunking, Extended Fabric, and Advanced Performance Monitor. The software bundle provides the SAN administrator with the necessary tools to proactively monitor the health and performance thresholds of the network while ensuring the highest levels of security, scalability, and manageability. Power Pack+ software enables infrastructure management through a single pane of glass.

HPE 8/8 Base SAN Switch and 8/8 SAN Switch family

Figure 6-10 HPE 8/8 Base SAN Switch

Features of HPE 8/8 Base SAN Switch (Figure 6-10) include

- 8 Gb/s
- 8–24 ports
- Smaller SAN installations
- Optional ISL Trunking
    - Up to 8 ISLs
- Optional Power Pack+
- One-year on-site warranty, NBD response

With the HPE 8/8 Base SAN Switch, you can consolidate data center connectivity for smaller companies. This switch offers 8 Gb/s connectivity and scalability from 8 to 24 FC ports. Other features include high availability and lossless networking between applications and data as well as between servers and storage networks.

The HPE 8/8 Base SAN Switch and the HPE 8/8 SAN Switch are targeted for small to medium-sized enterprises, and they are ideal for entry-level SANs and as edge switches for core-to-edge SAN environments. These switches provide SAN connectivity that simplifies IT management infrastructures, improves system performance, maximizes the value of virtual server deployments, and reduces overall storage costs.

Innovative hardware and software features make the HPE 8/8 SAN Switch easy to deploy, manage, and integrate into a wide range of IT environments. The 8 Gb/s FC 8/8 Base SAN Switch provides a simple, affordable, single-switch solution for a new, small SAN installation and connects to storage and servers only.

An optional Full Fabric SAN scaling license provides connectivity to other switches. The 8/8 SAN Switch has eight active ports and can be connected to multiple switches and scale to full fabric limits as defined in the *SAN Design Guide*.

Both new 8-port switches provide growth options with Ports on Demand, which is the ability to activate additional ports in 8-port increments to a maximum of 24 ports. This provides the flexibility to start small and grow storage networks in a nondisruptive manner. In addition, organizations can choose to initially deploy 4 Gb/s SFPs and upgrade to 8 Gb/s SFP+ when necessary.

The HPE 8/8 SAN Switch significantly increases performance and functionality for SANs at an entry-level price. This switch provides autosensing 1, 2, 4, and 8 Gb/s full-duplex performance to enable up to 128 Gb/s of uncongested throughput. Autosensing and speed-matching of data traffic provide interoperability with previous 1, 2, and 4 Gb/s devices.

To provide more targeted performance, enhanced ISL Trunking combines up to eight ISLs between a pair of switches into a single, logical, high-speed trunk that is capable of up to 64 Gb/s of throughput. The evolutionary design provides these capabilities while consuming less than 2.5 watts of power per port for exceptional power and cooling efficiency.

An optional enhanced Power Pack+ software bundle can be purchased separately. It includes Fabric Vision, Fabric Watch, ISL Trunking, Extended Fabric, and Advanced Performance Monitor. The software bundle provides the SAN administrator with the necessary tools to proactively monitor the health and performance thresholds of the network, while ensuring the highest levels of security, scalability, and manageability. Power Pack+ software enables infrastructure management through a single pane of glass.

H-Series: Entry-level and midrange switches

HPE SN6000 Stackable FC Switch

Features of the HPE SN6000 Stackable FC Switch (Figure 6-11) include

  - 8 Gb/s FC
  - 10/20 Gb/s stacking (ISL)
- 544 Gb/s bandwidth
- Complete SAN management system
- 8–20 ports (depending on the model)
  - Up to 120 ports per stack

Figure 6-11 HPE SN6000 Stackable FC Switch

- Aggregate switch bandwidth of 544 Gb/s
- Included management features:
    - SCM
    - QuickTools
    - Command line interface

The HPE SN6000 Stackable 8 Gb FC switch simplifies SAN deployment in small and medium-sized businesses and in enterprise environments with an affordable and easy-to-manage SAN solution.

This switch delivers scaling efficiency that keeps up with FC port growth as the network expands. Built with low-cost, high-performance scalability in mind, each SN6000 features up to 20 FC device ports enabled, supporting 8/4/2 Gb/s speeds, plus four dedicated high-speed 10/20 Gb/s stacking ports for linking or stacking SN6000 switches.

Heterogeneous fabric connection is simplified with the transparent router functionality that is standard in H-Series switches (the SN6000 and 8/20q models). This functionality provides seamless, nondisruptive interoperability and access to servers and storage on HPE B-Series and HPE C-Series SANs.

The economical SN6000 Stackable 8 Gb FC switch includes SCM software for complete end-to-end fabric management. With SCM software, you can easily provision storage and manage HBAs and switches from a single screen that displays a comprehensive topology view of the SAN. The SCM software integrates with HPE Virtual Connect Enterprise Manager (VCEM) environments to provide an end-to-end integrated solution for managing HPE BladeSystem c-Class solutions. This software also provides one management interface for solutions including VCEM, HPE P2000 MSA storage arrays or EVA storage solutions, and SN6000 switches.

No other 8 Gb FC switch in the industry offers the same scaling efficiencies as the HPE SN6000 Stackable FC switch. Adding more usable FC ports in the data center is full of challenges, but with the SN6000 switch, that process is as simple as stacking SN6000 switches. For example, you can begin with 20 device ports for an affordable solution and add new stacking switches for up to 120 usable device ports per stack (single manageable unit), or more than 500 ports in multistack topologies.

To achieve higher performance, 10 Gb/s stacking ports can be upgraded to 20 Gb/s speed at any time, nondisruptively, extending the solution lifespan while preserving current investments. Best of all, no matter how large a stacking SAN becomes, it is managed as one logical, end-to-end fabric with several availability features to prevent downtime.

B-Series: Midrange switches

HPE SN6000B FC Switch

Figure 6-12 HPE SN6000B FC Switch

Features of the HPE SN6000B FC Switch (Figure 6-12) include

- 16 Gb/s
- 24, 36, or 48 ports
- 1536 Gb/s bandwidth
- Power Pack+ optional or part of the product (depends on the Part Number)
    - Fabric Vision, Fabric Watch, Advanced Performance Monitor, and Extended Fabric
    - ISL Trunking, Integrated Routing, and Ports-on-Demand
- In-flight encryption
- Two integrated, redundant, hot swappable power supplies
- One- year on-site warranty, NBD response

Designed to enable better flexibility and investment protection, this switch scales from 24 to 48 ports, and supports 4, 8, 10, or 16 Gb/s speeds in an efficiently designed 1U package. It offers a wide range of performance for long-distance connectivity.

ISL compression provides the capability to compress all data in flight, which allows customers the choice of either moving more data or reducing bandwidth costs.

The Power Pack+ models include a set of tools that monitor the health and performance of the network, while also ensuring the highest levels of security, scalability, and manageability. Also included are Fabric Watch, Extended Fabric, ISL Trunking, Adaptive Networking, Server Application Optimization, and Advanced Performance Monitoring.

Note

Starting with Fabric OS (FOS) 7.4.0a, Fabric Watch and Advanced Performance Monitor will be obsolete. Fabric Watch settings can be converted to MAPS before upgrading, and Advanced Performance Monitor is replaced by Flow Vision. Both MAPS and Flow Vision require a Fabric Vision license.

A new configuration of this switch bundled with SFP transceivers is available.

Like the entry-level B-Series switch, the HPE SN6000B FC Switch meets the demands of hyperscale, private cloud storage environments by delivering market-leading 16 Gb FC technology and capabilities that support highly virtualized environments.

It also provides a simplified deployment process and a point-and-click user interface—making it both powerful and easy to use. The SN6000B Switch offers low-cost access to industry-leading SAN technology while providing "pay-as-you-grow" scalability to meet the needs of an evolving storage environment.

The SN6000B is available in six models:

- HPE SN6000B 16 Gb 48-port/48-port Active Power Pack+ FC Switch
- HPE SN6000B 16 Gb 48-port/24-port Active Power Pack+ FC Switch
- HPE SN6000B 16 Gb 48-port/48-port Active FC Switch
- HPE SN6000B 16 Gb 48-port/24-port Active FC Switch
- HPE SN6000B 16 Gb 48-port/24-port Active Port Side Air Intake FC Switch
- HPE StoreFabric SN6000B 16 Gb Bundled FC switch

C-Series: Midrange switches

HPE SN6000C 8 Gb FC Switch

Figure 6-13 HPE SN6000C 8 Gb FC Switch

Features of the HPE SN6000C 8 Gb FC Switch (Figure 6-13) include

- 8 Gb/s
- 16–48 ports
- 768 Gb/s of aggregate bandwidth
- Nonblocking architecture
- Virtual SANs (VSANs), Inter-VSAN Routing (IVR), QOS, Port Channels, NPIV, Cisco NX-OS, and Cisco Data Center Network Manager (DCNM)
- One-year on-site warranty, NBD response

The HPE SN6000C 8 Gb Fabric Switch (MDS 9148) is a high-performance, flexible, cost-effective platform providing high-density for storage networking deployments in small, medium-sized, and large enterprise environments.

This switch provides a highly available, secure, and easy-to-use storage networking solution in a compact 1U form factor. With the flexibility to expand from 16 to 48 ports in eight-port increments, the SN6000C switch offers the densities required to scale from an entry-level departmental switch to a top-of-the-rack (ToR) switch, to edge connectivity in enterprise SANs. It supports line-rate 8 Gb FC and an aggregate switch bandwidth of up to 384 Gb/s. The SN6000C Switch delivers a nonblocking architecture, with all 48 of the 1, 2, 4, or 8 Gb/s ports operating at line rate concurrently.

The switch supports innovative features such as VSANs, port channels, QoS, SAN security, SAN management, hardware-enforced zoning, and In-Service Software Upgrade (ISSU) at no additional cost.

The HPE SN6000C 8 Gb Fabric Switch supports the C-series Device Manager quick configuration wizard, which allows it to be deployed quickly and easily in networks of any size. Powered by C-Series MDS 9000 NX-OS Software, it includes advanced storage networking features and functions and is compatible with C-series SN8000C (MDS 9500) and SN8500C (MDS 9700) Series Multilayer Directors and C-series MDS 9100 and 9200 Series Multilayer Fabric Switches. This switch provides transparent, end-to-end service delivery in core-edge deployments.

HPE SN6010C 48-port 16 Gb FC Switch

Figure 6-14 HPE SN6010C 48-port 16 Gb FC Switch

Features of HPE SN6010C 48-port 16 Gb FC Switch (Figure 6-14) include

- 16 Gb/s
- 12–48 ports
- Nonblocking architecture
- Up to 768 Gb/s aggregate bandwidth
- VSANs, IVR, QoS, Port Channels, NPIV, Cisco NX-OS, and Cisco DCNM
- 1-1-1 warranty, NBD response

The HPE SN6010C 48-port 16 Gb Fabric Switch (MDS 9148S) is an affordable, high-performance, and flexible platform providing high-density, 16 Gb/s ports for storage networking deployments in small, medium-sized, and large enterprise environments. The SN6010C Switch offers outstanding value by providing high availability, security, and ease of use at a cost-effective price in a compact one-rack unit (1U) form factor.

With the flexibility to expand from 12 to 48 ports in 12-port increments, the SN6010C Switch offers the densities required to scale from an entry-level departmental switch to a ToR switch to edge connectivity in enterprise SANs. The SN6010C Switch delivers a nonblocking architecture, with all 48 16-Gb/s ports operating at line rate concurrently.

The HPE SN6010C 48-port 16 Gb Fabric Switch supports the C-Series Device Manager quick configuration wizard for quick and easy deployment in networks of any size. Powered by C-Series MDS 9000 NX-OS Software, it includes advanced storage networking features and functions and is compatible with C-Series SN8000C (MDS 9500) and SN8500C (MDS 9700) Series Multilayer Directors and C-series MDS 9100 and 9200 Series Multilayer Fabric Switches. This Switch provides transparent, end-to-end service delivery in core-edge deployments.

Embedded switches for HPE BladeSystem c-Class enclosures

An HPE BladeSystem solution can be equipped with various interconnect modules. FC switches for HPE BladeSystem c-Class enclosure (Figure 6-15) include the following:

Figure 6-15 Embedded switches for HPE BladeSystem c-Class enclosures

- Brocade 16 Gb SAN Switch for HPE BladeSystem c-Class

  The Brocade 16 Gb SAN Switch for HPE BladeSystem c-Class represents an easy-to-manage next-generation embedded FC switch with 16 Gb/s FC performance. It hot plugs into the back of the HPE BladeSystem c-Class Enclosure and doubles the network performance of the HPE BladeSystem c-Class solution. Additionally, it increases flexibility while reducing costs through rack consolidation, reduced cabling, and fewer SFP+ transceivers.

- Brocade 8 Gb SAN Switch for HPE BladeSystem c-Class

  The Brocade 8 Gb SAN Switch for HPE BladeSystem c-Class represents an embedded FC switch with 8 Gb/s FC performance. It hot plugs into the back of the HPE BladeSystem c-Class Enclosure, and its integrated design frees up rack space, enables shared power and cooling, and reduces cabling and SFP requirements.

- Cisco MDS 8 Gb Fabric Switch for HPE BladeSystem c-Class

  The Cisco MDS 8 Gb Fabric Switch for HPE BladeSystem c-Class integrates the FC switch within the HPE BladeSystem c-Class Enclosure. It reduces costs through consolidation while increasing flexibility and performance and simplifying management. The switch hot plugs into the back of the HPE BladeSystem c-Class Enclosure to free up rack space, and the integrated design reduces cabling and SFP requirements and further reduces the cost of power and cooling.

Example—Command line interface to FC switch

Following example shows how to use a command line user interface to an HPE FC switch.

There are different user interfaces to a FC switch, but the basic one is using a Secure Shell (SSH) client and a command line interface (CLI).

- To use the CLI, you need to open putty from the desktop (Figure 6-16).

Figure 6-16 Putty icon

- You must enter a host name or IP address (Figure 6-17).

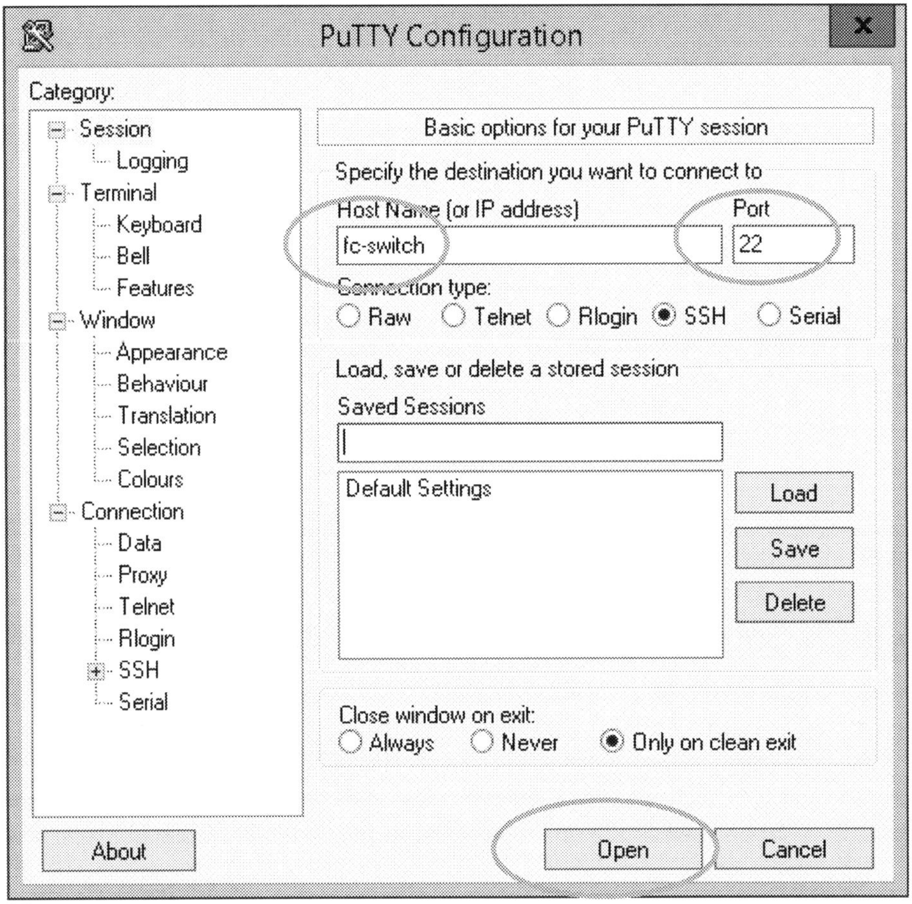

Figure 6-17 Host name or IP address

- FC switch credentials also must be entered to access the switch interface (Figure 6-18).

```
fc-switch - PuTTY
login as: hpadmin
hpadmin@fc-switch's password:
```

Figure 6-18 FC switch credentials

- The CLI to the FC switch contains a descriptive help system. To list all available commands, you can enter the `help | more` command (Figure 6-19).

```
fc-switch - PuTTY
login as: hpadmin
hpadmin@fc-switch's password:

-----------------------------------------------------------------
r169eva2sw1:hpadmin> help | more
```

Figure 6-19 To list all available commands

Press the Space key to move to the next page, and press the "q" key to stop the "more" pager.

- The `switchshow` command displays switch and port status information such as the name and ID of the switch, the World Wide Name (WWN) and mode of the switch, the number of ports, their states, and the FC speed (Figure 6-20).

```
fc-switch - PuTTY
r169eva2sw1:hpadmin> switchshow
switchName:     r169eva2sw1
switchType:     70.3
switchState:    Online
switchMode:     Native
switchRole:     Principal
switchDomain:   1
switchId:       fffc01
switchWwn:      10:00:00:05:33:ad:de:05
zoning:         ON (san_a)
switchBeacon:   OFF

Index Port Address Media Speed State     Proto
==============================================
  0   0   010000   cu    N4   Online      FC  F-Port  50:01:43:80:11:37:f6:a8
  1   1   010100   id    8G   Online      FC  F-Port  20:02:00:11:0a:02:54:fb
  2   2   010200   id    8G   Online      FC  F-Port  20:03:00:11:0a:02:54:fb
  3   3   010300   id    8G   Online      FC  F-Port  50:01:43:80:16:7d:61:1a
  4   4   010400   --    N8   No_Module   FC
  5   5   010500   --    N8   No_Module   FC
  6   6   010600   --    N8   No_Module   FC
  7   7   010700   --    N8   No_Module   FC
  8   8   010800   --    N8   No_Module   FC
  9   9   010900   --    N8   No_Module   FC
```

Figure 6-20 Displays of switch and port status information

- The `version` command displays the firmware version (Figure 6-21).

```
r169eva2sw1:hpadmin> version
Kernel:     2.6.14.2
Fabric OS:  v6.4.1a
Made on:    Tue Dec 14 02:36:38 2010
Flash:      Tue Nov 1 14:49:06 2011
BootProm:   4.6.6
r169eva2sw1:hpadmin>
```

Figure 6-21 Firmware version

- The `portshow` command displays the status and configuration parameters for FC ports and GbE ports. The `portshow 1` command can be used to verify the state of port 1 (the number 1 can be replaced with the appropriate port).

  Important information about the port is displayed, such as its protocol type (Protocol), WWNs of connected devices (portWwn of device(s) connected), and its speed (portSpeed) (Figure 6-22).

fc-switch - PuTTY

```
r169eva2sw1:hpadmin> portshow 1
portIndex:   1
portName:
portHealth: No Fabric Watch License

Authentication: None
portDisableReason: None
portCFlags: 0x1
portFlags: 0x20b03        PRESENT ACTIVE F_PORT G_PORT U_PORT NPIV LOGICAL_ONLINE
 LOGIN NOELP ACCEPT FLOGI
LocalSwcFlags: 0x0
portType:  18.0
portState: 1    Online
Protocol: FC
portPhys:  6    In_Sync         portScn:   32   F_Port
port generation number:    8
state transition count:    1

portId:    010100
portIfId:    43020015
portWwn:   20:01:00:05:33:ad:de:05
portWwn of device(s) connected:
        20:02:00:11:0a:02:54:fb
Distance:  normal
```

Figure 6-22 The status and configuration parameters for FC ports and GbE ports

- One of the basic configuration settings done at the FC switch level is zoning. The three common configuration options to configure are aliases, zones, and configurations (cfg; Figure 6-23). For example:
  - You can enter the `aliashow` command to display the configured aliases.

```
r169eva2sw1:hpadmin> alishow
Defined configuration:
 cfg:    san_a    cveva_zone; Radek_Windows_Bootable_zone1;
                  Spyros_ZoneA_windows_boot; IO_Windows_Boot_Student2_ZoneA;
                  SPM_5001438002A30010; IO_Windows_Data_Spyros_zoneA;
                  SPM_5001438002A30022
 zone:   IO_Windows_Boot_Student2_ZoneA
                  IO_Windows_Boot_Student2_p1; EVAC1_P1; EVAC2_P1
 zone:   IO_Windows_Data_Spyros_zoneA
                  IO_Windows_Data_Spyros_p1; EVAC1_P1; EVAC2_P1
 zone:   Radek_Windows_Bootable_zone1
                  Radek_Windows_Bootable_alias1; EVAC1_P1; EVAC2_P1
 zone:   SPM_5001438002A30010
                  50:01:43:80:02:A3:00:10; 50:01:43:80:11:37:F6:AC;
                  50:01:43:80:11:37:F6:A8
 zone:   SPM_5001438002A30022
                  50:01:43:80:02:A3:00:22; 50:01:43:80:11:37:F6:AC;
                  50:01:43:80:11:37:F6:A8
 zone:   Spyros_ZoneA_windows_boot
                  Spyros_IO_windows_boot_p1; EVAC1_P1; EVAC2_P1
 zone:   cveva_zone
                  CVEVA; EVAC1_P1; EVAC2_P1
 alias: CVEVA    50:01:43:80:16:7D:61:1A
 alias: EVAC1_P1
```

Figure 6-23 The three common configuration options to configure are aliases, zones, and configurations

- Or you can enter the `zoneshow` command to list all of the configured zones (Figure 6-24).

```
 zone:   IO_Windows_Boot_Student2_ZoneA
                  50:01:43:80:02:a3:00:0a
                  50:01:43:80:11:37:f6:a8
                  50:01:43:80:11:37:f6:ac
 zone:   IO_Windows_Data_Spyros_zoneA
                  50:01:43:80:02:a3:00:18
                  50:01:43:80:11:37:f6:a8
                  50:01:43:80:11:37:f6:ac
 zone:   Radek_Windows_Bootable_zone1
                  50:01:43:80:02:a3:00:02
                  50:01:43:80:11:37:f6:a8
                  50:01:43:80:11:37:f6:ac
 zone:   SPM_5001438002A30010
                  50:01:43:80:02:a3:00:10
                  50:01:43:80:11:37:f6:ac
                  50:01:43:80:11:37:f6:a8
 zone:   SPM_5001438002A30022
                  50:01:43:80:02:a3:00:22
                  50:01:43:80:11:37:f6:ac
                  50:01:43:80:11:37:f6:a8
 zone:   Spyros_ZoneA_windows_boot
                  50:01:43:80:02:a3:00:04
                  50:01:43:80:11:37:f6:a8
                  50:01:43:80:11:37:f6:ac
```

Figure 6-24 To list all of the configured zones

- Enter the `cfgshow` command to list the defined configurations.

HPE StoreFabric FC HBAs and CNAs

Figure 6-25 HPE StoreFabric SN1000Q 16 GB 2-port PCIe FC HBA

The HPE StoreFabric Host FC HBAs and CNAs bring high-performance host connectivity to SANs (Figure 6-25).

Differentiators:

- Host interface
  - FC
  - Ethernet
- Performance
  - 16 Gb/s FC
  - 8 Gb/s FC
  - 10 Gb/s Ethernet
- Number of ports
  - Single port
  - Dual port

- Model or hardware vendor
    - Emulex
    - QLogic
    - Brocade
- Supported servers

HPE FC HBAs

HPE StoreFabric HBAs enable more rapid storage and retrieval of critical information when using high-bandwidth cloud applications and storage-intensive applications such as backup, database transactions, and rich media.

Customers looking for a high-bandwidth solution to meet the requirements of increasing server virtualization and consolidation will benefit from HPE StoreFabric HBAs. These HBAs help to realize the full bandwidth potential of servers running I/O-intensive applications.

HPE StoreFabric HBAs enable more applications and VMs to run on a single server and port, resulting in reduced cabling and higher return on IT investment. Higher speed HBAs like the 16 Gb FC HBAs are backward compatible with 8 Gb and 4 Gb storage networks, and they will protect future investments.

To bring storage networking performance and efficiency to the next level, take advantage of the 16 Gb HBAs that deliver twice the I/O performance of 8 Gb FC HBAs.

Models:

- HPE StoreFabric SN1100E 16 Gb Single Port FC HBA
- HPE StoreFabric SN1100E 16 Gb Dual Port FC HBA
- HPE StoreFabric SN1000Q 16 Gb Single Port FC HBA
- HPE StoreFabric SN1000Q 16 Gb Dual Port FC HBA
- HPE 82B 8 Gb Dual Port PCIe FC HBA
- HPE 81B 8 Gb Single Port PCIe FC HBA
- HPE 82E 8 Gb Dual Port PCIe FC HBA
- HPE 81Q 8 Gb Single Port PCIe FC HBA
- HPE 81E 8 Gb Single Port PCIe FC HBA
- HPE 82Q 8 Gb Dual Port PCIe FC HBA

HPE CNAs

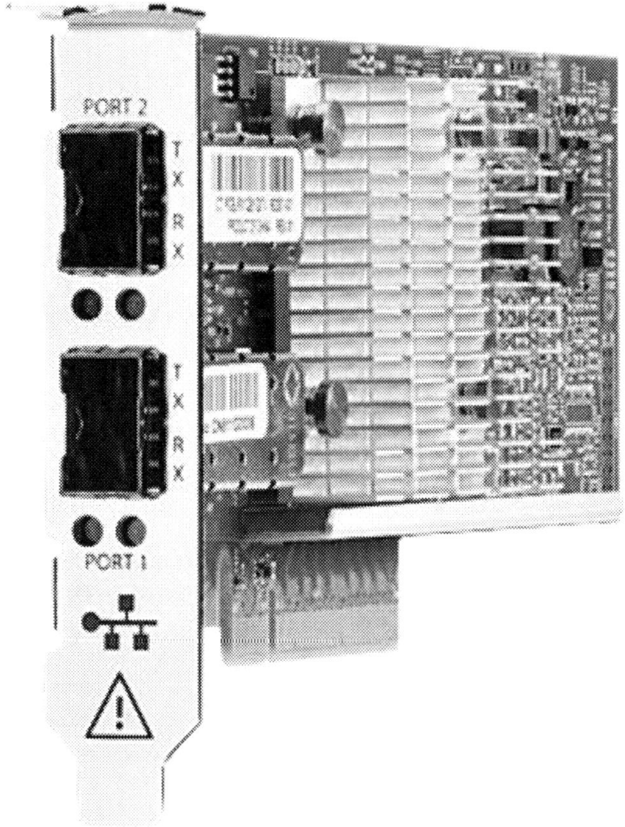

Figure 6-26 HPE StoreFabric CN1100R Dual Port CNA

The HPE CNAs are dual-port adapters that provide Ethernet, iSCSI, and FC connectivity over 10 GbE using both FCoE and CEE standards (Figure 6-26). By consolidating Ethernet, iSCSI, and FC on a converged adapter, HPE CNAs provide the following benefits:

- Reduce the number of separate adapters and cables required for data centers
- Reduce operational power and cooling costs
- Preserve the existing Ethernet and FC infrastructure

HPE CNAs have been thoroughly tested with HPE ProLiant servers and HPE ToR switches to deliver outstanding HPE FCoE solutions for the data center.

Models:

- HPE StoreFabric CN1200E 10 GbE CNA
- HPE StoreFabric CN1000Q Dual Port CNA

- HPE StoreFabric CN1100R Dual Port CNA
- HPE StoreFabric CN1100E Dual Port CNA

Cables, transceivers, and port licenses

The following storage networking accessories are available for many HPE StoreFabric products:

- Flexible fiber optic cable for data center designs
- Transceivers for networking and storage interfaces
- Cables, transceivers, and license upgrades

HPE StoreFabric multimode fiber cables (OM4) and active optical cables (AOC) offerings reduce data center costs with modern multimode systems that provide better signal integrity. The StoreFabric Converged SFP+ single transceiver allows flexible port capability without the replacement of SFPs.

- Flexible fiber optic cable for data center designs

Figure 6-27 Fiber optic cable

You can span cable across rows within the data center with HPE PremierFlex Fiber Optic Cables (Figure 6-27). The revolutionary PremierFlex cables use a new bendable fiber technology that significantly improves bend performance over existing fiber cables. This, combined with improvements in fiber clarity and bandwidth, allows HPE PremierFlex cables to transmit data over longer distances, at higher rates, with fewer transmission errors.

- Transceivers for networking and storage interfaces

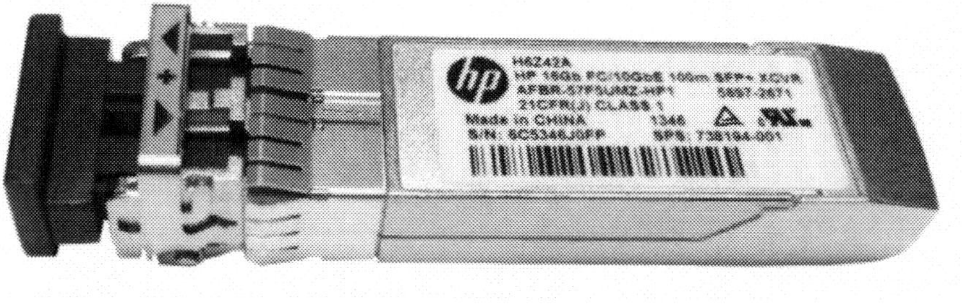

Figure 6-28 HPE FC transceiver

HPE FC transceivers offer maximum performance, reliability, and compatibility for HPE storage and computing products (Figure 6-28). HPE converged transceivers offer wire-once transceiver simplicity for networking and storage interfaces. You can eliminate the guesswork about which transceiver to use. HPE offers one universal transceiver for LAN and SAN traffic from 10 GbE, FCoE, iSCSI, and 16 Gb FC.

- License upgrades

HPE license upgrades provide the flexibility to tailor IT infrastructure to your business advantage.

Why choose HPE StoreFabric cables and transceivers?

Following are some key reasons why customers choose HPE StoreFabric cables and transceivers:

- **Flexible**—The new cables have up to 10x better bend performance than existing fiber cables.
- **Durable**—AOC offerings are lighter than copper cables and provide a better (lower) electromagnetic interference (EMI).
- **Reliable**—You can transmit data over longer distances, at higher rates, with fewer transmission errors.
- **Cost-effective**—Both FC and FCoE can be supported in a single SFP+ transceiver. The protocol is controlled with Electrically Erasable Programmable Read-Only Memory (EEPROM) settings.

Sizing SAN infrastructure products

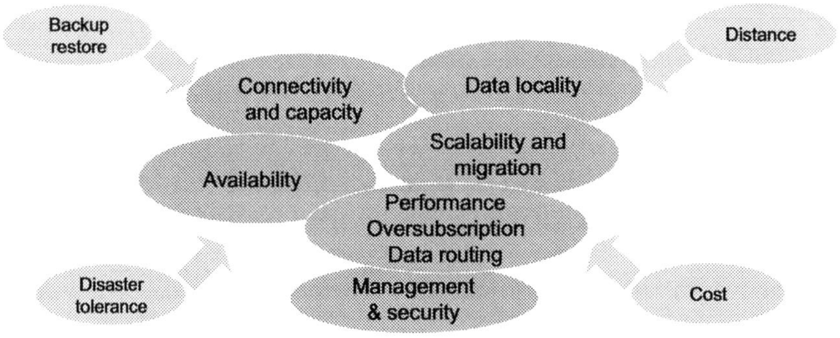

Figure 6-29 SAN architecture choices and considerations

SAN architecture choices and considerations

SAN architecture is the overall design, or structure, of a SAN solution (Figure 6-29). It includes one or more fabrics, each of which has a topology. Fabric topology refers to the layout of switches that form a fabric.

Distance or geographic layout

The physical layout of campuses, facilities, and the location of servers and storage within individual buildings can be a major factor where these physical issues dictate the outcome of the appropriate SAN topology design.

Such situations include the following:

- Running out of available F-ports on a switch
- Changing protocols to go greater than 100 km (asynchronous transfer mode or dense wavelength-division multiplexing)
- Traversing someone else's right-of-way and using a fiber owned by them
- Wanting distributed remote connectivity in a campus environment

---

Note

The SAN Design Reference Guide, Chapter 8, discusses supported SAN extension technologies.

Data locality

Data locality refers to I/O traffic that stays within (or is local to) the FC switch that connects local servers and storage arrays. I/O traffic between distant servers and local storage arrays must pass through two or more FC switches.

When the design has a minimal number of ISLs, the traffic is assumed to be local to the switch that connects the immediate servers and storage arrays. Changing the I/O traffic patterns by adding new server connections after the SAN has been implemented without adding ISLs can reduce the effective bandwidth of the SAN.

In the best case, data locality can be achieved when the path between a server and its storage does not leave the switch. The selection of the appropriate SAN topology should be based on the expected primary data locality need.

Design considerations

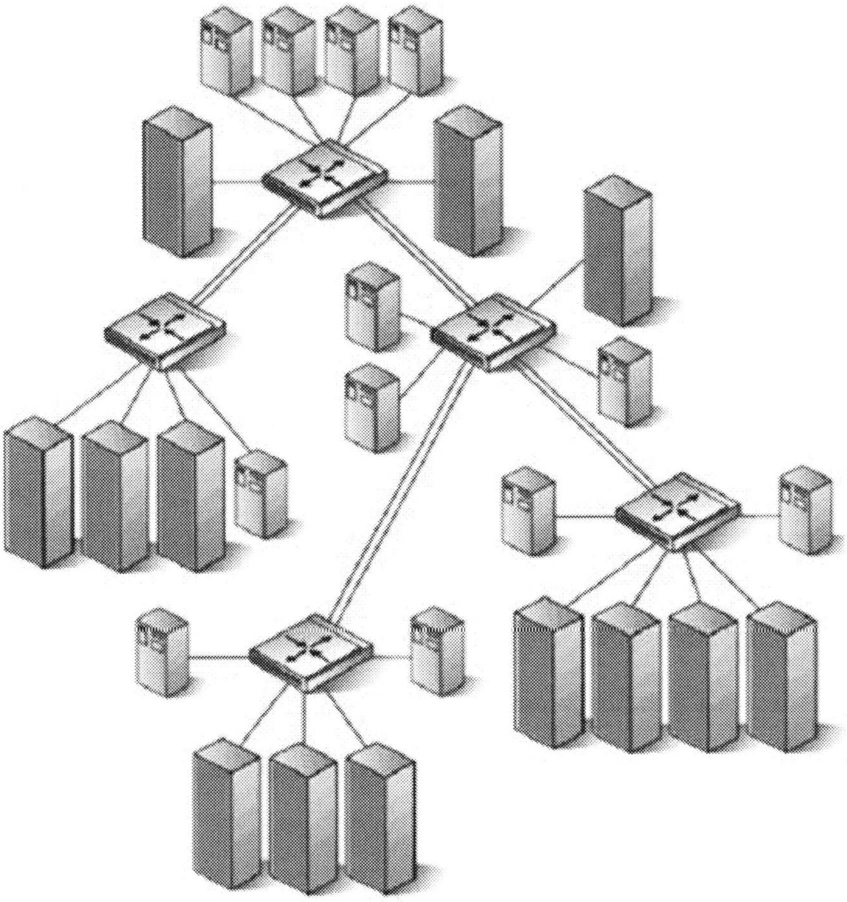

Figure 6-30 Example of SAN design

Consider these factors when designing a SAN (Figure 6-30):

- Applications and operating systems
- Availability
- Accessibility
- I/O profile
- Backup
- Security
- Migration
- Management

Connectivity and capacity

When determining connectivity and capacity needs for a customer environment, be sure to consider both their current and future needs. Choose a design that can be scaled or easily migrated to a topology design with more capacity, if required.

Connectivity

Connectivity is the total number of FC ports—the fabric port count—needed to connect servers and storage to the fabric. The ports available for server or storage connections are called user ports.

Connectivity needs directly affect the total number of FC switches required, and they are a major consideration when evaluating different SAN topology designs.

If the total number of ports required exceeds what is supported in a given topology, consider higher capacity topologies or deploying multiple independent SANs.

The design should provide adequate fabric performance by implementing the required number of ISLs.

Capacity

Storage capacity can be expanded by adding larger capacity disk devices or more disks or by deploying additional storage systems in the SAN.

Scalability and migration

SANs should be able to grow or scale incrementally over time as storage and connectivity needs increase. Customers want to scale their environment without affecting the operations of the environment already in place. Choose a design that will accommodate expected future growth and usage requirements.

Each topology has a migration path that provides configuration flexibility, expansion, and increased capabilities. You should consider aspects of scaling and migration when choosing a SAN design.

Availability

Data availability is a broad measure of how reliable a storage system is in routine operation. SAN implementations can provide different levels of data availability. In many cases, different levels can be mixed in the same SAN.

Examples:

- Fully redundant hardware—Redundant array controllers, power supplies, and HBAs
- Periodic backup
- Multipathing between servers and storage

- Replication configurations
- No single point of failure (NSPOF) designs, which provide the highest availability
    - Requires a SAN design that implements two or more separate fabrics and redundant components at all levels

Performance and application workloads

Consider the types of applications that will be used on the SAN relative to data access classification. Applications can usually be classified as high bandwidth or high throughput. The SAN should provide an adequate level of performance based on the application workload. In general, SAN topology designs with fewer switch hops between devices provide better performance because of the lower probability of oversubscription or congestion.

Oversubscription

To minimize oversubscription in the SAN design:

- Ensure there are an adequate number of ISLs between switches.
- Minimize the number of cases where many devices or ports are attempting to share a single switch port.

To improve data routing, minimize the number of hops between devices.

Backup

SAN-based backup, which is an enterprise backup solution, provides high bandwidth and centralized control for backup and restore operations. Your SAN design should provide adequate connectivity and bandwidth for backup.

Disaster tolerance

Consider the criticality of data in the event of unforeseen catastrophic site failures. You should also consider remote data replication requirements to ensure protection against site failures and full recovery of critical data.

Selected data can be copied to remote storage arrays automatically, providing recovery capabilities should a primary site interruption or possible loss occur.

By using multiple storage arrays, portions of the SAN can be configured for disaster tolerance, providing a common SAN with mixed data protection levels.

Disaster tolerance and disaster recovery require redundant locations.

Management

You can use management applications to centralize SAN management.

Security

Consider the customer's requirements for a secure environment. Ensure that each user can only access the data or storage devices the user is authorized to access. This is particularly critical in a heterogeneous platform environment.

SAN infrastructure performance factors

The following factors can decrease overall fabric-wide infrastructure performance:

- Congestion
- Blocking
- Fan-out
- Fan-in
- Oversubscription
    - Modifying ISL port routing
    - Ratios and locality

A SAN can be designed based on any of these factors:

- High performance
    - I/O per second
    - MB/s
- High availability
- Low cost

It is not possible to have all three factors, so you must determine what is most important to the customer and for the application.

When designing a SAN for performance, you first have to determine the performance requirements and the application I/O profile. Even if the SAN is already in production, administrators need to understand the application profile to determine how the SAN can be optimized or how performance can be maintained. Without this knowledge, any change or upgrade could result in lower performance.

Availability is another consideration, but be aware that designing or redesigning a SAN for availability will probably change the performance characteristics of the SAN.

A multiswitch fabric has two factors that decrease overall fabric-wide infrastructure performance:

- Latency through multiple switches (hops)
- Oversubscription or congestion of ISLs

Performance testing and measurement by HPE has shown switch latency to be less than 5% of the time lost resulting from the congestion of a full frame from another path. This finding implies that the number of switches and hops between devices is not a major factor affecting performance.

However, as devices send frames through more switches and hops, the chances increase that other traffic in the SAN might be routed over the same ISL/path. This routing can decrease performance because of oversubscription of a particular ISL or a path that is serving multiple devices. Oversubscription has been determined to be the largest contributing factor to reduced FC performance.

When devices must contend for the same ISL or path, the best result will be that each competing device will receive 1/n of the available bandwidth on the path, where *n* is the number of contending devices. HPE recommends following these guidelines when configuring a SAN:

- Whenever possible, devices that exchange the highest amount of data should be connected to the same FC switch.
- For high bandwidth, the number of application servers should be balanced with storage by using as much one-to-one access as possible.
- When devices exchanging data are on different switches, minimize the number of hops between those devices.
- For high-bandwidth (large transfer size) applications, configure a maximum of two active storage controller ports per ISL.
- For high-throughput (small transfer size) applications, configure a maximum of 20 active storage controller ports per ISL.
- For mixed applications, configure a maximum of four active storage controller ports per ISL.

Congestion

Congestion occurs when oversubscription is in place. That means, for example, more servers try to talk to the storage device over the same ISL. In this case, multiple servers are contending for bandwidth. Because the link has limited bandwidth, the servers will be throttled down to the total bandwidth.

A port, link, or fabric can become congested.

Blocking

Blocking means that the data does not get to its destination. This is different from congestion, in which the data will be delivered, but with a delay. Consider the example of more servers talking to the same storage device over the same ISL. In a blocking environment, whenever the new server tries to use the ISL (which is already in use), the server will be denied access and will have to wait until the ISL is free.

Fan-out

Fan-out is the ratio of server ports to a single storage port, and this ratio is important in the SAN design. For example, if the storage device has only one connection and six servers are connecting to it, the ratio is 6:1. In such an example, it might be reasonable to have fewer ISLs than the full capacity of the servers because the storage end can only handle limited bandwidth.

Fan-in

Fan-in is the ratio of storage ports to a single server port. This information is also important in the SAN design. For example, by reallocating the storage ports across the fabric, a bad ISL oversubscription ratio can be overcome.

Oversubscription

All HPE FC switches implement a nonblocking design. This means that any pair of ports can be active and transfer data without impacting the data transfer between another pair of ports. This feature should be carried throughout the design of the fabric itself.

Oversubscription or congestion can occur in a fabric with multiple switches when data from multiple sources must be sent to a single destination port, or when data must be sent across an ISL from multiple input ports. In situations where this occurs, the FC switches use fairness algorithms to ensure that all devices are serviced.

The switches will interleave frames from multiple devices, thus giving fractional bandwidth to all devices. If this situation occurs often, overall performance in the fabric will be reduced. Oversubscription can be minimized by ensuring that the fabric topology chosen for the design provides for an adequate number of ISLs between all switches and that it minimizes scenarios in which many devices or ports could attempt to access a single destination device or port.

Modifying ISL port routing

ISL oversubscription will not affect the majority of properly sized SAN fabrics, so this is not a routine maintenance task. Although it is technically easy to modify the routing paths, an inexperienced person should not attempt this task. This functionality should be reserved for custom SAN designs and should only be done by qualified SAN design engineers.

Ratio and locality

When designing a SAN, architects must understand the performance boundaries such as storage fan-out ratios and storage performance. Although any SAN device that connects to a SAN at 4 Gb/s is theoretically capable of 4 Gb/s, in reality, that device is most likely to provide much lower performance. If this device truly is capable of generating 4 Gb/s of I/O, the principles of locality should be applied or sufficient bandwidth should be provisioned for the ISLs.

A popular SAN application is storage consolidation in which many hosts share a storage device or port. Several popular storage vendors target a 6:1 fan-out ratio, which means that on average, six hosts are sharing a single storage port. If there were 32 storage ports in a fabric, architects would expect to find an average of 192 hosts. Even if every host requires 2 Gb/s or 4 Gb/s of bandwidth, the storage devices in the fabric are only capable of delivering 64 Gb/s (2 Gb/s ports) or 128 Gb/s (4 Gb/s ports), which equates to 6–12 MB/s per host.

While some ports in the fabric might require maximal bandwidth, not all ports require sustained maximal bandwidth, and rarely, if ever, do these ports require maximal bandwidth simultaneously.

SAN infrastructure availability factors

Several factors affect SAN data availability:

- Application software
- Server operating systems
- Server hardware
- SAN fabric infrastructure
- Primary and secondary storage
- Number of switches
- Number of ISLs
- Number of paths between a server or clustered servers and the fabric
- Number of storage controller paths in the fabric

Data availability in an IT environment is influenced by many factors, including the application software and operating systems in the servers, the server hardware, the SAN fabric infrastructure, and the primary and secondary storage. Operational parameters, including backup schedules and machine room procedures, as well as personnel issues and overall administrative practices, all make important contributions to the availability of data in an IT environment.

In some environments, adequate data availability is established by a routine backup procedure that is performed on a scheduled basis. Other cases might require online dynamic backup of primary data to a remote site. Some environments use clustered servers and redundant fabrics in the SAN design to achieve the data availability goals.

When selecting a SAN fabric topology, the number of FC switches and the number of ISLs between the switches have the largest effect on the data availability. The number of connections or paths between a given server or a group of clustered servers and the fabric, and the number of storage controller connections or paths into the fabric, also affect data availability.

From the perspective of SAN architecture and fabric topology design, fabric availability can be classified into at least four categories. The different categories offer a range of availability levels, from the most basic interconnect scheme with no redundancy, up to fully redundant NSPOF designs (Figure 6-31):

- Single connectivity fabric
- Single resilient fabric
- Single resilient fabric with multiple device paths
- Multiple fabrics and device paths

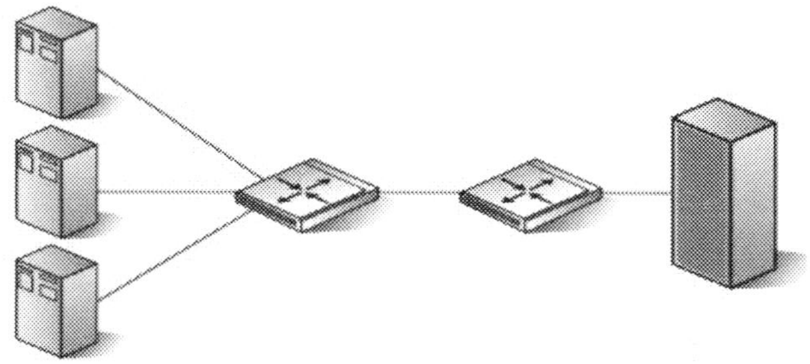

Figure 6-31 Example: Single connectivity fabric

Reference documents and sizing tools

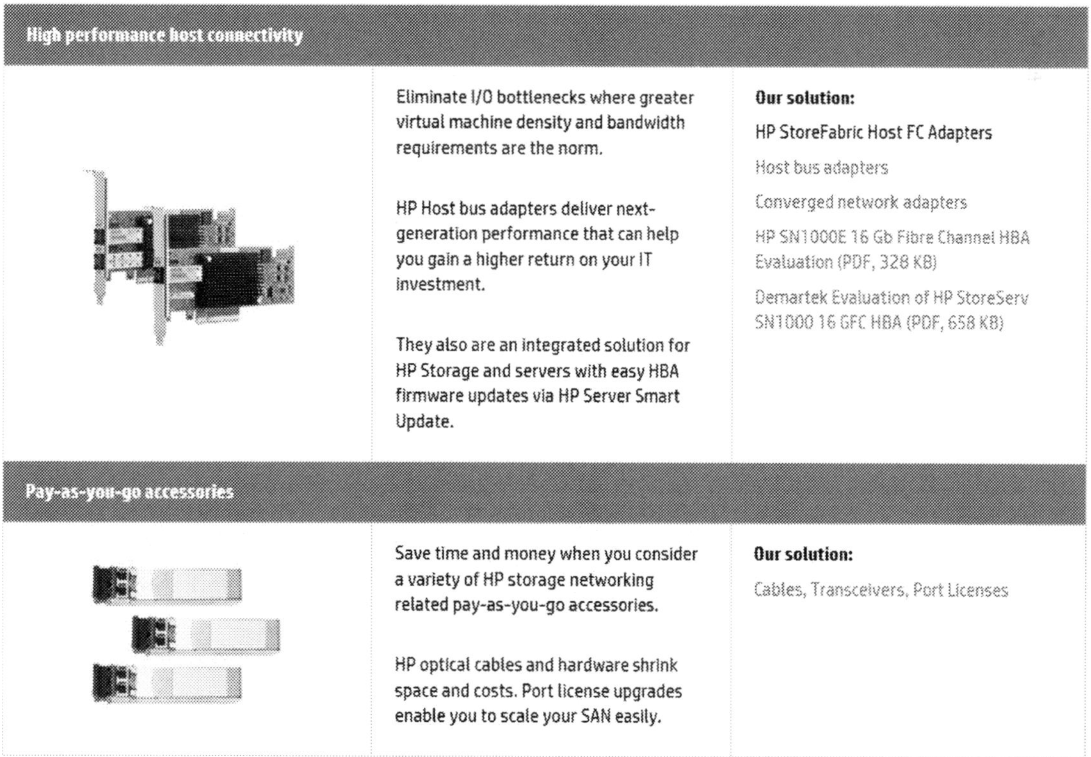

Figure 6-32 HPE website

Sizing tools are described in more detail in Chapter 8. You can use the following reference documents and sizing tools to select and configure SAN infrastructure components (Figure 6-32):

- *HPE QuickSpecs*
- *HPE Product Bulletin*
- HPE website: https://www.hpe.com/us/en/storage/networking.html.

Learning check

The following questions will help you to measure your understanding of the material presented in this chapter. Read all of the choices carefully because there might be more than one correct answer. Select or write the correct answer for each question.

1. Name at least three SAN infrastructure products.

   ....

   ....

   ....

   ....

2. SAS is the common protocol for connecting FC switches and servers.

   ☐ True

   ☐ False

3. Name at least two FC switch differentiators.

   ....

   ....

   ....

4. The host interface (FC versus Ethernet) is the main differentiator for HPE StoreFabric FC HBAs and CNAs.

   ☐ True

   ☐ False

5. What is the common HPE FC transceiver speed?

   ....

   ....

   ....

   ....

6. Congestion can limit the speed between servers and storage devices.

   ☐ True

   ☐ False

For answers, see Chapter 6 Answers on page 441.

7 HPE StoreOnce and Entry-Level Tape Libraries

OBJECTIVES

In this chapter, you should learn to:

- ✓ Describe the HPE backup and restore strategy
- ✓ Identify the products in the HPE StoreOnce backup portfolio, including their unique features and target customers
- ✓ Describe the HPE Data Protector Software
- ✓ Locate sizing tools and reference information to help you configure an HPE backup solution

INTRODUCTION

This chapter provides an overview of HPE backup products. It begins with the basics of a backup and restore strategy and defines the different types of backup and related terminology. The next section introduces HPE products in the backup category, particularly HPE StoreOnce products and HPE Data Protector, describing their unique features, capacity limits, performance figures, and target customers. The chapter then describes reference documents and tools you can use for sizing and configuration of a backup and restore solution.

HPE Backup and Restore Strategy

One of the most common causes of data loss is physical failure of the storage media. Normally, hard drives function for years without incident, but eventually they fail. It might happen gradually as more bad clusters accumulate until most of the drive is unusable, or it might happen suddenly, when the hard drive fails without warning.

What is backup?

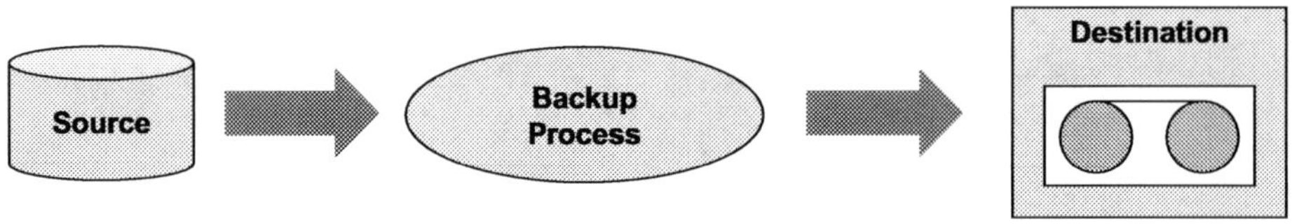

Figure 7-1 Backup process

A backup is a process that creates a copy of data on backup media (Figure 7-1). This copy is stored for future use in case the original is destroyed or corrupted.

What is restore?

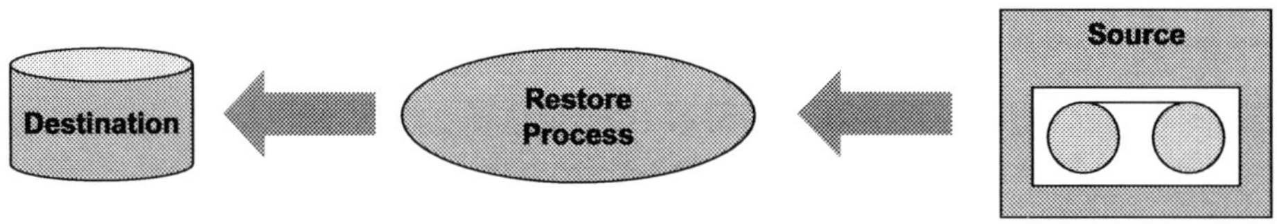

Figure 7-2 Restore process

A restore is a process that re-creates the original data from a backup copy (Figure 7-2). This process consists of the preparation and actual restoration of data and some postrestore actions that make that data ready for use. The source of the restored data is a backup copy.

A restore application is software that writes data to a destination device. The destination device is usually the disk to which the original data was written.

A restore cannot always replace all data. For example, power failures or spikes might cause data to be permanently lost. Many computer viruses will delete files on an infected machine.

Backing up the environment

During backups, data is transferred over the infrastructure, which is typically a storage area network (SAN) or LAN (Figure 7-3). The data comes from the systems to be backed up and is sent to the systems that have the backup media where the data is to be stored.

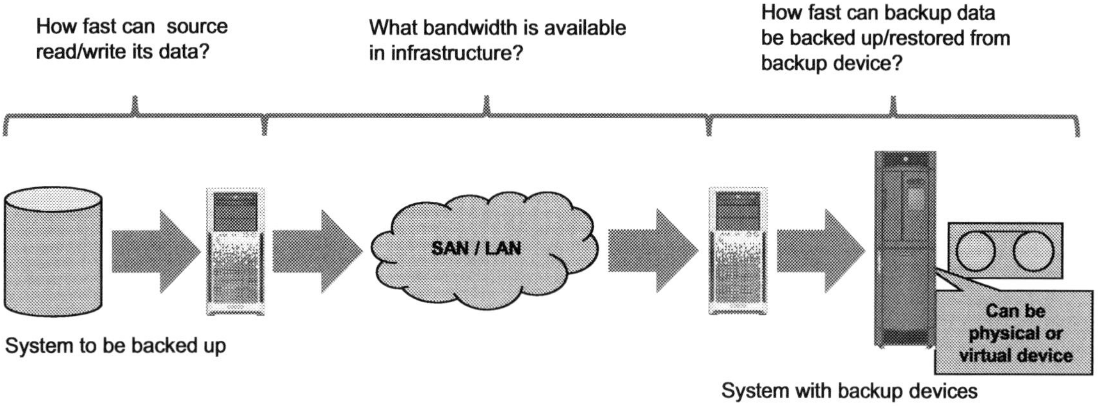

Figure 7-3 Backing up the environment

Backup configuration

Backups must be performed on a disciplined, scheduled regimen that has been carefully designed to address matters such as the following:

- The frequency and schedule for backups
- Whether to use partial or full backups
- Whether partial backups should be incremental or differential

The terms "full, partial, incremental," and "differential" describe the amount of information that is copied and backed up. All backups, whether full or partial, can be performed online or offline and are based on the status of an archive bit.

The archive bit is a file attribute that identifies whether the file has been backed up. If a file is changed after the last backup, the archive bit is cleared until the file is backed up again.

Full backup

A full backup is a complete backup of the entire server or client. A full server backup includes all volumes, directories, and files. A full client backup includes all drives, directories, and files.

The two types of full backups are

- **Normal backup**—Backs up all files on the hard drive and resets the archive bit
- **Copy backup**—Similar to a normal backup but does not reset the archive bit

Partial backup

A partial backup can be incremental or differential (Figure 7-4).

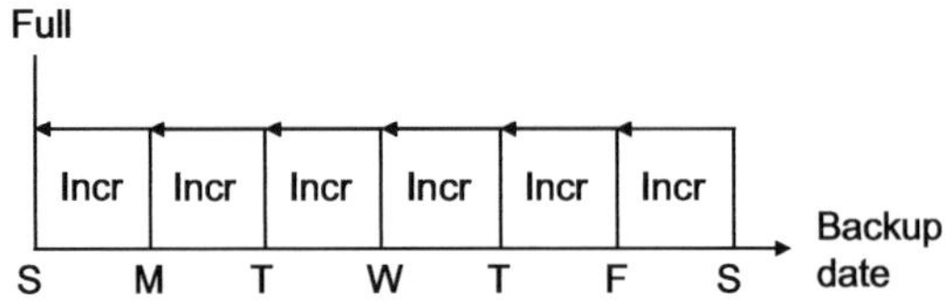

Figure 7-4 Incremental backup

- **Incremental**—Copies all files that were changed after the last backup, regardless of what kind of backup it was. Incremental backups back up only the data that changed since the last full backup, and they reset the archive bit.

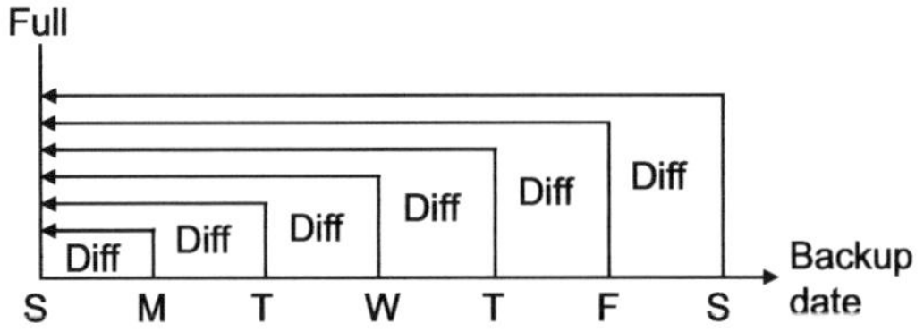

Figure 7-5 Differential backup

- **Differential**—Copies all files that were changed after the last complete backup. Differential backups back up all the data since the last full backup but do not reset the archive bit (Figure 7-5).

Verifying the backup

After a full system backup has been performed, it should be verified with the existing data on the system to ensure consistency. All backup software provides the capability to compare and verify the backed up copy with the original data on the system. Confirm that the verification process has completed successfully before proceeding with the upgrade.

A basic approach to improving performance is to reduce the amount of data to be backed up. Careful planning is necessary for full and incremental backups. Consider whether full backups of all the client systems could be performed at different times.

Recovery point and recovery time objectives

The recovery time objective (RTO) is the length of time that is acceptable for the business to be without a specific application. This is often associated with the maximum allowable or tolerable outage (Figure 7-6). The RTO is used when determining whether to use replication or backup to tape or disk.

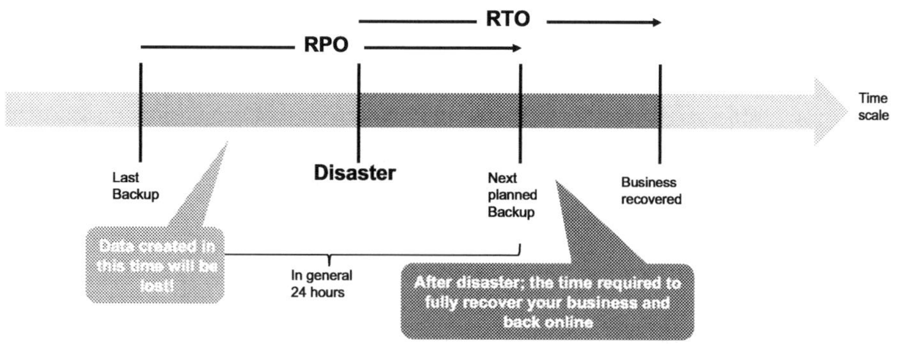

Figure 7-6 Recovery point and recovery time objectives

- If downtime is not an option, RTO would be zero. In this case, a clustered or a redundant infrastructure with replicated data off-site would be an ideal solution.
- If the RTO is measured in days, tape backup might be acceptable for a specific application.

The recovery point objective (RPO) is the point (before a failure or an outage) to which systems and data must be restored. The RPO dictates the allowable data loss and thus the type of data protection solution to be used.

Data deduplication can reduce the need for backup storage space by up to 20 times. Specific savings depend on the number of backups, the data retention policy, and how much the data changes.

The backup process and its definitions

It is important to understand that backing up is a process. As long as new data is being created and changes are being made, backups will need to be updated. Individuals and organizations with anywhere from one computer to thousands (or even millions) of computer systems all have requirements for protecting data. While the scale is different, the objectives and limitations are essentially the same. Likewise, those who perform backups need to know to what extent the processes were successful, regardless of scale.

Note

If you need more detailed information about HPE backup and restore solutions and concepts, see the *HPE Product Bulletin and the HPE Enterprise Backup Solution Design Guide.* Refer to the *HPE Enterprise Backup Solution Matrix* or *the HPE Single Point of Configuration Knowledge* (SPOCK) to determine whether products and firmware are compatible.

The effectiveness of recovery

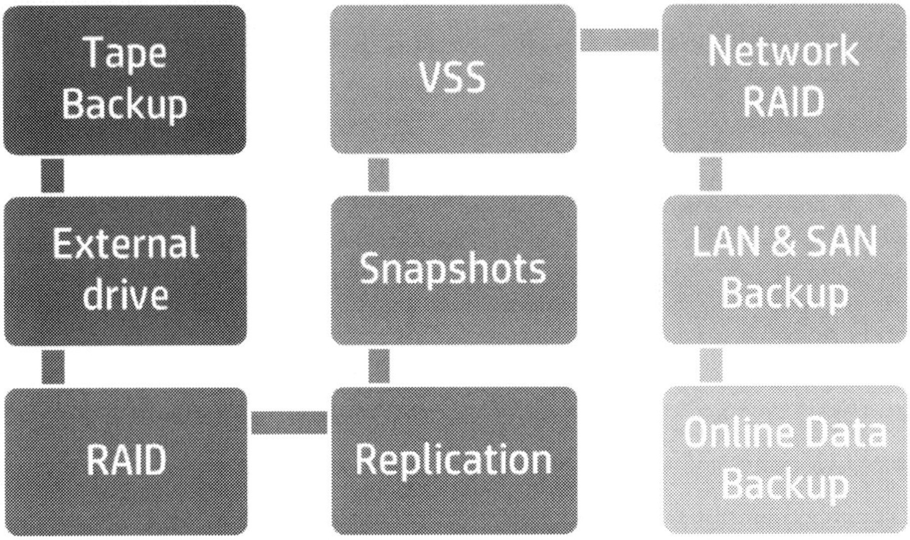

Figure 7-7 The effectiveness of recovery

As a leader in data protection, HPE delivers total business protection with simple, cost-effective, and easy-to-manage backup and recovery solutions, regardless of the size of the data center (Figure 7-7). HPE has solutions that can back up data three times faster and restore that data five times faster than competitive platforms. The HPE technologies and product portfolio can help to solve customer data recovery problems and provide business value.

HPE backup and recovery solutions provide these services:

- Backup modernization for customers struggling with antiquated backup plans and infrastructure
- Help at every stage of the planning and implementation of business protection plans
- Support for all major operating systems

HPE StoreOnce backup systems

The HPE StoreOnce backup system is a disk-based storage appliance for backing up host network servers or PCs to target devices on the appliance (Figure 7-8). These devices are configured as either network-attached storage (NAS), virtual tape library (VTL), or Catalyst targets for backup applications.

These appliances can be all NAS, all VTL, or any combination of NAS, VTL, and Catalyst devices. Similarly, the total number of backup targets can be split between NAS, VTL, or Catalyst devices. All

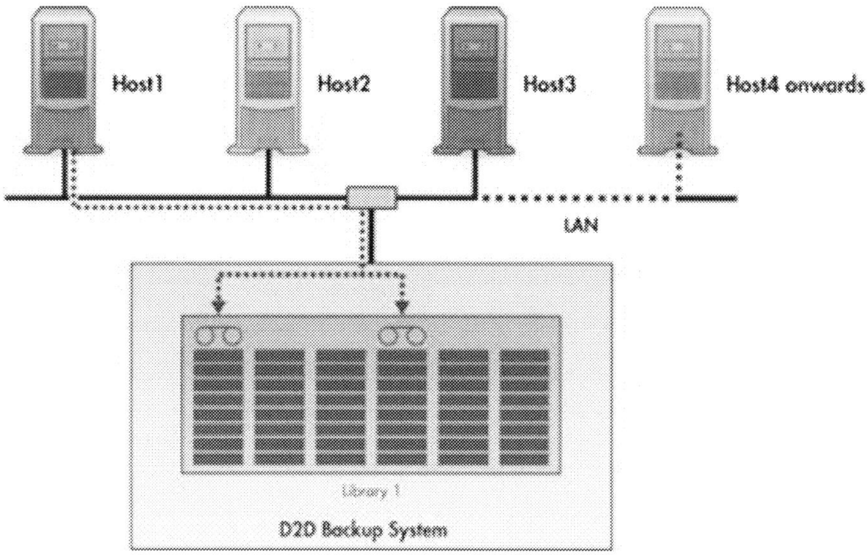

Figure 7-8 D2D Backup System example

HPE StoreOnce devices that are configured for NAS, VTL, or Catalyst automatically make use of HPE deduplication to ensure the most efficient and cost-effective use of disk space.

What is HPE StoreOnce?

Figure 7-9 HPE StoreOnce products

HPE StoreOnce is an HPE brand (Figure 7-9).

- **HPE StoreOnce backup system**—A backup appliance
- **HPE StoreOnce VSA**—Software-defined storage (SDS) through a virtual storage appliance (VSA)
- **HPE StoreOnce Deduplication**—Federated deduplication that defies limits
- **HPE StoreOnce Catalyst**—An interface to write and read from HPE StoreOnce devices

Why use an HPE StoreOnce (disk-based) backup target?

In disk solutions, data is backed up from an application server (disk) over a dedicated SAN to a disk-based system and from there to a traditional tape library. This procedure provides enhanced solutions for slow servers, single-file restores, and perishable data.

One of the particular benefits of the HPE Virtual Library System (VLS) and disk-to-disk (D2D) solutions is that they make a disk array look like a tape library to a backup server. Implementation requires no new software and no significant redesign of backup processes.

Customers who are familiar with tape backup environments can find these benefits from HPE solutions:

- Seamless integration into the existing backup infrastructure
- Stable, long-established backup application features such as media import and catalog regeneration
- A Fibre Channel (FC) interface or Internet Small Computer System Interface (iSCSI) for performance and flexibility
- NAS shares for simple integration into existing backup-to-disk environments
- Special-purpose and high-performance backup interfaces

HPE StoreOnce family

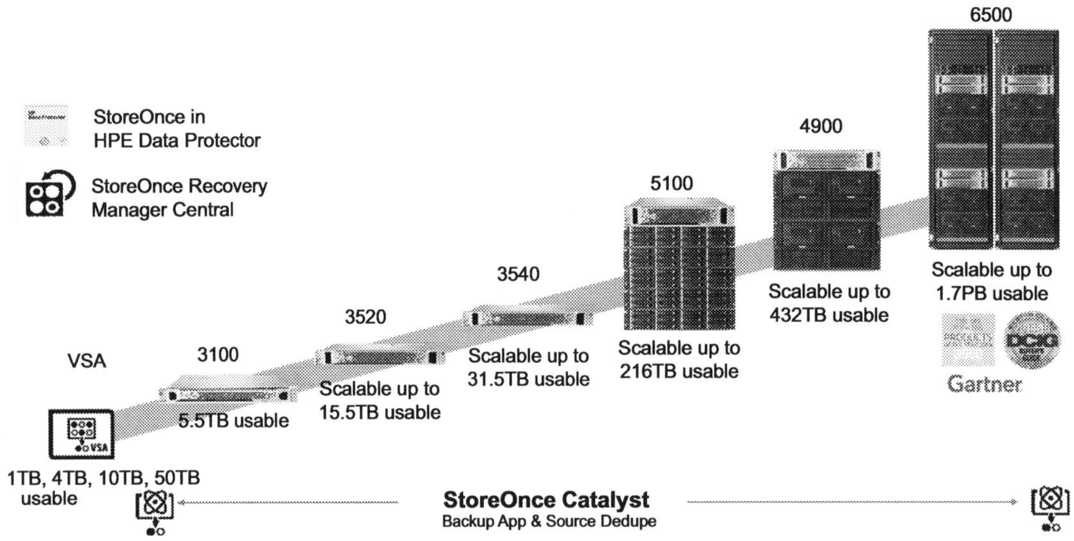

Figure 7-10 HPE StoreOnce family

Wherever customers need data protection, there is an HPE StoreOnce platform that fits (Figure 7-10).

- **HPE Data Protector**—Software-based deduplication
- **HPE StoreOnce VSA**—Virtual appliance with up to 50 TB usable capacity
- **HPE StoreOnce 3100**—Up to 5.5 TB usable capacity
- **HPE StoreOnce 3520**—Up to 15.5 TB usable capacity
- **HPE StoreOnce 3540**—Up to 31 TB usable capacity
- **HPE StoreOnce 4900**—Up to 432 TB usable capacity
- **HPE StoreOnce 5100**—Up to 216 TB usable capacity
- **HPE StoreOnce 6500**—Scale-up and scale-out data protection for enterprise data centers, with 72 TB to 1728 TB usable capacity

All new HPE StoreOnce products (3100, 3520, 3540, and 5100) are based on the proven HPE ProLiant Server Gen9 platform. The initial licenses for StoreOnce 3100, 3520, 3540, and 5100 are preinstalled at the factory.

Note

Not all products are suitable for small and medium-sized businesses.

HPE StoreOnce deduplication

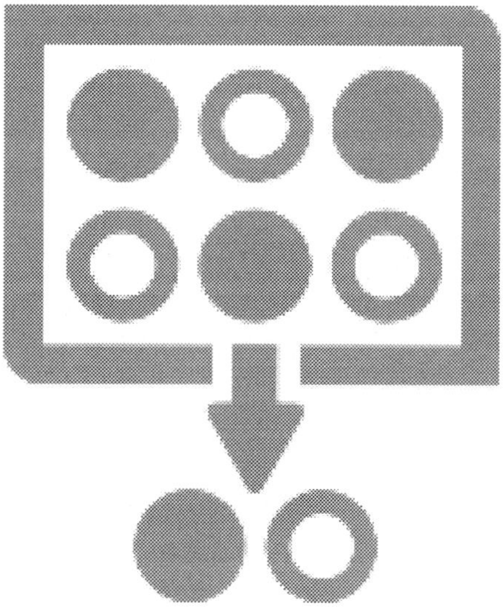

Figure 7-11 HPE StoreOnce deduplication icon

HPE StoreOnce deduplication reduces the disk space required to store backup data sets without impacting backup performance (Figure 7-11). Retaining more backup data on disk for a longer period of time enables greater data accessibility for the rapid restore of lost or corrupt files and reduced downtime.

Deduplication ratios are strongly influenced by two factors—the data change rate and the backup data retention period. Low data change rates and data retained for longer periods of time yield higher deduplication ratios.

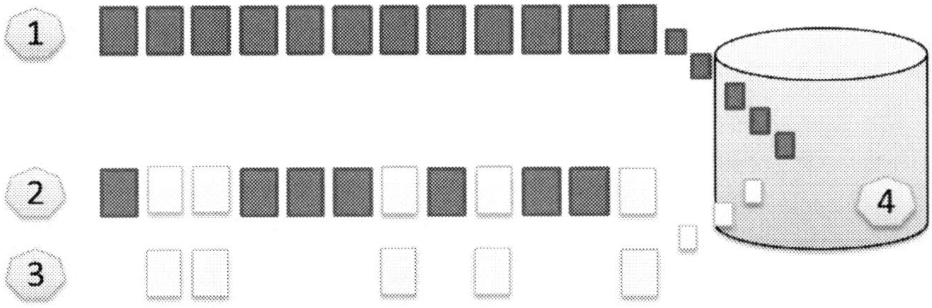

1 Data from the first backup stream are stored to disk
2 Duplicate data as well as unique data in a second backup stream are identified
3 Duplicate data in the second backup streams are eliminated
4 Unique Data in the second backup stream are stored to disk

Figure 7-12 HPE StoreOnce deduplication example

HPE StoreOnce deduplication software simplifies the deployment of deduplication technology across IT infrastructures (Figure 7-12). With explosive data growth driving IT sprawl, deduplication technology is quickly becoming a requirement for many customers to help reduce the capacity required to store information.

Traditional deduplication technologies tend to approach the problem from a fragmented perspective. This results in multiple deduplication methodologies being deployed, which adds to the management complexity of the infrastructure.

HPE StoreOnce technology is different from traditional deduplication technology. As a next-generation deduplication architecture, HPE StoreOnce deduplication software is not sold as stand-alone software, but rather, it is a portable engine that can be consistently embedded in multiple products. This architecture eliminates the complexity of first-generation deduplication. The HPE StoreOnce portfolio uses patented innovation and features designed by HPE Labs to maximize backup and restore performance while minimizing management and hardware overhead.

Deduplication works by examining the data stream as it arrives at the storage appliance, checking for blocks of data that are identical and eliminating redundant copies. If duplicate data is found, a pointer is established to the original set of data as opposed to storing, removing, or duplicating the redundant blocks.

The key advantage is that the data deduplication is being done at the block level, which removes far more redundant data than deduplication that is done at the file level, where only duplicate files are removed.

Data deduplication is especially powerful when it is applied to backup operations because most backup data sets have a lot of redundancy. The amount of redundancy depends on the type of data being backed up, the backup methodology, and the length of time the data is retained.

Deduplication can be implemented three ways:

- Client side
- Server side
- Target side

Benefits of deduplication

Customer benefits of deduplication include the following:

- **The ability to store dramatically more data online**—Retaining more backup data on disk for a longer duration enables greater data accessibility for the rapid restore of lost or corrupt files, and it reduces the impact of downtime on business productivity. Deduplication also provides savings in IT resources, physical space, and power requirements. Disk recovery of single files is faster than from tape.
- **Reduced investment in physical tape**—Deduplication also reduces the overhead of tape management by restricting the use of tape to the deep archiving and disaster recovery (DR) usage models.
- **A network-efficient way to replicate data off-site**—Deduplication can automate the DR process by enabling site-to-site replication at a lower cost. Because deduplication recognizes what data has changed at the block or byte level, replication becomes more intelligent, and it transfers only the changed data as opposed to the complete data set. This saves time and replication bandwidth, enabling better disaster tolerance without the need and operational costs associated with transporting data off-site on physical tape.
- **A broader range of available RPOs**—Data can be recovered from farther back in time, when compared to earlier backup sets, to better meet service level agreements (SLAs).

HPE StoreOnce replication

HPE StoreOnce deduplication is the technology enabler for HPE StoreOnce replication, which allows fully automated replication without rehydration. Optimized replication can use low-bandwidth WAN links to a DR site. This is also a cost-effective DR solution for data centers and remote offices and branch offices (ROBO).

HPE StoreOnce systems can be configured to replicate backup through the LAN or the WAN to other StoreOnce systems. In the case of and HPE StoreOnce Catalyst, two or more devices have a network connection and, using standard TCP/IP protocols, data is sent between systems.

HPE StoreOnce replication uses the fact that with deduplicated data stores, only new data "chunks" need to be sent over the WAN or the LAN after the initial synchronization. This process is often called bandwidth optimization, and it saves valuable bandwidth when replicating data.

HPE StoreOnce technology extensively checksums each data frame with an MD5 checksum. If an error is detected, the frame is retransmitted. This capability is especially important over WAN links because the error checking is less thorough than that used on LANs.

Disk-based backup systems (D2D)

Disk-based backup systems deliver leading price: performance, deduplicate backup data, and automatically consolidate multiple servers into a single backup process (Figure 7-13).

Figure 7-13 HPE StoreOnce 4900

The HPE D2D backup systems with HPE StoreOnce deduplication provide disk-based data protection for data centers and remote offices. These systems automate and consolidate the backup of multiple servers onto a single, rack-mountable device while improving reliability by reducing the errors caused by media handling.

All HPE D2D backup systems feature HPE StoreOnce deduplication software for efficient, longer-term data retention on disk and for enabling network-efficient replication. The software provides a cost-effective way of transmitting data off-site for DR purposes.

The D2D backup systems integrate seamlessly into current IT environments and offer the flexibility of using both VTL and NAS targets.

HPE StoreOnce Catalyst technology

HPE StoreOnce Catalyst technology enables backup applications to communicate directly with the HPE StoreOnce appliance. This capability allows advanced features such as deduplication at the backup server for bandwidth-optimized backups and replication to one or more StoreOnce appliances without the overhead of rehydration. The unique ability to perform some of the deduplication workload on the backup server, makes communication more bandwidth-efficient and faster.

The HPE Catalyst solutions use a standard network connection to a backup server, but they can communicate with a "client" application programming interface (API) that is integral to backup applications. Because a Catalyst system understands commands from backup software, it can move data without rehydration to other StoreOnce systems and remove expired data automatically.

The HPE StoreOnce Catalyst technology brings the HPE StoreOnce vision of a single, integrated enterprise-wide deduplication algorithm a step closer.

Comprehensive error checking enables HPE StoreOnce Catalyst systems to operate over WAN and LAN connections, even for international links. Now, data can really be moved offshore with no human intervention.

HPE StoreOnce Catalyst solutions are supported with HPE Data Protector, Symantec NetBackup with OST, Backup Exec, Oracle RMAN, and BridgeHead Healthcare Software. Oracle users can also back up directly to a Catalyst store, increasing deduplication efficiency.

HPE LTO Ultrium tape drives

Figure 7-14 Sixth-generation LTO tape devices

The family of HPE Linear Tape-Open (LTO) Ultrium tape drives is the premier line of backup devices. It is based on the LTO Ultrium format, which is an open standard (Figure 7-14).

With a well-defined eight-generation road map, HPE LTO Ultrium tape drives offer customers the best choice for investment protection. The rugged design is built on superior LTO technology to provide ultimate reliability and ease of use, even at 100% duty cycles.

HPE has added advanced features like hardware data encryption to create a new level of data protection. The tape drives employ dynamic data rate matching to adjust to the speed of the host, reducing wear and tear on both the drive and the media and increasing performance, even on slower hosts. The HPE ultra-durable, soft-load feature automatically positions the data cartridge together with the highly reliable HPE leader-capture mechanism to make backup and restore operations more reliable.

Support for HPE One-Button DR provides the easiest way to restore data in the event of a disaster. It provides quick recovery of the operating system, applications, and data from the latest full backup set.

This family of tape drives also features support for HPE Library and Tape Tools, which provide HPE TapeAssure health monitoring, diagnostics, and performance optimization. Included with the HPE LTO Ultrium tape drives is the option to download Yosemite Server Backup Basic, which is a backup and recovery software that provides a backup application to protect a single server. If data is critical to the business, protect it with an HPE LTO Ultrium tape drive.

The HPE LTO-6 drive supports the Reed-Solomon Cyclic Redundancy Check (CRC).

Sixth-generation LTO tape

The HPE Ultrium tape blades offer a complete data protection, DR, and archiving solution for HPE BladeSystem c-Class customers who need an integrated data protection solution. These half-height tape blades provide direct-attach data protection for the adjacent server and network backup protection for all data residing within the enclosure.

LTO-6, released in December 2012, was the sixth generation of LTO tape. It offers capacity for 2.5 TB of raw data (up to 6.25 TB with 2.5:1 compression) and a transfer rate of 160 MB/s, which translates to 400 MB/s compressed. The impressive increase in tape density was achieved by using a newly developed magnetic particle and other key advances in manufacturing technology.

New features in LTO-7 tape

Figure 7-15 HPE StoreEver LTO-7 Ultrium 15000 device

LTO-7 is the seventh generation of LTO tapes, with 6 TB of raw data, and it was introduced in 2015 (Figure 7-15).

HPE LTO-7 drives have added support for the Append-only mode, as defined in the T10 draft of the SSC-4 (SCSI-3 Stream Commands) standard. In this mode, the default behavior of the drive is unchanged, but when a WRITE command is received, the drive processes it as normal and writes data at the current logical position (subject to any applicable constraints such as write once, read many [WORM] media). Any previous data after that logical position will be overwritten. This is why the new mode is sometimes referred to as "Overwrite allowed" mode.

In Append-only mode, write operations are only allowed when the current logical position is one of the following:

- Beginning of the partition (BOP) with nothing written to tape
- BOP with only file marks before the end of data (EOD)
- Between BOP and EOD with only file marks between this position and EOD
- EOD

Write operations will complete normally in these cases, but in all other cases (where data would be overwritten) the WRITE command will fail with CHECK CONDITION status, with the sense key set to DATA PROTECT, and with the additional sense code set to OPERATOR SELECTED WRITE PROTECT.

The HPE LTO-7 drives support the CRC32C (Castagnoli) polynomial, which on some systems might be considerably faster than the Reed-Solomon polynomial.

Library and Tape Tools

HPE Library and Tape Tools (L&TT) is a free, downloadable, robust diagnostic tool for all HPE tape storage and magneto-optical storage products. Provided for a wide range of users, L&TT is ideal for customers who want to verify their installation, ensure product reliability, perform their own diagnostics, and achieve faster resolution of tape device issues.

- L&TT performs firmware upgrades, verification of device operation, failure analysis, and a range of utility functions.
- Performance tools assist in troubleshooting bottlenecks, and system configuration checks warn of common host issues.

L&TT also provides seamless integration with HPE Support by generating and emailing test results and support tickets.

Note

HPE Support requires the use of L&TT to troubleshoot most device issues, so HPE recommends pulling a support ticket and running the device assessment test before calling.

HPE BURA initiative

BURA stands for backup, recovery, and archive. The accelerating rate of change demands a new approach to BURA.

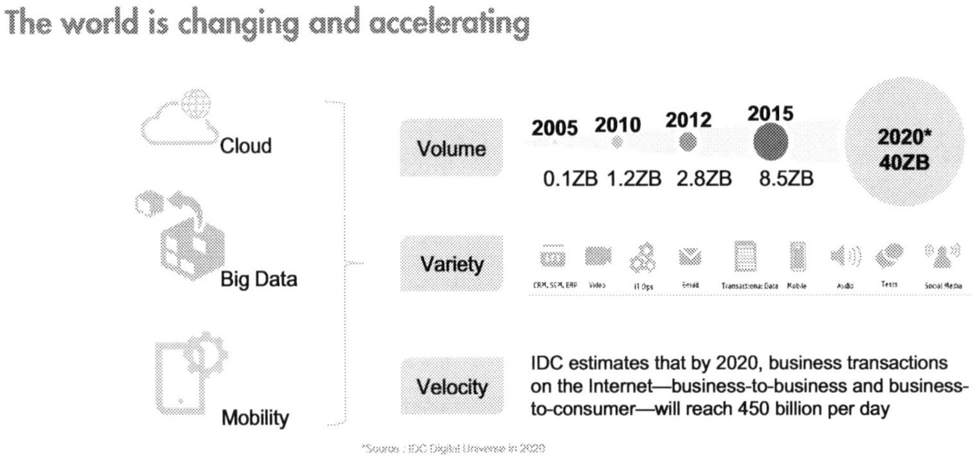

Figure 7-16 Ways the world is changing and accelerating

Data is growing at an unprecedented rate, with one statistic saying that 90% of all the data ever created was created in the past 12 months (Figure 7-16).

The IDC Digital Universe survey estimates a massive 40 zettabytes (40,000 exabytes or 40 million terabytes) of data will be created by 2020, up from almost 3 zettabytes in 2012. This represents a more than 13-fold increase in 8 years.

The variety of data is changing too, driven partially by the growth of mobile devices and the ability to communicate with anyone, anytime, anywhere. It is estimated that there will be more than 450 billion business transactions over the Web every day.

This is shaping how IT is evolving. A new era of IT is rapidly taking shape, from one era where silos of data were once used to a new era where data must be flexibly stored and be agile enough to move around organizations.

Some trends are emerging. They include cloud (both private and public), and software-defined storage and software-defined data centers, which are now industry buzzwords and trends. These trends are rapidly gathering momentum.

There is also massive data growth of both structured and unstructured data, and more importantly, businesses want to extract value from that data to help them make better decisions in all areas of their organization (including targeted advertising). With so many large online retailers and the huge growth in mobile devices, there are many challenges that organizations face in managing, storing, and protecting that data.

Most organizations today see data growth in the area of 25%–50% per annum, and in many cases, they are seeing it double every year.

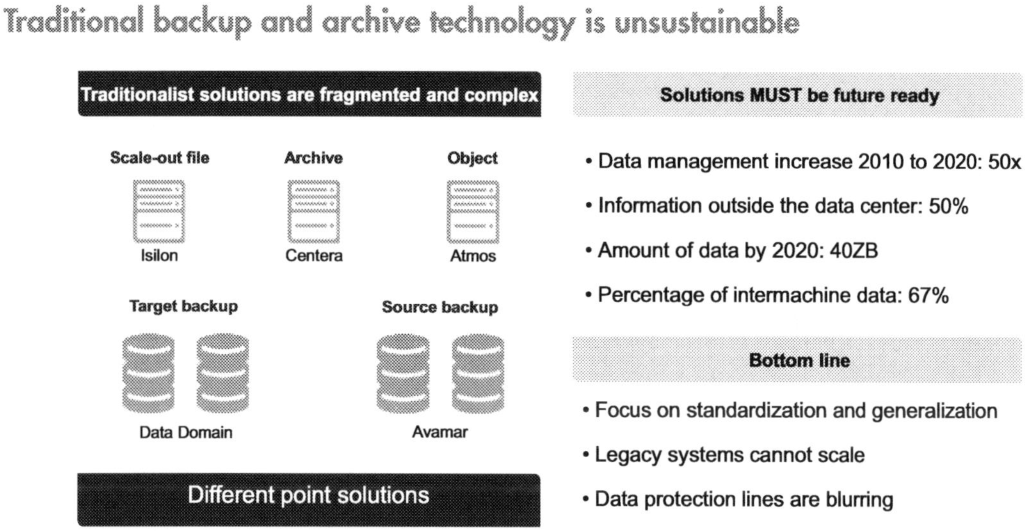

Figure 7-17 Ways that traditional backup and archive are unsustainable

Many storage vendors continue to design and produce storage architectures that have done an excellent job in storing, protecting, and retaining data. However, these designs will not support the changes that are going to happen in IT over the next 5–10 years (Figure 7-17).

If you consider the architectures on the left side of the graphic, you will see that they are siloed and limited in terms of their ability to scale capacity and performance. When the businesses outgrow them, they will need a "forklift upgrade" to create a brand new and costly solution.

The traditional backup architectures have been good solutions to date, but they are based on what HPE sometimes refers to as "Deduplication 1.0" architectures. These architectures are siloed, homogeneous deduplication engines that are focused on specific and limiting use cases—one use case for data centers, another use case for remote offices, and so on.

The current backup and recovery solutions are fragmented and complex, and they are built for specific traditional infrastructures. The challenge now is that IT organizations have to be dynamic and agile enough to grow with the business and to enable the business to respond to the market.

Enormous data growth is predicted over the next few years, with 50% of that data typically residing outside of the traditional boundaries of the main data center. Fifty percent data growth would equate to 40 zettabytes of data existing in the world. Yet if you analyze this, you would find that approximately 33% of that data could be distilled into usable and valuable information. The rest is normal communications between people or infrastructure components (source: a Forrester report). So why would someone make that 67% (which equates to just over 26 zettabytes of data) part of their backup strategy? The answer from HPE is that you should not do so.

The bottom line is that for organizations to succeed, they have to focus on how they will standardize all aspects of their data center. Just as importantly, they must generalize their resources. This means that by using fewer specialized staff, you need a smaller premium to pay for them.

Another issue driving the changes to a more flexible architecture is that legacy systems cannot scale, even to the point of getting to 33% of the market. By addressing a converged approach, HPE can help organizations scale much more easily.

As part of standardization, what typically was traditional data protection might have been an afterthought. However, it no longer can be an afterthought.

It is equally important to protect data residing on primary storage and in regional and branch offices. Today, many primary applications, databases, and so on, have backup integrated into them.

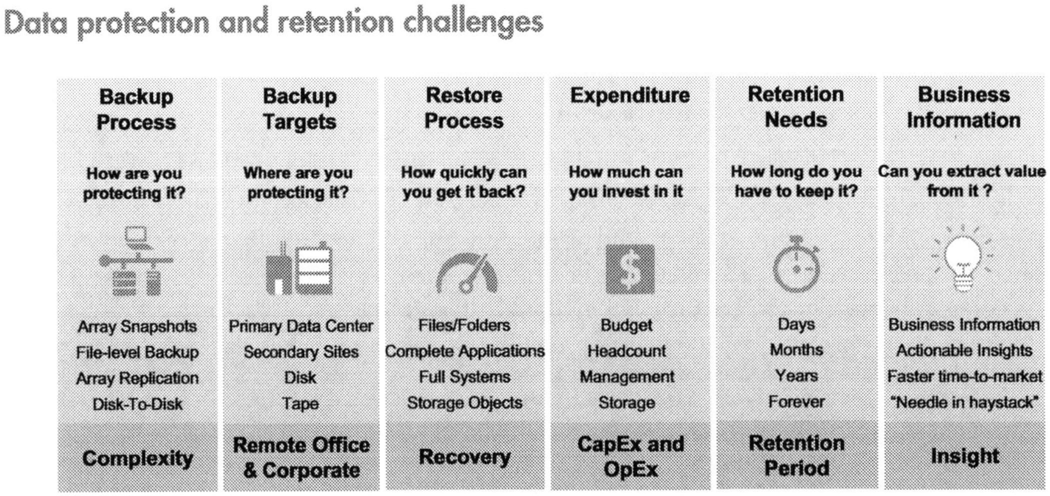

Figure 7-18 Data protection and retention challenges

The basic challenges of data protection and retention have remained the same; however, they are becoming increasingly more complex with the evolution of the IT infrastructure (Figure 7-18).

HPE BURA solutions

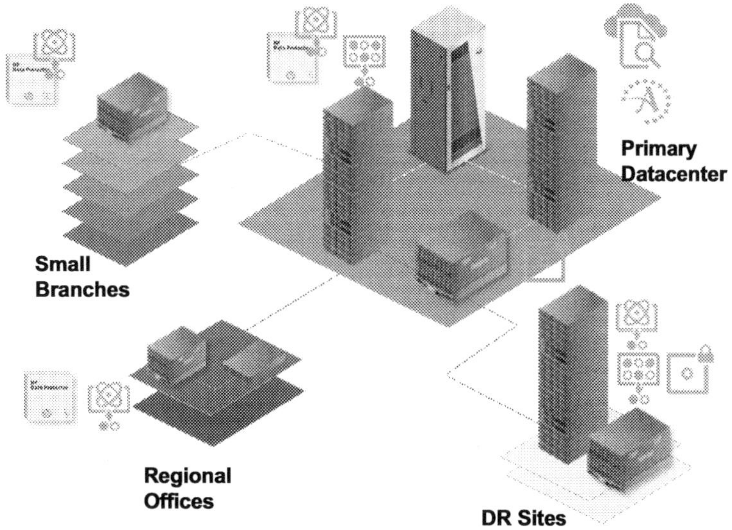

Figure 7-19 HPE BURA configuration

Standardized infrastructure—Edge to core

With HPE BURA solutions, the primary data center contains the following components (Figure 7-19):

- Standardized hardware
- Simplified operations and automation
- Less resource overhead
- Intelligent archiving

DR sites contain these components:

- Standardized hardware
- Array-level capabilities
- Faster failover and failback operations

Remote offices contain these components:

- Standardized hardware
- Less or no resource overhead
- Shared deduplicated data, resulting in less network traffic

These solutions can look complex, but there is a fairly standard deployment model for many organizations. In Figure 7-19, the main data center in the center is dark gray, the regional offices are purple, and the smaller branch offices are blue. One or more DR sites are light grey. This is a typical model for an organization.

BURA solutions are not bound to the primary data center. They cannot be because 50% of an organization's data typically resides outside that data center. For this reason, an ideal solution must encompass any place that data is generated for the business, and it also must consider where the organization uses and relies on the data.

All organizations evolve and often merge. When customers try to integrate a new company, it brings new systems and new processes. For example, a large company might have 14 backup processes and solutions because it did not consolidate data protection processes that came with the acquired companies.

The environment might not be that complex, but if customers apply the BURA perspective to their approach to a data protection strategy, they can assure the speed at which they want to protect and restore their data. The value customers can extract from that data could be immense, while the BURA solution reduces cost and risk.

For example, the data on an HPE 3PAR infrastructure could be moved to archiving devices, an HPE scale-out archiving platform, and an HPE Autonomy Consolidated Archive, to provide lower cost and longer-term retention of data. This solution would also provide incredibly fast extraction of value from that data for business intelligence purposes.

HPE StoreOnce deduplication solutions can be deployed with Data Protector Software for the smallest remote sites (containing just application servers that can be virtualized or nonvirtualized) to the largest data center or DR site. This would be a single, completely compatible backup solution.

All of this data can be offloaded to tape for low-cost DR, compliance, and archival purposes.

HPE BURA solutions can be configured and managed from central data center, reducing branch office staffing costs, reducing space, reducing network traffic, and, more importantly, reducing costs and risk.

Before this ideal approach can be achieved, another topic must be addressed: the steps to classify the information that an organization generates. This classification will help to devise a successful data protection and retention strategy.

HPE StoreOnce backup

HPE StoreOnce backup solutions help with following customer concerns:

- Does data growth leave you struggling with complex, distributed, and costly data protection?
- Is some of your data not being protected because backup windows are not long enough or backup jobs are failing?

HPE StoreOnce backup provides disk-based, deduplicating, backup systems. You can use StoreOnce deduplication, which is available in a range of scalable dedicated appliances and flexible virtual appliances, with your choice of backup and recovery software to deliver robust enterprise-wide data protection.

HPE StoreOnce backup systems reduce the amount of backup data customers need to store by up to 95%, and with the scale-out architecture, customers can retain up to 34 petabytes of data in a single pool.

These backup systems provide automated backup and DR operations with all of the features that can be expected from disk backup, together with secure data retention with built-in data encryption for Data at Rest and Data in Flight. The secure erase functionality can be use if disks are lost, stolen, or discarded.

Note

At this time, Data in Flight encryption with IPSec is only supported on StoreOnce Catalyst systems. In addition, write performance might be impacted, but performance improvements should be seen after the first ingest.

HPE StoreOnce Catalyst systems deliver industry-leading backup speeds of up to 139 TB/hr, enabling customers to meet shrinking backup windows, plus Federated Deduplication across the enterprise, which lets customers choose where to deduplicate data. Federated Catalyst enables Catalyst stores to span nodes, simplifying backup management and optimizing the available storage in large environments.

Note

Actual HPE StoreOnce Catalyst system performance depends on the configuration data set type, the compression levels, the number of data streams, the number of devices emulated, and the number of concurrent tasks such as housekeeping or replication.

The HPE StoreOnce portfolio of backup systems provides a choice of powerful dedicated appliances for larger offices and data centers, flexible virtual appliances for highly virtualized or smaller and remote offices, and when customers do not want to use a dedicated deduplication appliance, Data Protector Software with StoreOnce Federated Deduplication and StoreOnce Catalyst. The HPE StoreOnce deduplication technology makes managing the movement of data across the enterprise easier.

StoreOnce backup solutions seamlessly integrate with current backup applications and provide flexible integration into FC and iSCSI SANs or into GbE and 10 GbE, virtualized, and other environments. Catalyst over FC provides all the independent software vendors (ISV) control and source-side deduplication benefits of the current StoreOnce Catalyst system, but over a FC fabric.

HPE StoreOnce VSA

Figure 7-20 StoreOnce VSA icon

The HPE StoreOnce VSA offering extends the deployment options for StoreOnce solutions with the agility and flexibility of a virtual appliance, eliminating the need to install dedicated hardware (Figure 7-20). This appliance provides a flexible and a cost-effective backup target for virtualized server environments.

HPE StoreOnce VSA can be used as part of a pure software-defined data protection solution or in conjunction with StoreOnce purpose-built appliances. Operation and integration with backup software is the same for the StoreOnce VSA appliance and the HPE purpose-built appliances. HPE StoreOnce VSA is an extension to a single, federated deduplication architecture and is a key component of HPE's emerging software-defined storage portfolio.

HPE StoreOnce VSA is HPE StoreOnce delivered as a VMware virtual appliance, with a single license that enables up to 50 TB of usable capacity (1000 TB logical equivalent with >20:1 data deduplication). VSA delivers a cost-effective and comprehensive data solution for IT-as-s-Service (ITaaS) and Backup-as-a-Service (BaaS) and for enterprise ROBOs, and it brings centralized management of all backup and copy activities.

HPE StoreOnce VSA target audiences

Typical target infrastructures that are suitable for HPE StoreOnce VSA implementations include

- Server virtualization
- Common backup apps such as Data Protector, Veeam, or Symantec
- Catalyst, VTL (iSCSI), and Common Internet File System (CIFS)

Typical target customers for HPE StoreOnce VSA solutions include

- Service providers looking to provide BaaS or DR-as-a-Service (DRaaS)
- Enterprises with remote and branch offices
- Small and medium-sized businesses

HPE StoreOnce VSA features and integration

HPE StoreOnce VSA features include

- Hardware agnostic
- 1 TB, 4 TB, 10 TB, or 50 TB
- Protocol support: StoreOnce Catalyst, CIFS, and VTL iSCSI
- Bulk deployment capability for large volume of remote offices
- Federated Deduplication and Copy/Replication compatible with StoreOnce hardware appliances
- 60-day instant-on trial license
    - The HPE StoreOnce VSA license enables the use of StoreOnce Catalyst, VTL, and NAS replication and includes three years of HPE support.

Application support includes

- The appliance runs on VMware, Microsoft Hyper-V, or KVM.
- Reporting of capacity utilization, deduplication ratios, and so on, is through StoreOnce Enterprise Manager.
- It is supported by HPE Data Protector, Symantec NetBackup, BackupExec (VTL iSCSI only), and Veeam. For other backup ISVs, submit a Deal Exception Request.

StoreOnce VSA specifications

Table 7-1 VSA specifications

Form factor	Virtual appliance (VMware, Hyper-V, KVM)
Total usable capacity	4 TB, 10 TB, and 50 TB
Replication fan-in/fan-out (appliance)	8/2
Write performance (maximum aggregated data transfer rate using VTL)	2.4 TB/hr
Read performance (maximum aggregated data transfer rate using VTL)	1.8 TB/hr
StoreOnce Catalyst performance (maximum aggregated data transfer rate)	6 TB/hr
Device interfaces	2 ×1 GbE vNICs (minimum)
Number of StoreOnce Catalyst, VTLs and NAS backup targets (combined)	16
Maximum number of cartridges emulated	8192

Capacity scalability

HPE StoreOnce VSA solutions are enabled with a single license. You can scale up to 4 TB, 10 TB, or 50 TB of usable capacity (upgrades are available; Table 7-1).To most efficiently use your appliance resources, HPE recommends provisioning of the required capacity. Scale-up is done by purchasing licenses and adding additional virtual disks (vDisks) as needed. The HPE StoreOnce VSA appliance is enabled by a three-year or a five-year license to use (LTU).

Note

The HPE StoreOnce VSA upgrade process is fully described in the user manual.

Capacity upgrades are nondisruptive; however, adding more memory and processor resources requires restarting the appliance, so always schedule these upgrades outside of backup times.

HPE StoreOnce 3100 System

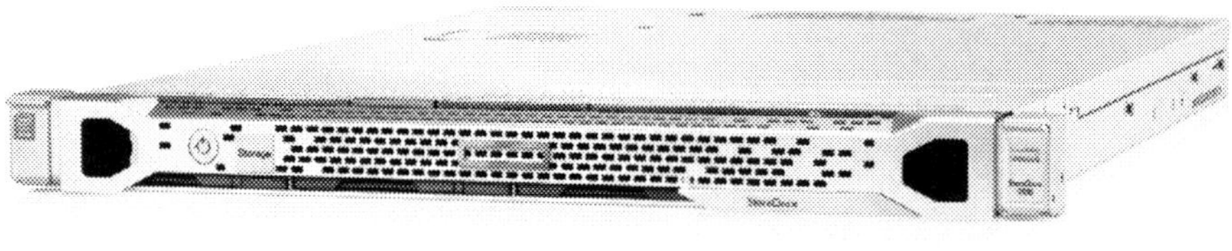

Figure 7-21 HPE StoreOnce 3100 System (with security bezel)

The HPE StoreOnce 3100 System delivers entry-level, disk-based, backup, and DR that is ideal for smaller remote or branch offices and data centers (Figure 7-21). This 1U backup system offers 5.5 TB of usable capacity (8 TB RAW) and speeds of up to 6.4 TB per hour with StoreOnce Catalyst technology, allowing a full 25.6 TB of backup to be completed in just 4 hours.

HPE StoreOnce 3100 System specifications

Table 7-2 System specifications

Form factor	1U rack
Total capacity (raw)	8 TB
Total capacity (usable)	5.5 TB
Write performance (max)	1.4 TB/hr
Catalyst performance (max)	6.4 TB/hr
Max fan-in/backup targets	8

Note

Actual usable capacity for customer data storage depends on the drive formatting, the log file and metadata size, and the housekeeping backlog. Actual performance is dependent on the configuration, the data set type, the compression levels, the number of data streams, the number of devices emulated, and the number of concurrent tasks such as housekeeping or replication and storage configuration (Table 7-2).

HPE StoreOnce 3520 System

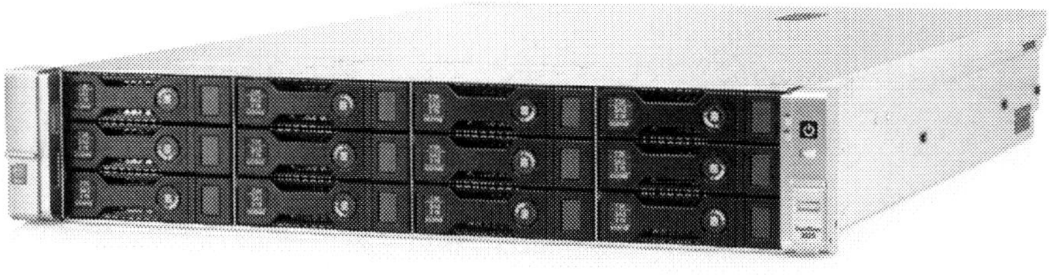

Figure 7-22 HPE StoreOnce 3520 System (without security bezel)

The HPE StoreOnce 3520 System is designed for small and medium-sized data centers and as a replication target device for remote and branch offices (Figure 7-22). The StoreOnce 3520 System delivers a scalable 2U solution from 7.5 TB to 15.5 TB of usable capacity (12 TB to 24 TB RAW) using an upgrade license. You can meet backup windows, with speeds of up to 12.7 TB/hr using the HPE StoreOnce Catalyst technology for the protection of up to 50.8 TB of data in a 4-hour window.

The StoreOnce 3520 System comes fully populated with twelve 2 TB disks. However, initial working capacity is 12 TB RAW; a capacity upgrade license is required to activate the additional disk capacity.

Note

Actual usable capacity for customer data storage depends on the drive formatting, the log file and metadata size, and the housekeeping backlog.
Actual performance is dependent on the configuration, the data set type, the compression levels, the number of data streams, the number of devices emulated, and the number of concurrent tasks such as housekeeping or replication and storage configuration.

HPE StoreOnce 3520 System specifications

Table 7-3 System specifications for HPE StoreOnce 3520

Form factor	2U rack
Total raw capacity	24 TB
Total usable capacity	15.5 TB
Maximum write performance	4.6 TB/hr
Maximum Catalyst performance	12.7 TB/hr
Maximum fan-in/backup targets	24

HPE StoreOnce 3540 System

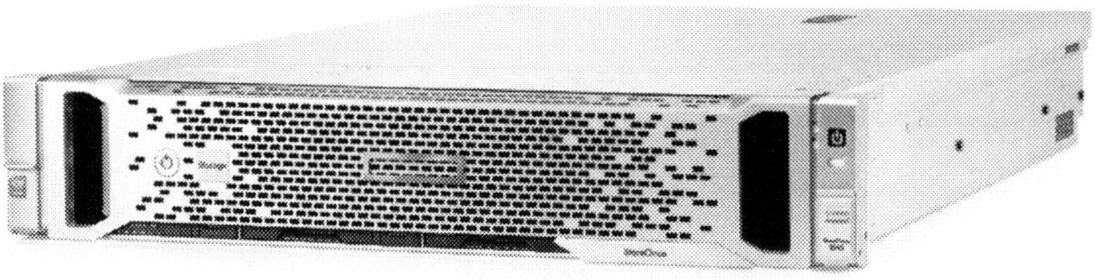

Figure 7-23 HPE StoreOnce 3540 System (with security bezel)

The HPE StoreOnce 3540 System is designed for small and medium-sized data centers and as a replication target device for up to 24 remote and branch offices (Figure 7-23; Table 7-3). This system delivers a scalable 2U solution, from 16 TB to 31.5 TB of usable capacity (24 TB to 48 TB RAW) using a simple and cost-effective capacity upgrade. The system offers speeds of up to 12.7 TB per hour using HPE StoreOnce Catalyst technology for the protection of up to 50.8 TB of data in a 4-hour window.

The StoreOnce 3540 comes fully populated with twelve 4 TB disks. However, initial working capacity is 24 TB RAW; a capacity upgrade license is required to activate the additional disk capacity.

HPE StoreOnce 3540 System specifications

Table 7-4 System specifications for HPE StoreOnce 3540

Form factor	2U rack
Total raw capacity	Up to 48 TB
Total usable capacity	Up to 31.5 TB
Maximum write performance	4.6 TB/hr
Maximum Catalyst performance	12.7 TB/hr
Maximum fan-in/backup targets	24

Note

Actual usable capacity for customer data storage depends on the drive formatting, the log file and metadata size, and the housekeeping backlog.

Actual performance is dependent on the configuration, the data set type, the compression levels, the number of data streams, the number of devices emulated, and the number of concurrent tasks such as housekeeping or replication and storage configuration (Table 7-4).

HPE StoreOnce 4900 System

Figure 7-24 HPE StoreOnce 4900 System

HPE StoreOnce 4900 System delivers cost-effective, scalable, disk-based backup with deduplication for longer-term on-site data retention and off-site DR for large data centers or regional offices. It also provides a replication target device for up to 50 remote or branch offices (Figure 7-24, Table 7-5).

The StoreOnce 4900 System delivers a scalable 7U to 12U solution from 36 TB to 432 TB of usable capacity (60 TB to 560 TB RAW) and speeds of up to 22 TB/hr with HPE StoreOnce Catalyst technology for the protection of 74 TB of data in a 4-hour window.

Note

Actual usable capacity for customer data storage depends upon drive formatting, log file and metadata size, and housekeeping backlog.

HPE StoreOnce 4900 specifications

Table 7-5 Specifications for HPE StoreOnce 4900

Form factor	7U–12U rack (+1U support shelf)
Total raw capacity	560 TB
Total usable capacity	432 TB
Maximum write performance	8.5 TB/hr
Maximum Catalyst performance	22 TB/hr
Maximum fan-in/backup targets	50

HPE StoreOnce 5100 System

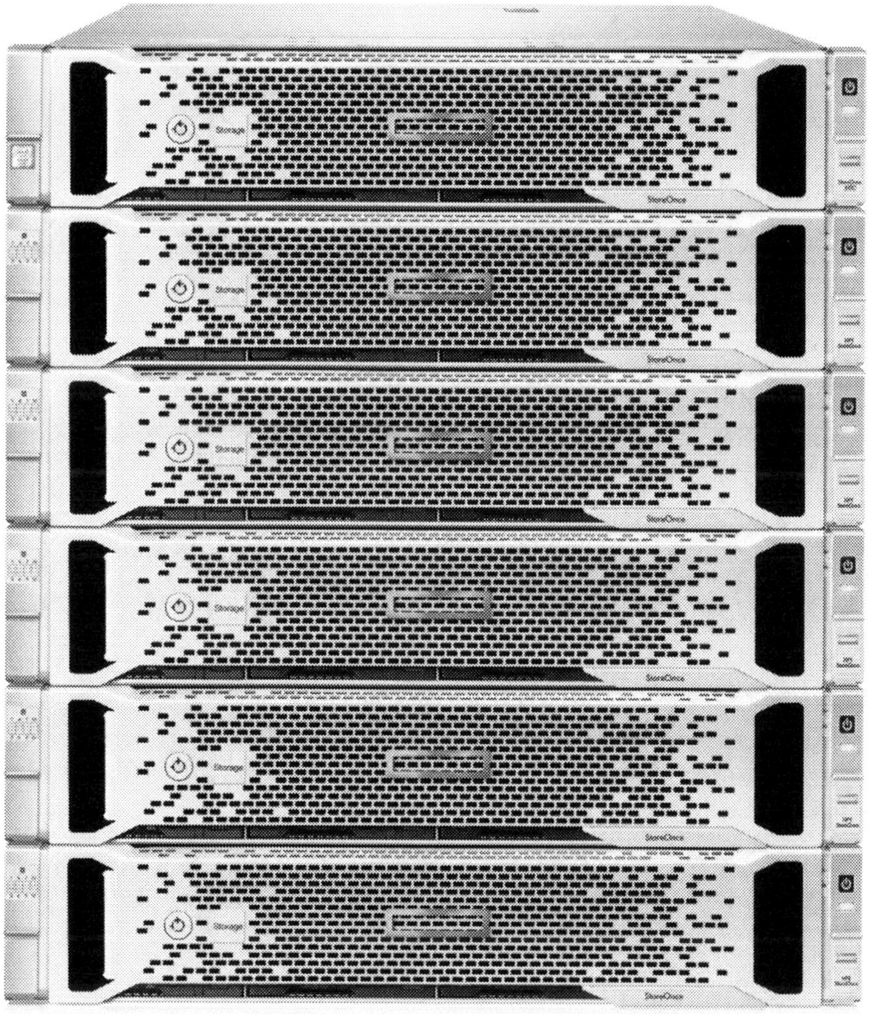

Figure 7-25 HPE StoreOnce 5100 System and HPE StoreOnce 5100 Capacity Upgrade Kits

The HPE StoreOnce 5100 System delivers cost-effective, scalable, disk-based backup with deduplication for longer-term, on-site data retention and off-site DR for larger data centers or regional offices. It also provides a replication target device for up to 32 remote or branch offices (Figure 7-25, Table 7-7).

This system delivers a scalable 2U to 12U solution from 36 TB to 216 TB of usable capacity (48 TB to 288 TB RAW) and speeds of up to 26.7 TB per hour with HPE StoreOnce Catalyst technology for the protection of more than 106.8 TB of data in a 4-hour window.

The HPE StoreOnce 5100 48 TB System upgrade kit is a D3650 base enclosure offering an additional 48 TB of RAW disk storage.

HPE StoreOnce 5100 System specifications

Table 7-6 System specifications for HPE StoreOnce 5100

Form factor	2U–12U rack
Total raw capacity	48 TB– 288 TB
Total usable capacity	36 TB–216 TB
Maximum write performance	13.8 TB/hr
Maximum Catalyst performance	26.7 TB/hr
Maximum fan-in/backup targets	32

Note

Actual usable capacity for customer data storage depends on the drive formatting, the log file and metadata size, and the housekeeping backlog.
Actual performance is dependent on the configuration, the data set type, the compression levels, the number of data streams, the number of devices emulated, and the number of concurrent tasks such as housekeeping or replication and storage configuration.

HPE StoreOnce Recovery Manager Central

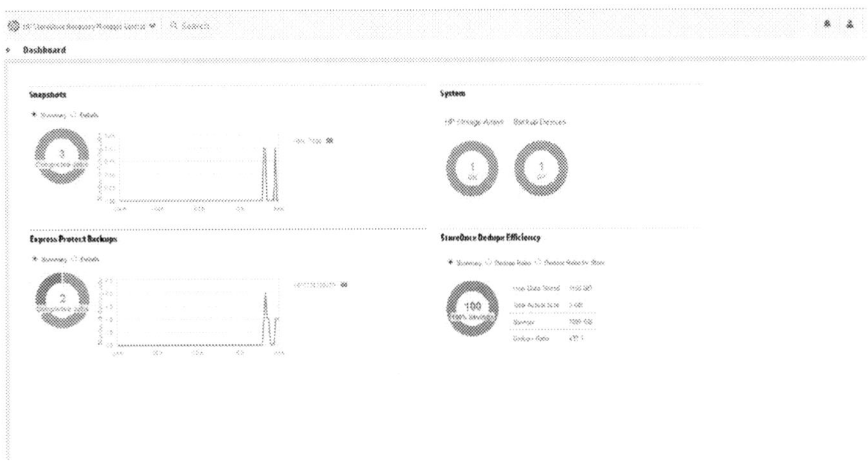

Figure 7-26 HPE StoreOnce Recovery Manager Central screen

The HPE StoreOnce Recovery Manager Central (RMC) software can be used to protect business-critical applications without impacting performance (Figure 7-26). This automated, nonintrusive software combines the simplicity and performance of snapshots with the reliability and cost-effective retention of deduplicated backups. RMC offers a simple, efficient, and fast backup and recovery solution that augments traditional backup applications rather than replacing them.

RMC integrates HPE 3PAR StoreServ primary storage and HPE StoreOnce backup for converged data protection that delivers assured recovery of application-consistent recovery points with flexible recovery options. This integration also provides a converged availability and flat backup service that augments traditional backup approaches.

HPE StoreOnce RMC 1.0 enables customers to back up their VMware virtual machines VMs and data stores, using the HPE 3PAR Virtual Copy (VC) functionality. The VC snaps can be maintained on the 3PAR system, or the data on them can be migrated to a StoreOnce appliance for long-term storage, providing a single point of backup and recovery management. RMC 1.0 can also manage crash-consistent VC backups of all other data stored on HPE 3PAR virtual volumes (VVs).

Features and benefits of RMC include

- Single snapshot and replication-enabling architecture
- Integrated ISV partnership ecosystem
- Open and modular solution
- Virtual appliance deployment
- VSS provider, scripts, and third-party agents
- Support for multiple customer workflows
- Support for the HPE 3PAR and HPE MSA Recovery Manager install base
    - Provides backup and recovery for existing applications
- Extensibility to support other applications, databases, and solutions
- RESTful APIs/OpenStack®

HPE StoreOnce Recovery Manager Central for VMware

HPE StoreOnce Recovery Manager Central provides integration with specific hypervisor or application environments through plug-ins.

The HPE StoreOnce RMC for VMware software enables RMC customers to protect VMware Virtual Machine Disks (VMDKs) and data stores using application-consistent snapshots for rapid online recovery. VMware administrators can create hundreds of VM-aware, consistent snapshots and initiate rapid online recovery directly from within the VMware vCenter Server virtualization management console.

In addition, the Express Protect feature (enabled by the StoreOnce Catalyst license) offers a second tier of data protection by facilitating direct backup from 3PAR StoreServ to StoreOnce. Backups to StoreOnce are self-contained volumes, deduplicated to save space, and they can be used to recover back to the original or to a different 3PAR StoreServ array, even if the original base volume is lost. VMware administrators can access all of this functionality from within the familiar VMware vCenter management console.

HPE Data Protector

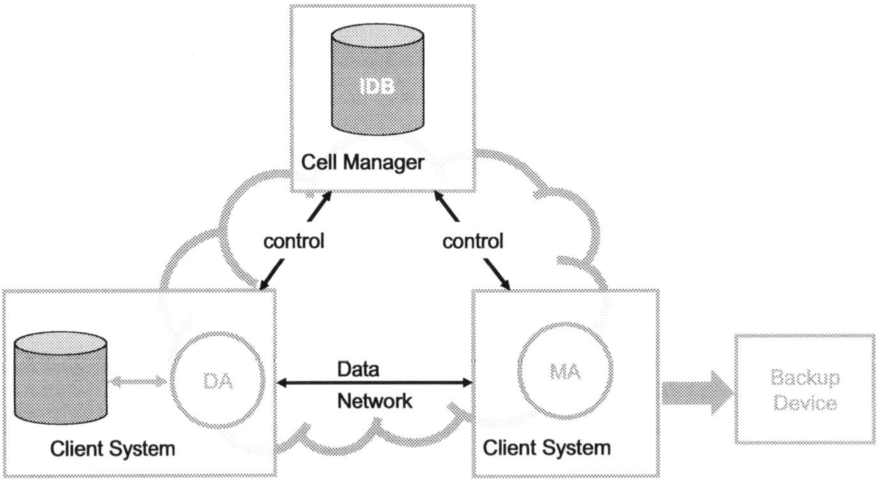

Figure 7-27 HPE Data Protector Software solution example

HPE Data Protector is the industry's first adaptive backup and recovery solution that uses an intelligent approach to data protection and management (Figure 7-27). Advanced integration, with applications and infrastructure, and operational analytics to optimize backup and recovery process help customers to improve business continuity and boost uptime within their IT environments.

In the IT arena, shifting data center requirements are forcing organizations to reassess backup and recovery processes, strategies, and infrastructure to

- Keep pace with increased data volumes, diverse data formats, and complexity
- Provide advanced integration with cloud and virtualized platforms for comprehensive backup and rapid recovery
- Take advantage of new business models such as ITaaS and Bring Your Own Device (BYOD)
- Manage and protect data according to tightening internal policies and always-on 24×7 service level objectives

HPE Data Protector Software is helping organizations of every size address these challenges with software that offers advanced data protection capabilities such as HPE StoreOnce federated deduplication, integrated cloud backup, Enhanced Automated DR (EADR), Zero Downtime Backup, Instant Recovery, advanced virtual server protection, and mission-critical application recovery down to the exact second. HPE Data Protector provides better transparency, predictability, and business resiliency within IT environments.

With HPE Data Protector Software, customers can centrally manage and protect critical data scattered across remote sites and data centers in physical, virtual, and cloud infrastructures. Today, more than 45,000 customers worldwide use HPE Data Protector Software to safeguard their business information.

A backup session can be started interactively by an operator using the Data Protector user interface or unattended using the Data Protector Scheduler. The Backup Session Manager process starts Media Agents and Disk Agents, controls the session, and stores generated messages to the internal database (IDB). Data is read by the Disk Agent and sent to a Media Agent, which saves it to media.

Key product highlights

Advanced analytics for smarter backup and recovery

Optimizing the backup and recovery process with an approach that also addresses future needs requires the ability to derive value and insight from backup-related operational information. HPE backup solutions like Data Protector Software integrate advanced analytics that deliver operational intelligence to enable IT to gain better utilization of backup resources, make informed decisions related to future capacity and infrastructure needs, and proactively address current and potential issues in the backup environment.

HPE StoreOnce federated deduplication

HPE StoreOnce is a patented, deduplication technology that is deployed with HPE Data Protector Software and HPE Storage systems. Today, only HPE can deliver the same deduplication algorithm that scales from a software-only footprint to a scale-out, purpose-built appliance, enabling efficient data replication and storage.

Cloud-based backup

Through integrated cloud protection, HPE Data Protector extends backup to the cloud, blending the two worlds of on-premises and secure hosted backup into one offering. The integrated cloud capability brings the power and simplicity of cloud-based backup to the traditional data center. This integration leverages the 14 global cloud data centers of HPE Autonomy, managing more than 50 petabytes of customer information worldwide.

Integrated Enhanced Automated DR

HPE Data Protector Software delivers centralized system recovery (bare metal recovery) to virtual or physical servers (from P2V or V2P) from a single backup—at no additional cost—to streamline the DR process. Unique to HPE Data Protector is the ability to create a DR image from any existing file system or image backup, including object copies, without creating a separate special backup for system recovery. With a single click, you can initiate the DR process and automatically rebuild an entire system and the partitioning.

Advanced application protection

Data Protector Software offers advanced functionality to simplify and automate application protection and recovery. You can automate the entire recovery process by rolling forward the application transaction logs and recovering applications to an exact point in time. The advanced snapshot functionality of Data Protector protects mission-critical applications without impacting application performance, and it provides automated recovery for applications and data in seconds to meet even the most demanding recovery objectives.

Comprehensive virtualization support

HPE Data Protector Software provides comprehensive virtual server protection by supporting all hypervisors. With VMware and Hyper-V, Data Protector Software offers advanced hypervisor integration and array snapshot management for the most demanding virtualized applications. For VMware, Data Protector Software also offers One-Touch protection, automatic protection of new VMs, single item recovery, and native best-in-class integration for vCloud Director environments.

Centralized management

HPE Data Protector Software offers a single console to centrally manage data protection for physical and virtual servers across on-premises, hybrid, and cloud infrastructures. The product offers best-in-class integration with HPE Storage, from tape automation to array snapshots, and replication from a centralized console without scripting, to support any RTO or RPO.

HPE Data Protector architecture

Key components of the HPE Data Protector architecture include

- **Cell Manager**—The Cell Manager is the main system that centrally controls and manages the cell. It contains a database, runs the core software, and runs the backup and restore sessions, writing session information to the database.
- **Client systems**—Client systems to be backed up must have the Data Protector disk agent, also called the backup agent, installed. To back up online database integrations, install the application

agent. The disk agent reads or writes data from a disk on the system and sends or receives data from a media agent. The disk agent is also installed on the Cell Manager, enabling customers to back up data from the Cell Manager, including the Data Protector configuration and the IDB.

Note
The term "disk agent" is used for both agents.

- **Systems with backup devices**—Client systems with connected backup devices must have a Data Protector media agent installed. Such systems are called backup servers or media servers. Customers can connect a backup device to any system, not only to the Cell Manager. The media agent reads and writes data from and to media in the device and sends or receives data from the disk agent.
- **Systems with a user interface**—You can manage Data Protector from any system on the network on which the Data Protector GUI and CLI are installed. As a result, the Cell Manager system can run in a data center, and Data Protector Software can be managed from a desktop system.
- **Installation Server**—The Installation Server holds a repository of the Data Protector Software packages for a specific architecture. The Cell Manager is by default also an installation server. At least two Installation Servers are needed for a mixed environment, one for UNIX and one for Microsoft Windows.
- **Internal database**—The IDB, which is located on the Cell Manager, holds information about which data is backed up; the media it resides on; the result of backup, restore, object copy, object consolidation, object verification, and media management sessions; and which devices and libraries are configured.

The information stored in the DIB enables the following:

- **Backup management**—Manage configuration and scheduling
- **Monitoring**—Check running backup sessions
- **Reporting**—Verify the result of backup sessions
- **Media management**—Perform tasks such as allocating media during backup, object copy, and object consolidation sessions; tracking media management operations and media attributes; grouping media in different media pools; and tracking media locations in tape libraries
- **Encryption/decryption management**—Allocate encryption keys for encrypted backup or copy sessions, and supply the decryption key when required to restore encrypted backup objects
- **Fast and convenient restore**—Browse files and directories

An essential part of the internal database configuration is configuring the backup of the IDB itself. Regular IDB backup is the most important preparation for recovery in the event of a disaster. The IDB recovery is essential for the restoration of other backed-up data if the Cell Manager is struck by a disaster.

Media management

HPE Data Protector Software provides powerful media management, which enables you to easily and efficiently manage large numbers of media in your environment in the following ways:

- Grouping media into logical groups, called media pools, enables you to think about large sets of media without having to worry about each medium individually.
- You can keep track of all media and the status of each medium, the data protection expiration time, the availability of media for backup, and a catalog of what has been backed up to each medium.
- If HPE Data Protector Software controls enough media in the library devices, you can use automated operation to run the backup sessions without operator intervention.
- Automated media rotation policies allow media selection for backups to be performed automatically.
- The software can recognize and support barcodes on large library devices and provides barcode support for silo devices.
- The software provides recognition, tracking, viewing, and handling of media used by Data Protector Software in large library devices and in silo devices.
- Information about the media is available in a central place and can be shared among several Data Protector cells.

Media pools

Data Protector Software uses media pools to manage large numbers of media. A media pool is a logical collection of media of the same physical type and with common usage policies (properties). Usage is based on the data on the media. The structure and number of pools, as well as which pool contains what type of data on its media, depend entirely on customer preferences.

When a device is configured, a default media pool is specified. This media pool is used if no other media pool is defined in the backup specification.

Backup devices

Data Protector defines and models each device as a physical device with its own usage properties, such as the default pool. This device concept is used because it enables customers to easily and flexibly configure devices and use them in conjunction with backup specifications. The definitions of the devices are stored in the Data Protector Media Management Database.

Reporting

Data Protector reporting capabilities include a notification function that enables you to forward events to HPE Support or to HPE or third-party reporting or management tools such as HPE Storage Essentials and IBM Tivoli.

Data Protector Reports provide various details about the backup environment. For example, you can check the status of the last backup, object copy, object consolidation or object verification; check which systems in the network are not configured for backup; check on the consumption of media in media pools; check the status of devices; and more.

You can also configure reports and report groups using the Data Protector GUI or any web browser with Java support.

- Parameters enable you to customize reports.
- Report groups enable you to manage reports easily, to schedule the reports in the report group, and to define the criteria for grouping the reports in report groups.

Data Protector Reporter software

For customers who require advanced reporting capabilities, HPE provides Data Protector Reporter software. This customer-installable software delivers centralized, automated reporting to optimize operations and infrastructure.

Data Protector Reporter software features a powerful reporting engine that drives global, multisite backup and restore analysis. This capability enables IT staff to reduce the risk of data loss by easily identifying and troubleshooting issues such as failed backup clients, performance issues, and poor drive and media utilization. Data Protector Reporter software features more than 35 reports, including SLA and performance reporting.

This optional licensed module also provides enhanced, powerful, and customized ad hoc query and flexible analysis reporting. Collectively, these reports help you to optimize operations to meet SLAs, optimize capacity and performance to improve cost, plan for growth, and track compliance.

Disk backup

It has become more important that the time required for data backup and restore be reduced to a minimum so that it does not interrupt the daily operation of company applications.

Many applications and databases frequently make small changes to existing files or produce many new files containing business-critical data throughout the working day. These files must be backed up immediately to guarantee this data will not be lost. This requirement necessitates a fast medium that can store large amounts of data and work without interrupting ongoing processes.

Disk-based storage media have become increasingly affordable in recent years. At the same time, the storage capacity of disks has increased dramatically, leading to the availability of low-cost, high-performance single disks and disk arrays for storing data.

Disk (*D2D*) backup is increasingly important. In the past, tape storage was the favored medium for backup and restore because of its price and effectiveness in meeting DR requirements. Today, more businesses are augmenting their tape storage backup solutions with faster disk-based backup solutions that ensure faster data backup and recovery.

Disk-based devices

Data Protector Software supports the following disk-based devices:

- **Stand-alone file device**—A stand-alone file device is a file in a specified directory in which data is backed up instead of being written to a tape. This device saves data in the form of files. Each of these files is the equivalent of a slot in a tape device. The stand-alone file device is useful for smaller backups.
- **File jukebox device**—A jukebox is a library device. It can contain either optical or file media. If the device is used to contain file media, it is known as a file jukebox device. The file jukebox device is the logical equivalent of a tape stack; it saves data in the form of files, and each of these files is the equivalent of a slot in a tape device. The size of the slots is defined by the user during the initial device configuration. This device is configured manually, and its properties can be altered while it is being used. The file jukebox device writes to disk instead of tape.
- **File library device**—A file library device resides in a directory on an internal or external hard disk drive that has been defined by the user. A file library device consists of a set of directories. When a backup is made to the device, files are automatically created in these directories. The files contained in the file library directories are called file depots.

Customers need to determine the policies for retaining data and storage media:

- **Data protection**—Data Protector Software allows each customer to specify the amount of time that data on their media is protected from being overwritten by Data Protector. Customers can specify the protection duration in absolute or relative dates, and in different parts of the Data Protector interface.

If customers do not change the Data Protection backup option when configuring their backup policies, their data is permanently protected, and the number of media needed for backup grows constantly.

- **Catalog protection**—Data Protector Software saves information about backed up data in the IDB. Each time a backup is performed, the IDB grows with the number and the size of the backups. Catalog protection tells Data Protector Software how long the information about backed up data should be available to users browsing data during a restore.

After the catalog protection period has expired, the Data Protector Software will overwrite this information in the IDB (not on the media) during one of the subsequent backups. The administrator can specify the protection duration using absolute or relative dates. If the Catalog Protection backup option is not changed when the backup is configured, the information about backed up data has the same protection duration as the data protection setting.

Follow these guidelines when planning to back up large amounts of data:

- Consider using a library device.
- Consider backing up to a disk-based device. In addition to other benefits, backup to disk (B2D) reduces the time needed for backups and enables the use of advanced backup strategies such as synthetic backup and disk staging.

  - Synthetic backup is a backup solution that eliminates the need to run regular full backups. Instead, incremental backups are run and are subsequently merged with the full backup into a new, synthetic full backup. This can be repeated indefinitely, with no need to run a full backup again. In terms of restore speed, such a backup is equivalent to a conventional full backup.
  - The concept of disk staging is based on backing up data in several stages to improve the performance of backups and restores, reduce costs of storing the backed up data, and increase the data availability and accessibility for restore. The backup stages consist of backing up data to media of one type and later moving it to media of a different type. The data is backed up to media with high performance and accessibility, but limited capacity (for example, system disks). These backups are usually kept accessible for restore for the period of time when a need to restore is the most probable. After a certain period of time, the data is moved to media with lower performance and accessibility, but high capacity for storage, using the object copy functionality.
- After you determine the type of media to be used, determine how to group that media into media pools, and how to place objects on the media. Define the backup policies for how media are to be used.
- Decide whether to store the media at a safe place (such as a vault) and for how long the media should be retained. Consider duplicating backed up data during or after the backup for this purpose.
- Determine the rights of users who can administer and operate the storage products.

Compression

If data is compressed on a disk, the Windows operating system decompresses the data before sending it across the network. This reduces the backup speed and uses CPU resources.

Disk performance

All data that Data Protector Software backs up resides on disks in the system. Therefore, the performance of disks directly influences the backup performance. A disk is essentially a sequential device; it can be read or written to, but it cannot be read and written to at the same time. Also, reading and writing is performed one stream of data at a time. Data Protector Software backs up file systems sequentially to reduce disk head movements. It also restores files sequentially.

Sometimes this is not visible because the operating system stores the most frequently used data in the cache memory.

Disk fragmentation

Data on a disk is not kept in the logical order visible to users when they browse the files and directories. It is fragmented in small blocks all over the physical disk. Therefore, to read or write a file, a disk head must move around the whole disk area. Backups are most efficient for large files with little fragmentation.

Note

The extent to which this is true differs from one operating system to another.

Disk image backups

Data Protector Software also allows customers to back up UNIX disks as disk images. With a disk image backup, a complete image of the disk is backed up without tracking the file system structure. The disk head moves linearly across the surface. Thus a disk image backup can be considerably faster than a file system backup.

Disk agent performance on Windows systems can be improved by enabling asynchronous reading. Asynchronous reading improves performance of the disk agent when backing up data on disk arrays, especially if large files are being backed up.

Note

HPE recommends that customers perform test backups to establish whether asynchronous reading will improve performance in a specific environment and to determine the optimum asynchronous reading settings.

SAN performance

If large volumes of data need to be backed up in one session, the time needed to transfer the data over a connection (LAN, local, or SAN) to a backup device becomes significant.

Data Protector B2D device concepts

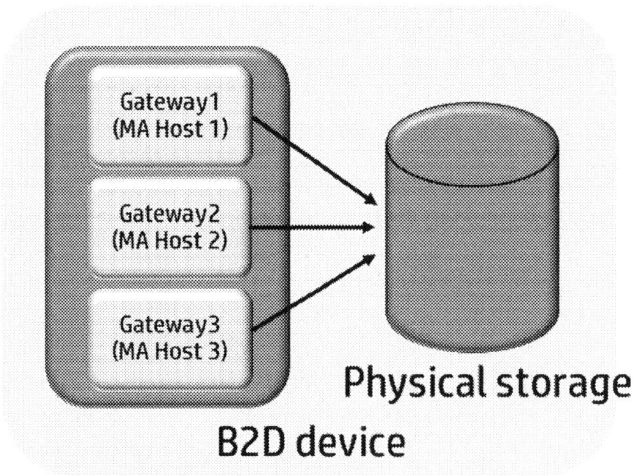

Figure 7-28 B2D device concepts

HPE Data Protector Software offers a new B2D device type to back up data to physical disk storage (Figure 7-28). This B2D device supports multihost configurations, which means that a single physical storage device can be accessed through multiple gateways.

Each gateway represents a Data Protector client with the Media Agent component installed. A B2D device is a logical device and consists of gateways and a store.

The physical storage can also be divided into individual stores representing specific storage sections, similar to partitioning a hard disk.

If customers want to use the new B2D functionality to support HPE StoreOnce software deduplication and HPE StoreOnce backup systems, they need to activate a separate B2D license.

HPE Data Protector licensing model

HPE Data Protector Software supports two licensing schemes.

- **Traditional licensing based on features and backup targets**—Available for all versions of HPE Data Protector Software
- **Capacity-based licensing**—Available with HPE Data Protector 7.01 and later

Traditional licensing

Data Protector Software primarily supports three different backup targets, snapshots, disk, and tape. Depending on the target used, the customer would license one or all of them because the target types can be combined. In addition, the customer can control the backup performance by parallelizing their use of targets.

The product structure is modular and offers a lot of flexibility. Customers can order the license that provides the Data Protector Software functionality that best meets the specific requirements of environment. The Data Protector Software 7.0 product structure and licensing consist of three main categories:

- **Starter Packs**—A management server (Cell Manager) is supported on HP-UX, Windows, and Linux.
- **Backup targets**—For example, tape drive licenses, referred to as Drive Extensions, are licensed per drive. Advanced B2D and Zero Downtime Backup are licensed by capacity.

- **Data Protector functional extensions**—The functional extension licenses are required once per instance (system, library, or terabyte) for the following functions:
    - Online backup of databases and applications
    - Manager-of-Managers (MoM) functionality
    - Using libraries with more than 60 media slots
    - Encryption
    - Instant Recovery
    - Network Data Management Protocol (NDMP)
    - Granular Recovery Extension

The UNIX product licenses operate on the UNIX, Windows, and Novell NetWare platforms, providing the functionality regardless of the platform, while the Windows product licenses operate only on the Windows, Novell NetWare, and Linux platforms.

Passwords are bound to the Cell Manager and are valid for the entire Data Protector cell.

Clients do not require any license for file system or disk image backups (Figure 7-29).

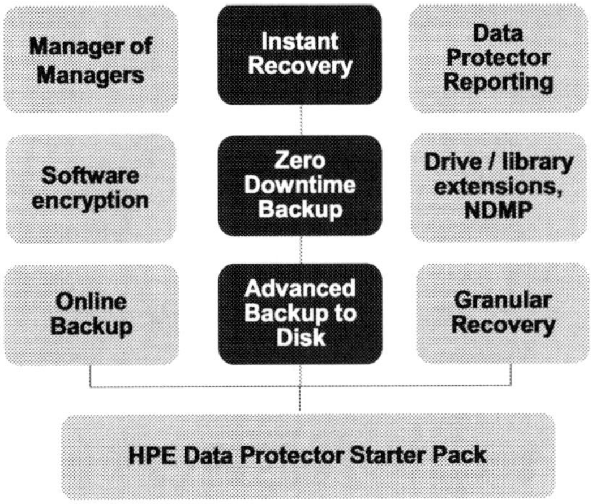

Figure 7-29 Traditional licensing

Capacity-based licensing

The capacity-based product structure was introduced in the Data Protector 7 Update 1 (DP 7.01) release. It is based on the volume of primary data protected by the HPE Data Protector instance. The capacity is measured in “front-end terabytes,” or front-end TB. The total amount of front-end TB is defined as the aggregate amount of source data from the largest full backup per backup specification under retention.

The following features are included in the capacity-based license:

- Cell Managers and MoM
- Tape drives and libraries
- Online Backup and Granular Recovery Extensions
- Zero Downtime Backup and Instant Recovery
- Advanced B2D and NDMP

Software encryption is **not** included with capacity-based licensing and must be ordered separately.

The complementary products that are sold separately are

- Distribution Media (DVD sets)
- Media Operations
    - HPE Data Protector Media Operations is only available for DP 7.0x.
- IDOL Server
    - This is required to take advantage of the inclusive IDOL integration.
- HPE Cloud Backup "protection plan" to use the integrated Cloud Backup
- HPE Backup Navigator
- HPE Data Protector Extended Online Backup
- HPE Data Protector Zero Downtime Backup (ZDB) for third-party storage arrays
- HPE Data Protector Management Pack
- HPE Data Protector Smart Plug-in for HPE Operations Manager

The Traditional and Capacity product structures can be used by the same customer, but they cannot be combined in the same Cell Manager or MoM environment. The complementary products listed are the exceptions to this rule because these licenses can be combined with both the Data Protector Traditional and capacity-based licensing methods. Migration from a traditional product structure to a capacity-based product structure is supported. Both licensing models are valid for any size of environment.

Note

Additional license authorizations and restrictions applicable to software products are found at http://www.hp.com/go/SWLicensing.

Third-party backup solutions—CommVault

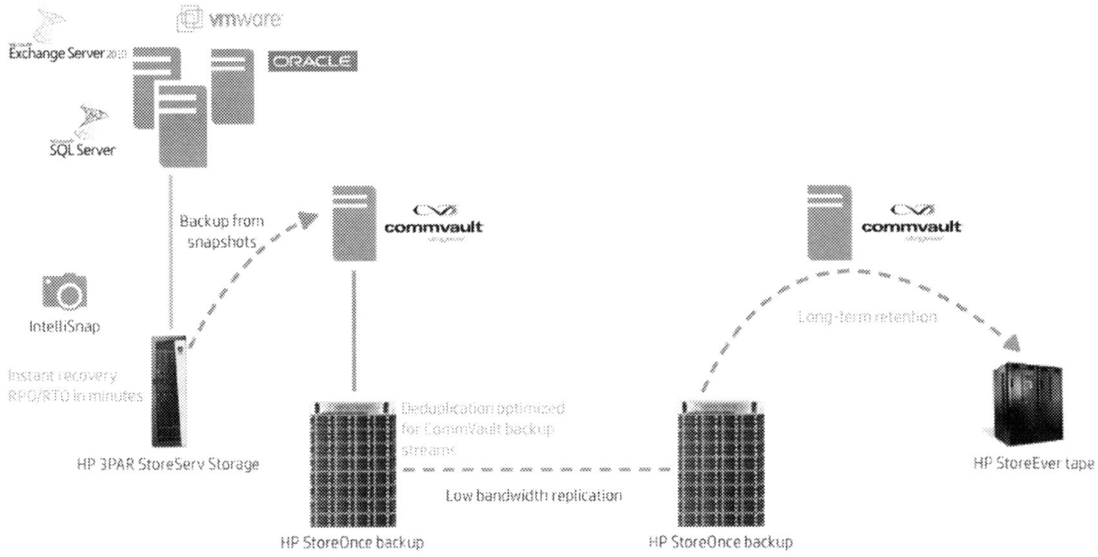

Figure 7-30 CommVault implementation example

Many enterprise IT organizations are challenged by accelerating data growth and increasing reliance on their data (Figure 7-30). A solution to this challenge requires fundamentally rethinking how the data is protected and accessed. HPE StoreOnce backup and CommVault Simpana deliver an integrated, end-to-end data protection and rapid recoverability solution that enhances business continuity and resiliency.

As businesses strive to stay competitive, 24×7 operations are the new norm, and the tolerance for downtime is rapidly diminishing. SLAs and uptime requirements are stringent, forcing organizations to modernize their protection and retention processes to align with their business objectives. At the same time, rapid data growth, expanding virtual server deployments, and budgetary pressures are further complicating today's legacy backup and recovery methods. These dynamics, compounded by demands for faster, easier recovery, are putting an increasing strain on already beleaguered data protection processes—and slowing down IT transformation initiatives. The combination of HPE 3PAR StoreServ, HPE StoreOnce backup system, and CommVault Simpana software delivers Tier 1 application resiliency, reliable data protection, enhanced data recovery, and complete life-cycle data management to improve operational efficiencies.

The CommVault Simpana IntelliSnap technology unites complete data protection life cycle management with primary and secondary tiers of HPE storage. The integration of IntelliSnap with the native HPE 3PAR StoreServ snapshot engine provides consistent point-in-time recovery copies for enterprise applications while connecting hardware snapshots to the rest of the data protection process, without the need for complex scripting. From the hardware snapshot, IntelliSnap indexes the contents, and selectively vaults data to the HPE StoreOnce backup device for longer-term retention.

Using the HPE 3PAR StoreServ Priority Optimization capability, storage administrators can control and balance the distribution of array resources across other workloads that leverage snapshots. Snapshots of the production database that are used for analysis, backup, and development purposes can be run with an appropriate I/O priority by using the Quality of Service feature so that the production applications are not affected during these auxiliary operations.

Third-party backup solutions—Veeam

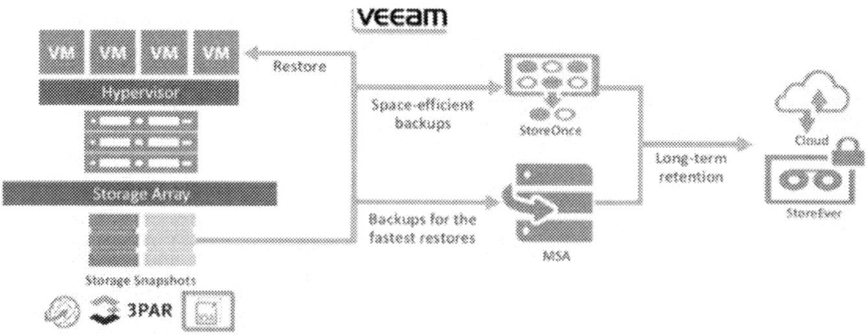

Figure 7-31 Veeam

Always-on is a requirement for most businesses today (Figure 7-31). Together, Veeam and HPE solutions help these businesses to meet their availability objectives and future-proof their data centers.

The combined Veeam and HPE solutions can help businesses meet always-on business goals by providing data protection without limits. Businesses can improve application availability, meet RTOs and RPOs, and enable fast, reliable DR capabilities.

Backup and recovery are at the core of all enterprise data protection strategies. Today's virtual environments present new challenges for protecting and recovering data. Legacy backup tools were built for physical servers, and when applied to a virtual environment, they can cause long backup times and performance constraints and have limited recovery capabilities. Because of this, many organizations are unable to meet the backup window for their VMs.

One key factor is that legacy backup tools typically rely on agents running inside the VM to accomplish backup and recovery operations. These agents running inside VMs can create resource problems. They use guest CPU cycles and consume network bandwidth and memory. These agents also add to the complexity of maintaining software and increase support costs. In addition, agent-based technology does not work at all when VMs are powered off, unavailable, or newly provisioned.

Adding further complexity, most organizations are using multiple hypervisors, and in many cases, each hypervisor might use a different backup solution, which results in increased management demands.

Veeam uses agentless technology that is hypervisor-agnostic and designed to protect VMs. The integration of Veeam and HPE storage solutions lets customers create application-consistent backups from HPE StoreServ and HPE StoreVirtual snapshots for fast and efficient data protection.

Veeam works with HPE StoreServ, StoreVirtual, StoreOnce, MSA, and StoreEver technologies to provide a complete data protection strategy.

Sizing HPE StoreOnce solutions

HPE StoreOnce backup device emulation options

The HPE StoreOnce backup systems offer flexibility with NAS (NFS and CIFS), iSCSI, and FC virtual tape libraries (VTL) and StoreOnce Catalyst targets for backup applications. Customers can easily integrate StoreOnce backup into an existing IT environment with minimal disruption. Note that not all products support all targets.

Table 7-7 Integrate StoreOnce backup into an existing IT environment with minimal disruption

StoreOnce Model	iSCSI VTL	FC VTL	CIFS share (TCP/IP)	NFS share (TCP/IP)	Catalyst Store (TCP/IP)
VSA	X		X		X
3100	X	X	X	X	X
3520	X	X	X	X	X
3540	X	X	X	X	X
5100	X	X	X	X	X
4900	X	X	X	X	X
6500		X	X	X	X

HPE StoreOnce performance and sizing guidelines

HPE StoreOnce products differ in scalability and performance counters.

Table 7-8 StoreOnce model differs in scalability and performance counters

StoreOnce model	VSA	3100	3520	3540	5100	4900
Total usable capacity	50 TB	5.5 TB	15.5 TB	31.5 TB	216 TB	432 TB
Total raw capacity		8 TB	24 TB	48 TB	288 TB	560 TB
Read performance	1.8 TB/hr	1.4 GB/hr	4.1 TB/hr	4.1 TB/hr	14.2 TB/hr	10.1 TB/hr
Write performance	2.4 TB/hr	1.4 TB/hr	4.6 TB/hr	4.6 TB/hr	13.8 TB/hr	8.5 TB/hr
Catalyst performance	6 TB/hr	6.4 TB/hr	12.7 TB/hr	12.7 TB/hr	26.7 TB/hr	22 TB/hr
Number of backup targets	16	8	24	24	32	50

Sizing considerations

To correctly select an HPE StoreOnce system, evaluate the following considerations:

- The total size of backup jobs and the anticipated room for growth
- Full versus partial backups
- The type of connectivity such as for servers and storage
- The backup windows and recovery speed requirements
- The deduplication ratios, considering the types of documents
- The requirements for installation support and enhancements to warranty services

Reference documents and sizing tools

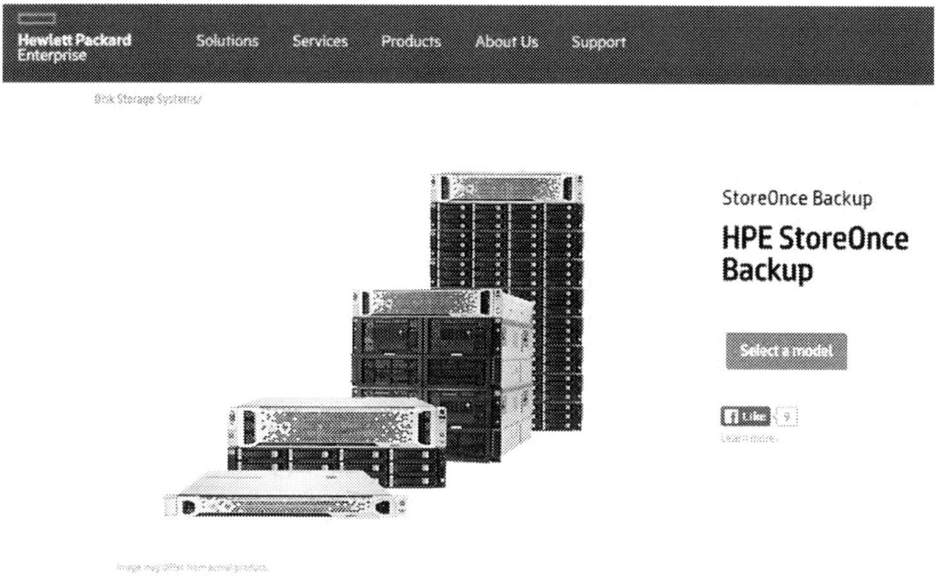

Figure 7-32 HPE website

Sizing tools are described in more detail in Chapter 8 (Figure 7-32). You can use the following reference documents and sizing tools to select and configure the correct HPE StoreOnce storage system:

- *HPE QuickSpecs*
- *HPE Product Bulletin*

- HPE website:
  - https://www.hpe.com/us/en/storage/storeonce.html
- *HPE Backup, Recovery, and Archive (BURA) Solutions design guide*
- HPE Storage Sizer
- NinjaProtected Tool

Example—Using the HPE Storage Sizing Tool to configure StoreOnce storage

This example shows how to configure an HPE StoreOnce solution based on the following customer description:

- A new backup infrastructure is needed.
- The old solution will be decommissioned because of high maintenance costs.
- The customer estimates 15 TB of mixed raw data and prefers the iSCSI protocol for backups.
- The customer anticipates 10% yearly storage growth.
- A compression ratio of 4:1 is estimated based on previous backups; the daily change rate is 1%.
- The solution must be calculated for the next 5 years.
- There will be daily incremental backups, and the data retention period is 4 weeks.
- The backup window for a full backup is 20 hours and 12 hours for incremental backups.

Following are key points for using the sizing tool.

- You can start the HPE Storage Sizing Tool from the icon on the desktop (Figure 7-33).

Figure 7-33 Storage Sizing Tool icon

- The Backup Calculators section in the left pane contains the sizers for backup solutions (Figure 7-34).

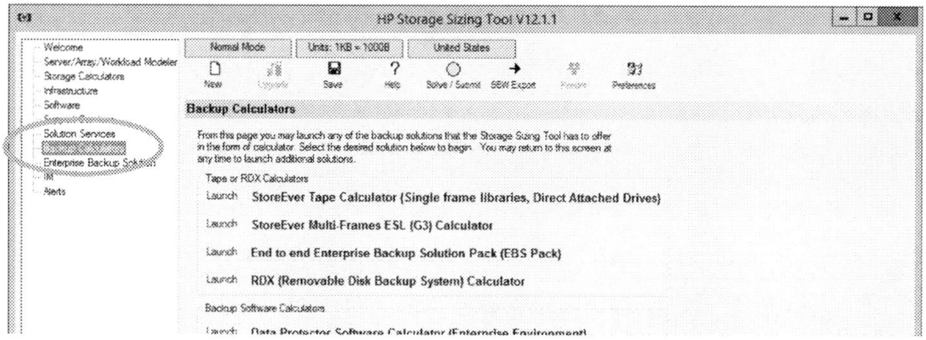

Figure 7-34 Backup Calculators

- You can start the HPE StoreOnce Calculator from the Launch link (Figure 7-35).

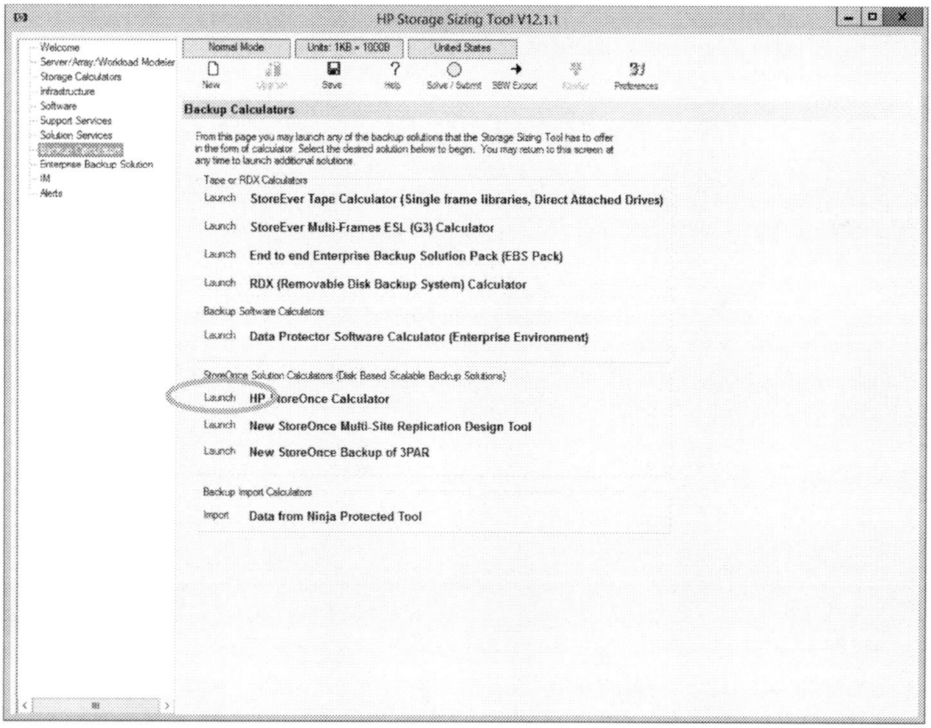

Figure 7-35 Launch link

- The StoreOnce calculator can work in two modes, Basic and Advanced (Figure 7-36).

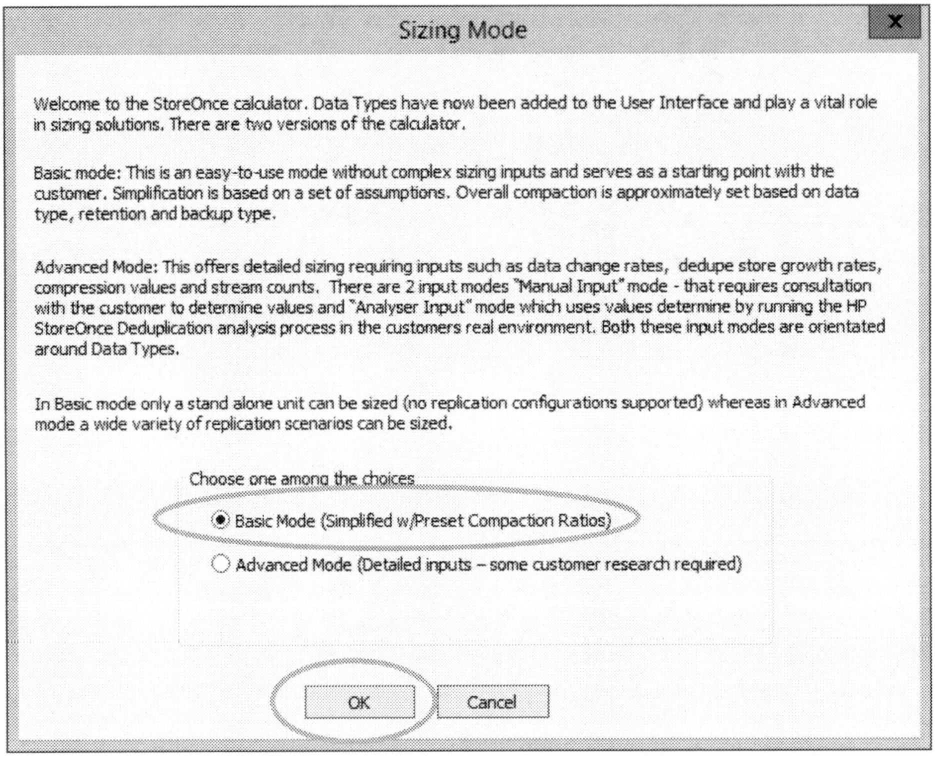

Figure 7-36 Calculator works in two modes, Basic and Advanced

- Backup jobs can be configured with information gathered from the customer. Following is a StoreOnce configuration example (Figure 7-37):

Job Size:	**15000 GB**
Full Backup Window	**20 hours**
Data Type:	**Mixed (catch all)**
Yearly Growth:	**10%**
Years to size for:	**5 Years**
Daily Backups in a week:	**5**
Backup Type:	**Incrementals and Fulls**
Retention In Weeks:	**4**

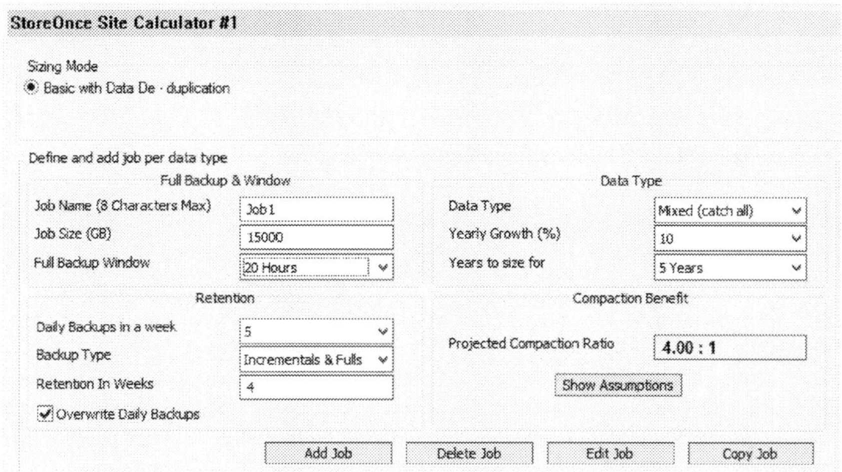

Figure 7-37 StoreOnce Site Calculator

- For the Basic mode, the sizing tool makes several assumptions. For example, it assumes an incremental backup size of 20%, which is too much for the customer scenario in this example (Figure 7-38).

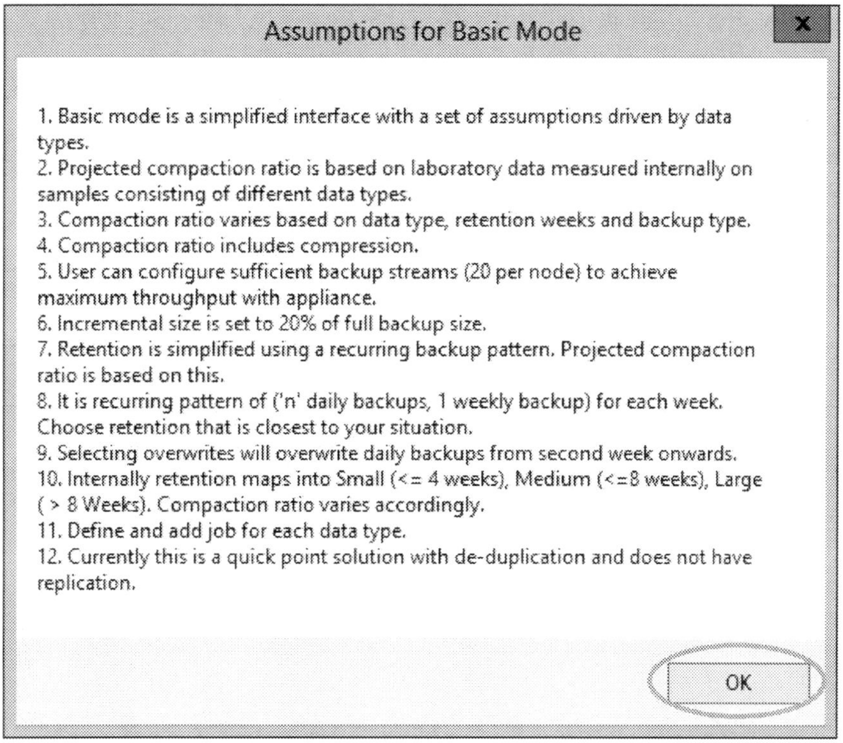

Figure 7-38 Assumptions for Basic Mode

- You can send a job to the calculator with the Add Job button (7-39).

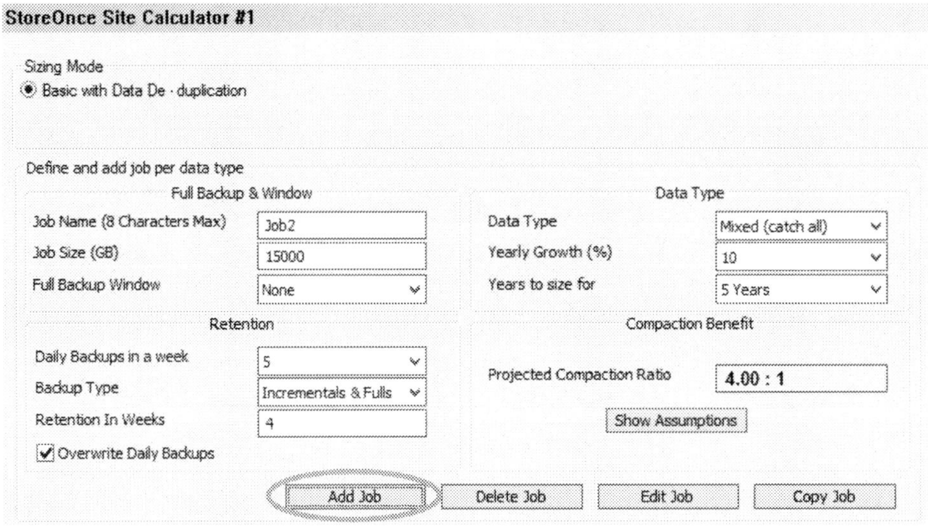

Figure 7-39 Add Job button

- The Solve/Submit button will create a configuration based on user input (Figure 7-40).

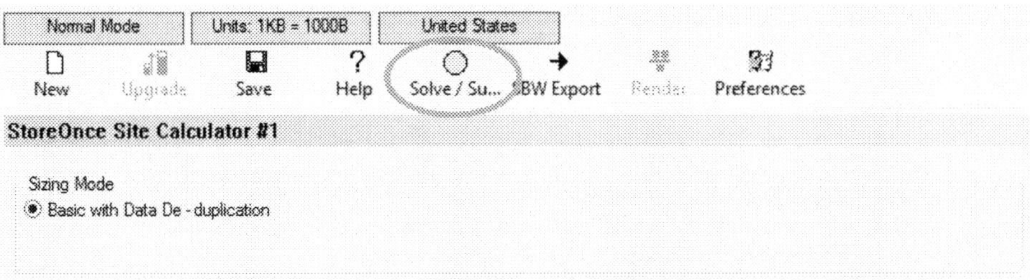

Figure 7-40 The Solve/Submit button

- When the sizing tool has finished making its calculations, Internet Explorer is started and Microsoft Excel automatically opens the file with results (Figure 7-41).

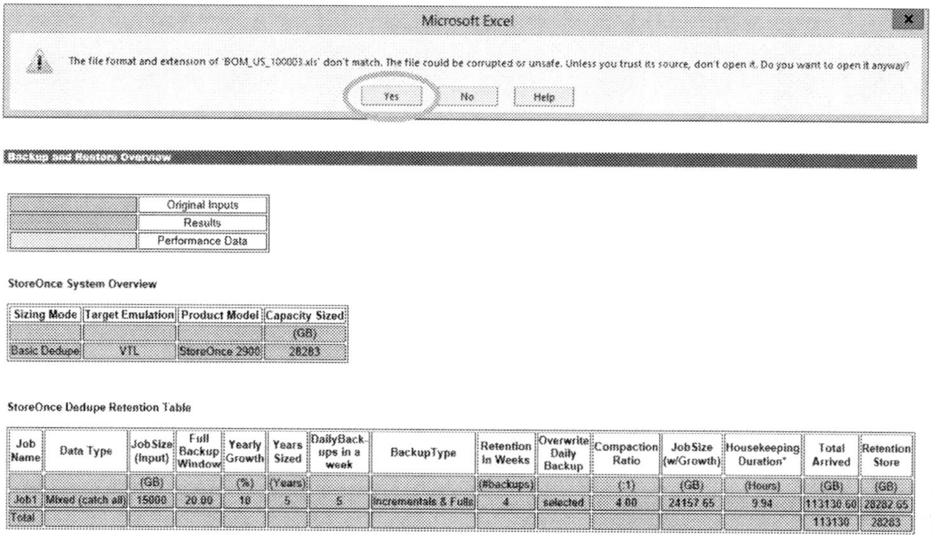

Figure 7-41 Sizing tool making its calculations

This is an example of the results in Excel (Figure 7-42).

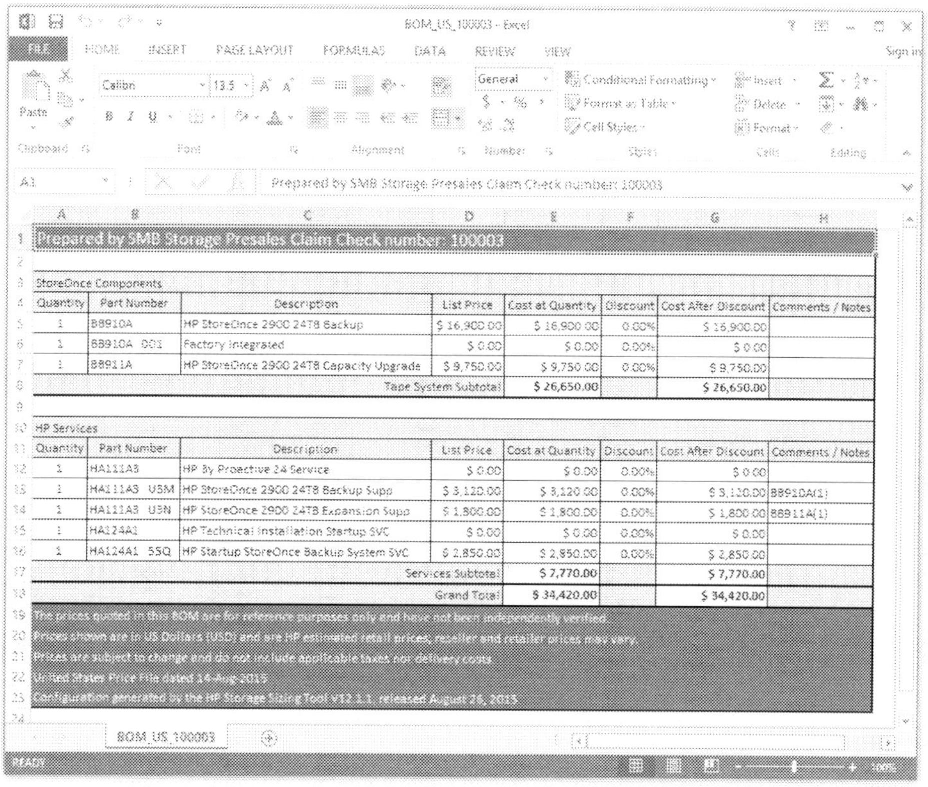

Figure 7-42 An example of the results in Excel

Example—Demonstrating the HPE Product Bulletin user interface

The following example shows you key points for using the HPE Product Bulletin to quickly address customer questions.

- You can start the HPE Product Bulletin from the icon on the desktop (Figure 7-43).

Figure 7-43 Product Bulletin icon

- To find the maximum usable capacity of HPE StoreOnce 2900 (Figure 7-44) before deduplication, you can enter storeonce in the search field and then click the first occurrence of "HPE StoreOnce Backup."

Figure 7-44 To find the maximum usable capacity of HPE StoreOnce 2900

- You can find the request value by scrolling down to the "HPE StoreOnce Backup Models" section or using the Find Text field to search for a phrase (Figure 7-45) like "before dedupe."

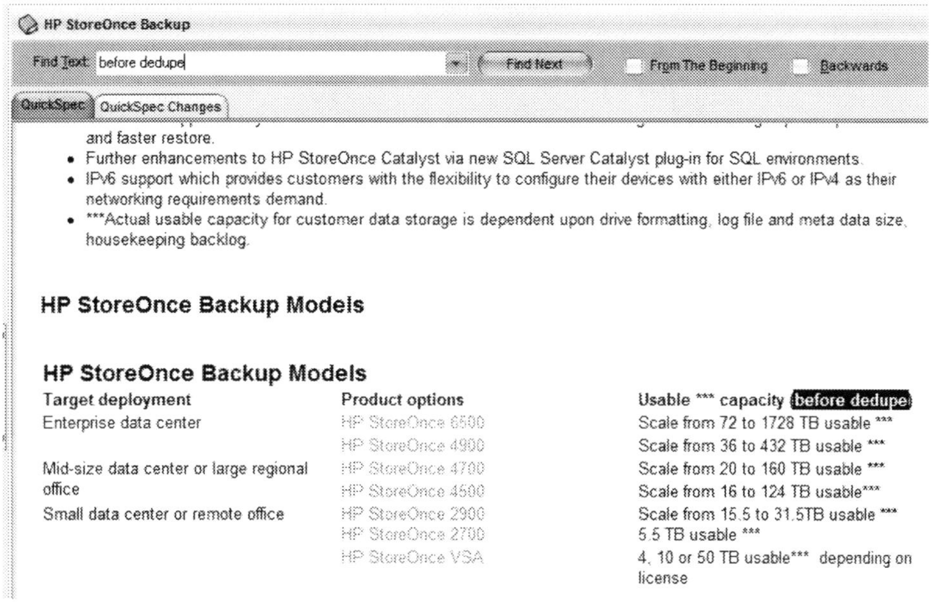

Figure 7-45 To search for a phrase like "before dedupe"

- Additional questions can be answered easily using the information in the HPE Product Bulletin, such as "What are the form factor and size of HPE StoreOnce 2900 Backup system?" or "What is the maximum HPE StoreOnce 2900 Backup write performance when using the HPE StoreOnce Catalyst?"

- Tape drives are also included in the database for the HPE Product Bulletin (Figure 7-46).

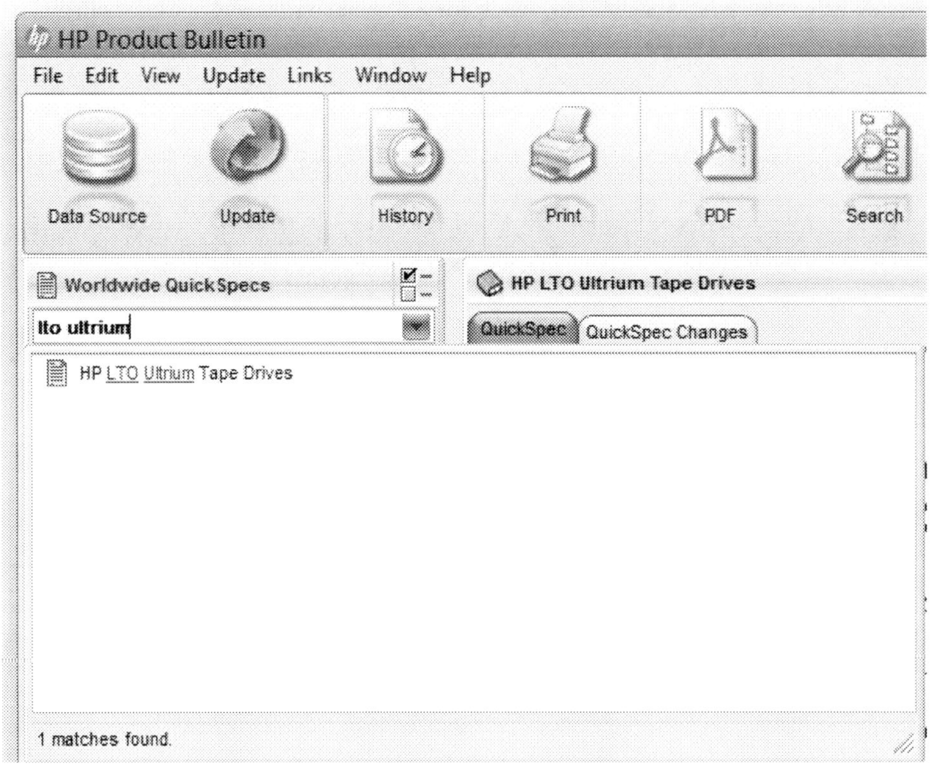

Figure 7-46 HPE Product Bulletin

For example, you can use the HPE Product Bulletin to display and compare the list prices of drives and data cartridges (Figure 7-47).

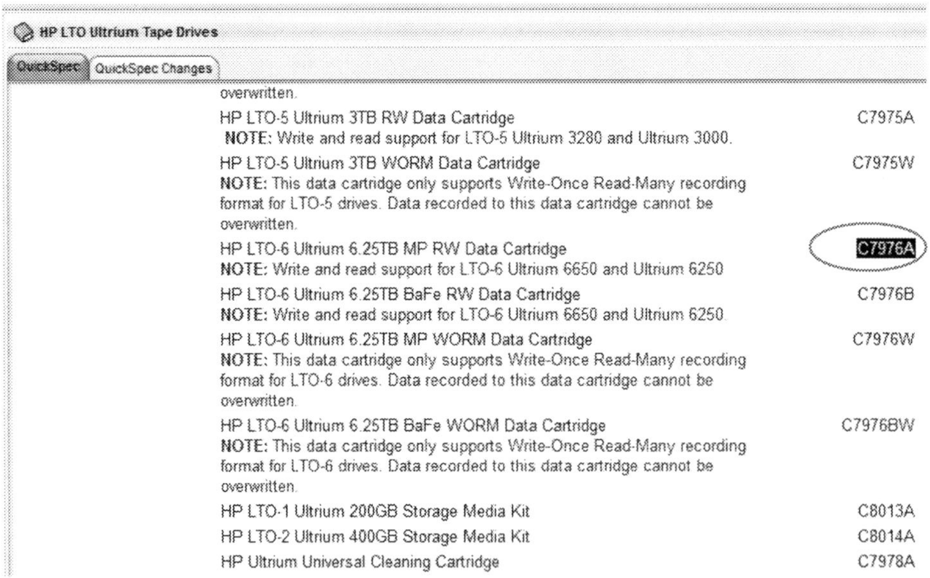

Figure 7-47 HPE Product Bulletin to display and compare the list prices of drives and data cartridges

- The Internet Price button in the top ribbon will show the list price of a product (Figure 7-48).

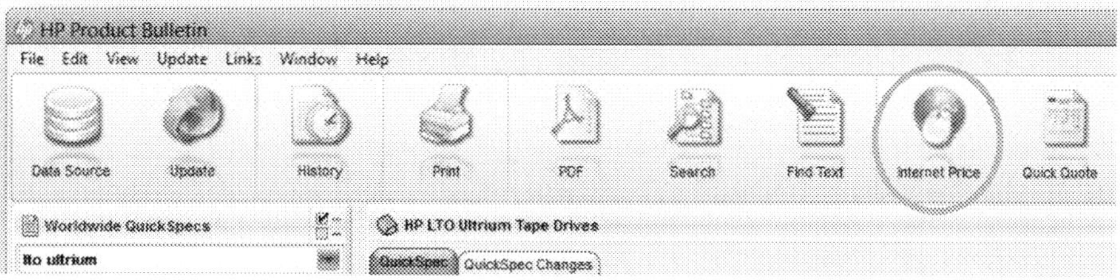

Figure 7-48 The list price of a product

- The part number is automatically filled in the Internet Price form (Figure 7-49).

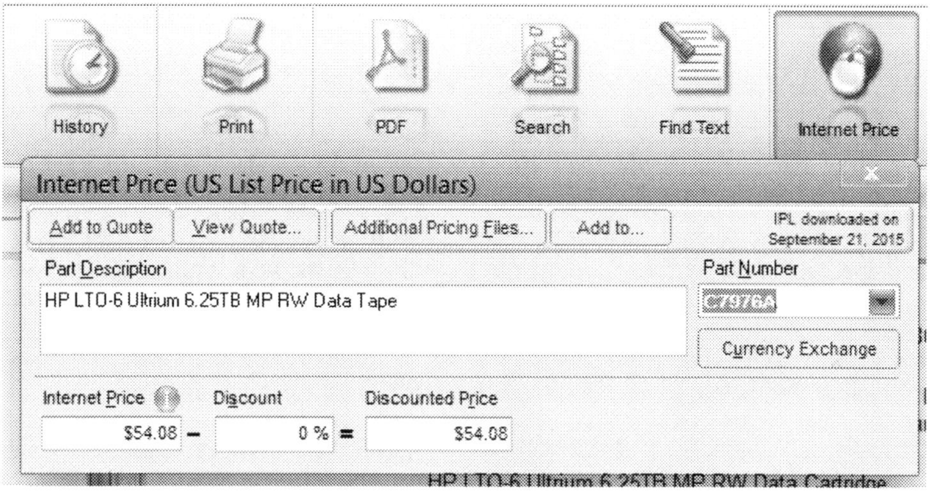

Figure 7-49 Internet Price form

NinjaProtected Tool and Get Protected Guarantee

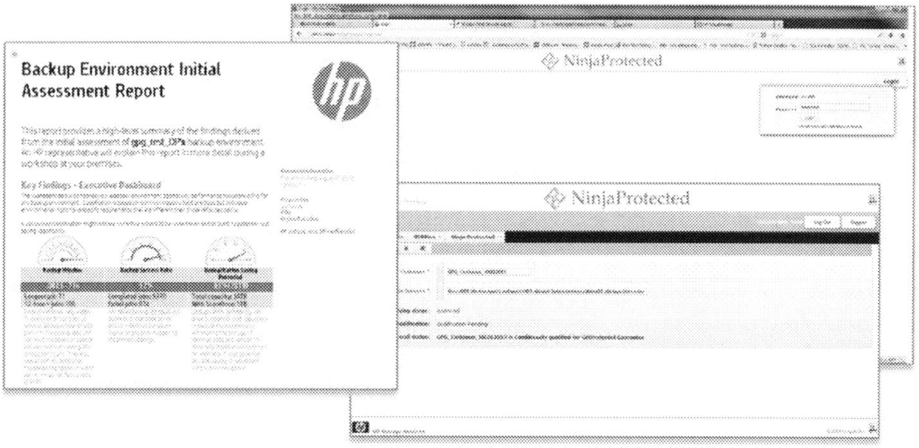

Figure 7-50 Screens from the NinjaProtected Tool

NinjaProtected is the new name for what used to be called SizerAdviser. The Get Protected Guarantee (GTG) is a marketing program being offered worldwide (Figure 7-50).

The NinjaProtected Tool is responsible for managing customer information, analyzing and parsing it, and generating useful reports from that data. This engine supports two reporting processes for HPE products:

- **The Assessment and Analysis Report**—This report is synonymous with what the SizerAdviser had been delivering for almost two years.
- **The Get Protected Guarantee program report**—This report works in conjunction with the Assessment and Analysis Report and supports the Get Protected Guarantee marketing program. It also delivers immediately useful information about the customer's backup environment to assist in driving BURA conversations at any level.

In Figure 7-50, you can see screens from the NinjaProtected Tool web interface. In the top-right corner is the login screen for the tool. This is the login page regardless of which process a user intends to follow, for example, to generate an Assessment and Analysis Report or a GPG program report.

At the bottom of the graphic is the qualification screen for a GPG customer process, showing the customer name, their backup media servers, and whether this customer has any backup jobs that qualify for the GPG program. In this example, the customer seems to have at least one qualifying backup job, but the "Qualification Pending" status indicates that the job must still be confirmed by running a follow-on script in their environment.

There are many pages and tabs in this tool to assist you with the process of gaining and managing the information needed to help a customer with their BURA assessment and analysis needs.

On the left side of the graphic, you can see what an actual Get Protected Report generated by the NinjaProtected Tool looks like. This is a multipage report intended for sharing with the customer. It shows how well their current backup environment is performing (the middle gauge), how much capacity it is using (the left-hand gauge), and most importantly, how much we can potentially improve it using HPE StoreOnce technology (the gauge on the right side of report).

There are additional pages of information appropriate for both chief information officers and initial engagements with backup administrators.

Note
You can access the NinjaProtected Tool at http://www.hp.com/go/ninjaprotected.

What are the NinjaProtected Tool deliverables?

One deliverable is a workshop where you can present results and recommendations on next steps to a customer. A high-level summary report about backup environment status can be presented to customers at any time. The report contains information such as daily data backed up, throughput data, total capacity used for backup retention, and data types being backed up. General client information a top-10 backup job summary are also included,

Why use the NinjaProtected Tool?

With the NinjaProtected Tool, valuable and relevant insights into the current backup environment can be received, and issues in the current environment can be uncovered, such as backup windows and failed backups. The NinjaProtected Tool can help you to finding opportunities for cost reduction (retention times) for the customer and identify potential deduplication-driven benefits.

Learning check

The following questions will help you to measure your understanding of the material presented in this chapter. Read all the choices carefully because there might be more than one correct answer. Select or write the correct answer for each question.

1. Name at least two different types of backup.

   ....

   ....

   ....

   ....

   ....

2. Incremental backups back up only the data that changed since the last full backup and reset the archive bit.

   ☐ True

   ☐ False

3. HPE StoreOnce VSA is HPE StoreOnce storage delivered as a VMware virtual appliance.

   ☐ True

   ☐ False

4. What is the typical write performance of an HPE StoreOnce backup system? (Consider StoreOnce 3100 to StoreOnce 4900 models without Catalyst technology.)

   a. 1.3 TB/hr to 8.5 TB/hr

   b. 1.5 GB/hr to 12 GB/hr

   c. 15 TB/hr to 32.5 TB/hr

   d. 120 GB/hr to 844 GB/hr

5. HPE Data Protector Software can be licensed based on the capacity needed.

   ☐ True

   ☐ False

6. Name at least three possible backup targets for HPE StoreOnce backup systems.

   ....

   ....

   ....

   ....

For answers, see Chapter 7 Answers on page 442.

8 Tools and Reference Materials

OBJECTIVES

In this chapter, you should learn to locate and use the following HPE tools and reference materials:

- ✓ HPE Tools website
- ✓ HPE storage area network (SAN) Design Reference Guide
- ✓ HPE Single Point of Configuration Knowledge (SPOCK)
- ✓ HPE Storage Sizing Tool
- ✓ Sales Builder for Windows (SBW)
- ✓ VisioCafe stencils for SAN design
- ✓ HPE Simple Configurator solutions
- ✓ HPE Product Bulletin
- ✓ Alinean return on investment (ROI) and total cost of ownership (TCO) calculators
- ✓ Other sizing and planning tools
    - ✓ SAN Designer
    - ✓ SAN Visibility
    - ✓ Storage Product Selector
    - ✓ Storage Security Self-Assessment Tool
- ✓ Solution Demo Portal
- ✓ Ninja tools
    - ✓ HPE NinjaSTARS for 3PAR
    - ✓ NinjaProtected Tool and Get Protected Guarantee
- ✓ DD Analyzer Tool for HPE Storage Presales

INTRODUCTION

This chapter provides an overview of various sizing tools and reference documents that HPE provides to help you with presales tasks.

HPE storage tools and resources

Figure 8-1 HPE offerings

HPE provides a variety of tools, resources, and other support to help technical presales professionals (Figure 8-1). The information available includes the following:

- **White papers**—Address key topics and technologies relevant to a particular technology or HPE product. Some HPE white papers are suitable for sharing with customers; look for the restriction information in the document.
- **Solution briefs**—Provide key information about an HPE product, service, or supporting activity and are typically suitable for customer viewing.
- **Customer presentations**—Provide information about HPE solutions or activities. These slide sets are designed for customer viewing.
- **Internal and external portals**—Provide information about HPE and its products, services, events, and so on. The portals are designed for HPE corporate customers and channel partners. These portals often contain special tools and utilities that can help facilitate the process of sizing and configuring a solution.

HPE Tools

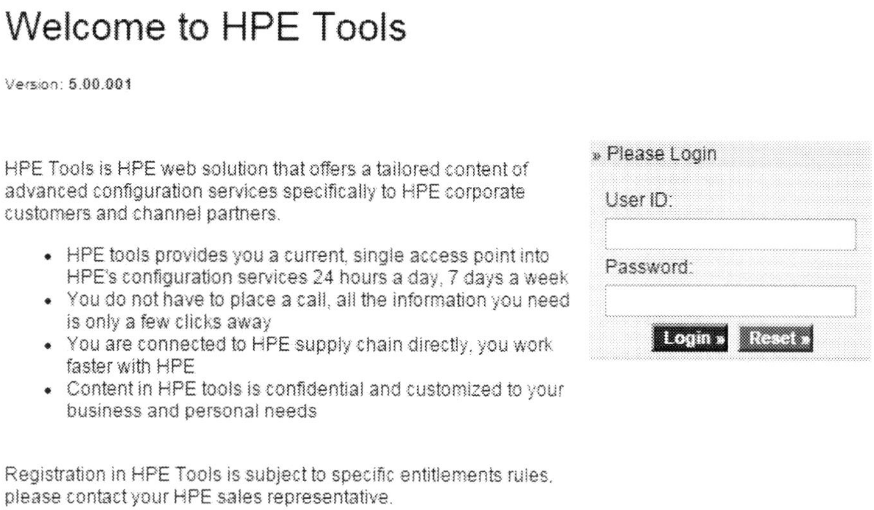

Figure 8-2 HPE Tools website

HPE provides a variety of tools that offer tailored content of advanced configuration services specifically to HPE corporate customers and channel partners (Figure 8-2). HPE is deploying a license-based security mechanism against unauthorized usage to protect many tools and applications used by HPE employees and certified HPE partners.

When a specific tool requires a license for use, a utility named the *HPE Sizing Tool License Manager* must be used. This utility applies licenses to use (LTUs) for individual applications (or several at one time) to a computer.

The HPE Sizing Tool License Manager is included with current HPE Enterprise Security Services (ESS) Alliances software tools, so there is no need to download a new application if an HPE tool is already installed.

Note

To access the HPE Tools website, visit: https://www.tools.hp.com/.

Obtaining and applying a license through the HPE Partner Ready Portal

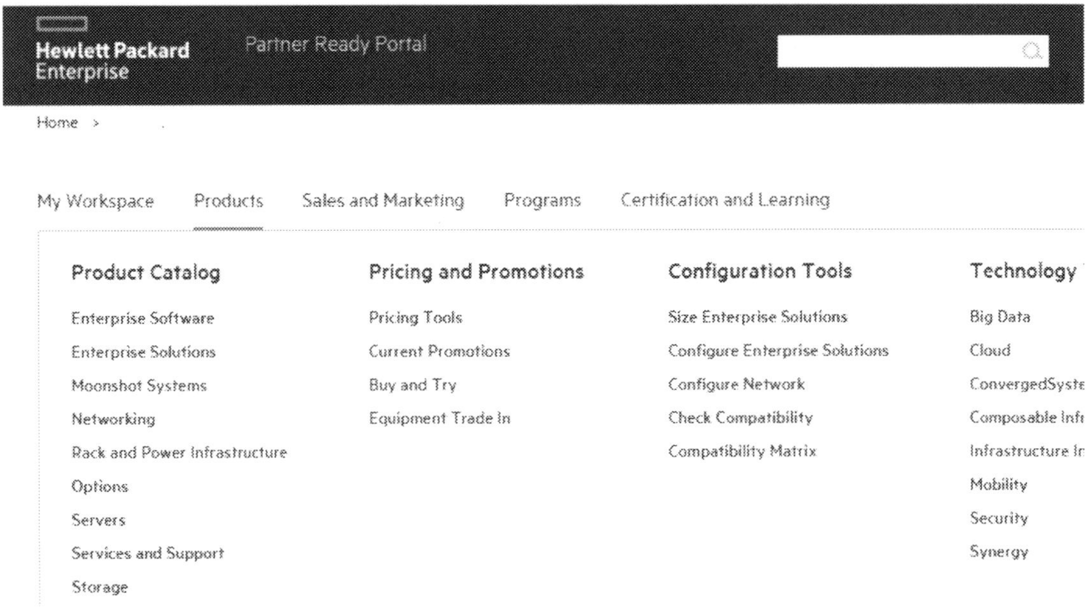

Figure 8-3 HPE Partner Ready Portal options

When a tool starts using the new security mechanism, you might need to obtain and apply a license to use that application (Figure 8-3). Some applications do not require a license, and sometimes a license for other applications is based on the level of affiliation with HPE. For example, the Storage Sizing Tool has migrated to the licensing model. Access to the tool differs according to your HPE affiliation:

- **HPE employees**—HPE employees should obtain their first-time LTU from the HPE Employee Page. After the initial application of their license, HPE employees will automatically receive an extension to that license as long as they periodically connect to the HPE intranet (office or VPN).
- **HPE Certified Partners**—HPE Certified Partners should get new licenses from a page within their respective partner portal.

Note

The procedure to obtain the license is changing at this time and is region specific.

You can search your respective portal using "Storage Sizing Tool" as your search term.

All other users must have an HPE Sales Representative act as a sponsor. You can start a licensing request by using the Request License feature of the License Manager as of Version 7.5.0. After providing the necessary information, and identifying an HPE employee as a sponsor, you can get a sponsor to complete your request and send you a license directly.

Applying a license

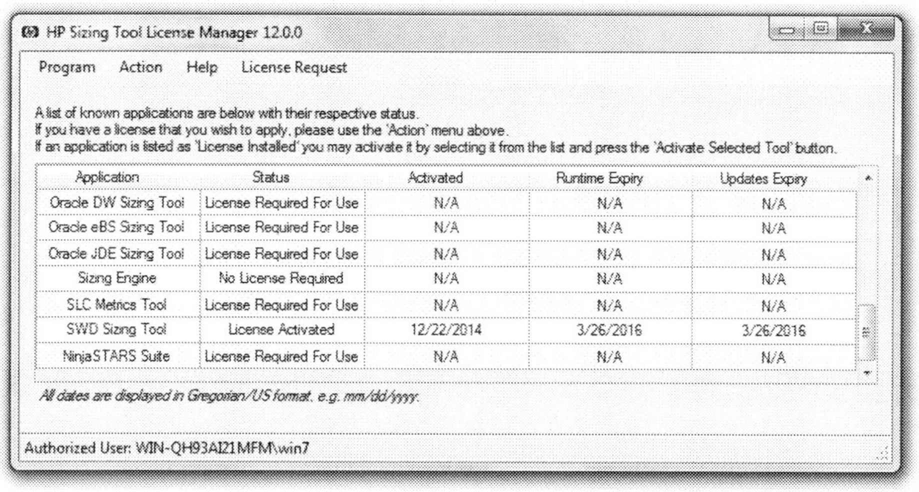

Figure 8-4 Licensing Storage Sizer screen

To apply a license, follow the steps (Figure 8-4):

1. Launch the HPE Sizing Tool License Manager in one of two ways:
    - Click **Start → Programs → Hewlett Packard Enterprise → HPE Sizing Tool License Manager**. Then right-click the HPE icon in the system tray and select **Open License Manager**.
    - Select **Action → Apply New License Data**.
2. Navigate to the directory containing the license, and select the license file.
    - For licenses that grant access, the affected tools will have their status changed to "License Installed" (assuming you had never installed it before).
    - For licenses that increase access time, the affected tools will have their expiration dates changed accordingly.
3. If this is the first license you have received for a tool, you will need to activate that tool.

Highlight the line representing the tool and click the **Install/Activate Tool** button. The Status Line for the tool will change to "License Activated," and the expiration dates will display.

Factors to consider for sizing a solution

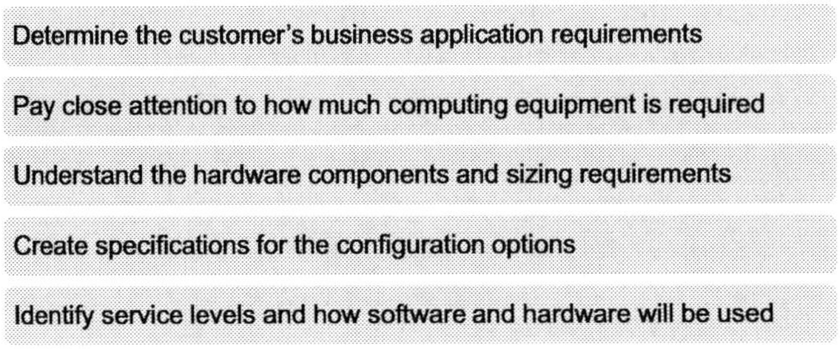

Figure 8-5 Factors for sizing a solution

Customers want a computing environment to help them stay ahead of the competition. Sizing an IT solution correctly is critical to meeting this objective (Figure 8-5). To recommend the proper hardware and software configuration and services, it is important to uncover the customer's business application requirements.

Pay close attention to the amount of computing equipment required to support the business applications. If overconfigured, the solution does not return good value for the investment. If underconfigured, the solution does not provide adequate service or scalability.

Although several hardware configurations might apply, understanding the hardware components and their sizing requirements enables you to recommend an accurate and beneficial product configuration. Also, consider the price:performance ratio, the customer's budget, and the time frame compelling the purchase.

After determining the server platform, create specifications for the configuration options by identifying how the software and hardware will be used and an acceptable level of performance. For example, determine the number of processors, the amount of memory, the I/O bandwidth, and the type and amount of storage to be configured.

HPE provides a variety of tools and resources to help you size and configure a customer solution. One of the best places to find online information about HPE products and solutions is the HPE website.

Note

For a complete list of available sizing tools, see the Size Enterprise Solutions section of the HPE partner portal:
http://h17007.www1.hp.com/us/en/enterprise/converged-infrastructure/info-library/index.aspx?type=20#.Vh-1lfmqpBc.

HPE SAN Design Reference Guide

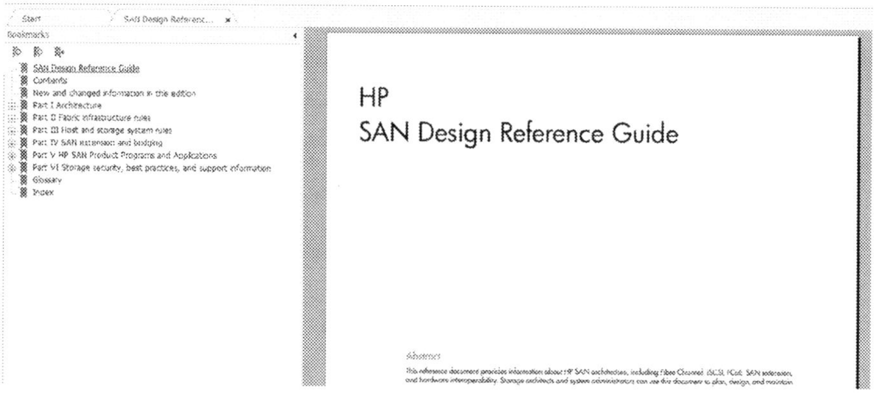

Figure 8-6 HPE SAN Design Reference Guide

The *HPE SAN Design Reference Guide* provides information on SAN design philosophy and considerations (Figure 8-5). The guide includes design rules concerning SAN topologies and supported configurations. In addition to sections on security and management, the guide also details HPE Best Practices regarding planning, upgrading, and migrating SANs.

This guide can help you to become more familiar with the components of a SAN when you are designing a SAN to meet customer needs, including the following:

- Architecture
- Configurations
- Implementation
- New technologies

Note

To access the *HPE SAN Design Reference Guide*, visit: http://www.hp.com/go/sandesignguide

---

HPE StoreFabric Storage Networking software and tools provide critical information to optimize the value of a storage network. Blueprints, SAN analysis tools, and SAN availability and management tools, as well as the *HPE SAN Design Reference Guide*, are available online.

Note

For information about storage networking software and tools, visit: https://www.hpe.com/us/en/storage/networking.html.

HPE Single Point of Configuration Knowledge

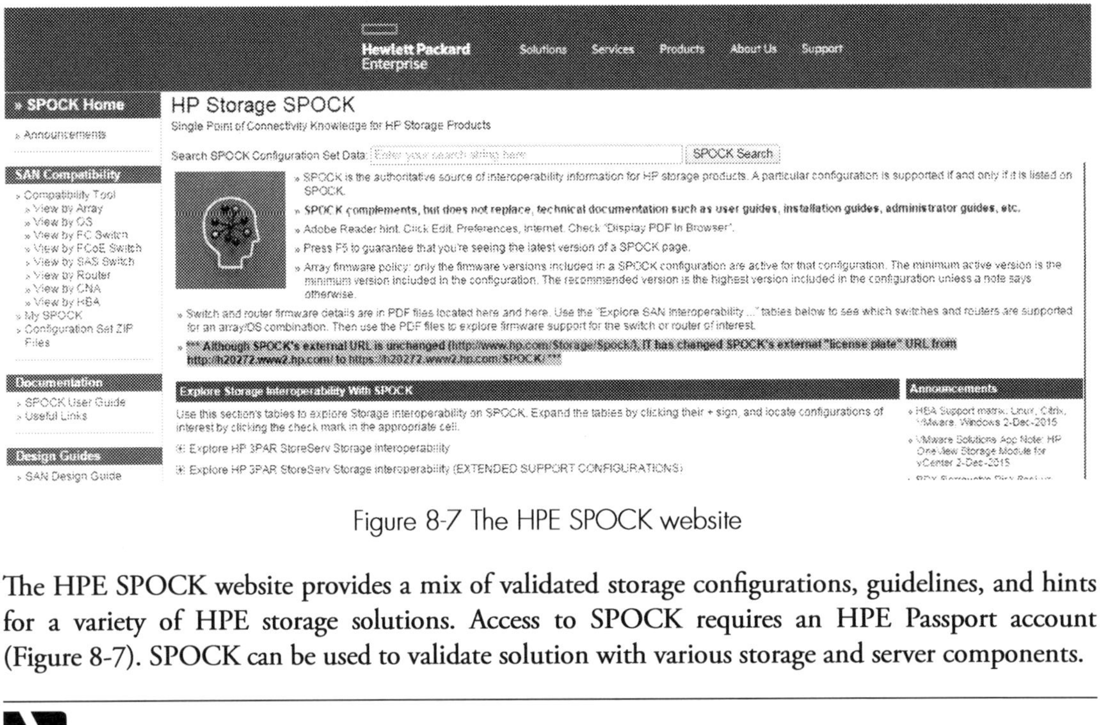

Figure 8-7 The HPE SPOCK website

The HPE SPOCK website provides a mix of validated storage configurations, guidelines, and hints for a variety of HPE storage solutions. Access to SPOCK requires an HPE Passport account (Figure 8-7). SPOCK can be used to validate solution with various storage and server components.

Note

You can access SPOCK at: http://www.hp.com/Storage/Spock.

You must have an HPE Passport account to access SPOCK from this link.

HPE Storage Sizing Tool

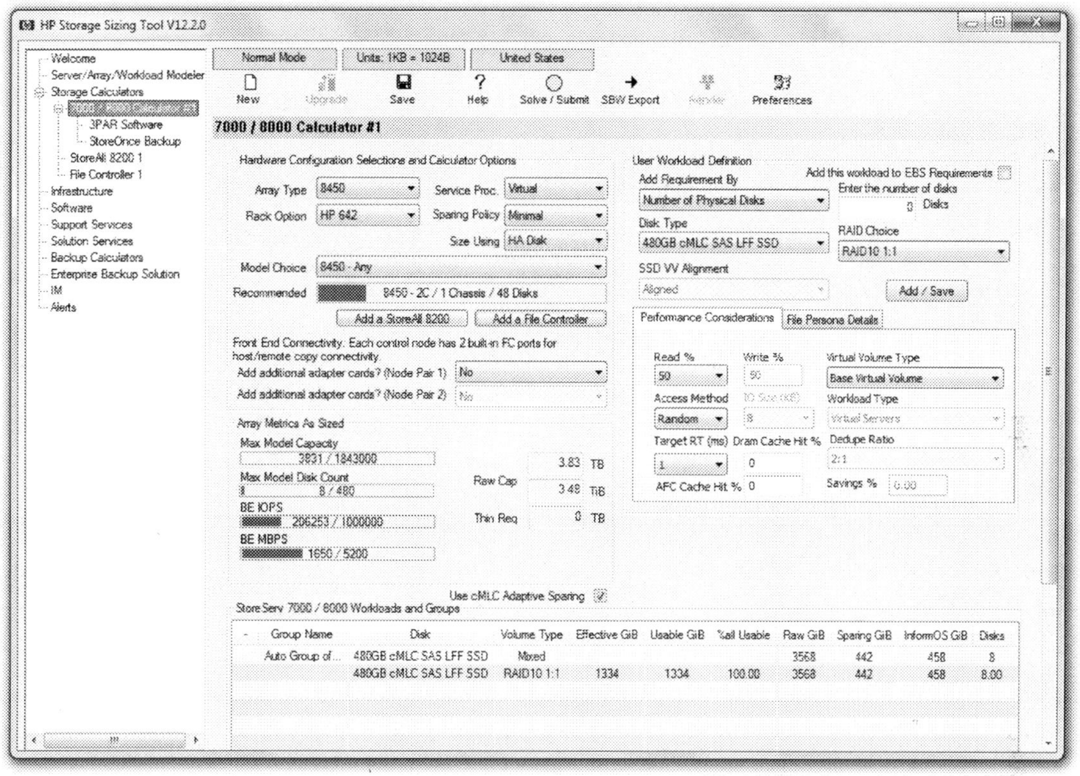

Figure 8-8 HPE Storage Sizing Tool

The HPE Storage Sizing Tool is a downloadable sizing tool that enables you to work with your customers to design a storage infrastructure to meet their needs (Figure 8-8). Whether they need to satisfy performance requirements with specific metrics, business requirements such as server consolidation, or simply pure capacity requirements, you can define that information in the Storage Sizing Tool.

Because the tool applies all the HPE SAN design rules, it gives you a valid, supported storage infrastructure to meet those requirements. The Sizing Tool is helpful when you are not sure what combination of products will satisfy your customers' requirements.

As you try different scenarios, a helpful wizard interface guides you through the process of sizing a SAN by asking a series of questions about the proposed configuration. The wizard is intended for use by those without much experience using the Sizing Tool.

The HPE Sizing Tool offers these benefits:

- Simplifies the process of designing a storage solution
- Applies storage design rules, licensing, and services rules
- Generates output in a valid, supported configuration that can be imported directly into SBW for quotation
- Provides localized parts and pricing for the Americas, APJ, and EMEA regions
- Brings new product information or functionality to you through Smart Update Technology
- Encompasses the entire HPE Storage family
- Will be updated at every new product launch as part of the New Product Initiative (NPI) process
- Adds new functionality based on user input, annual surveys, and quarterly focus groups

Important

The HPE Storage Sizing Tool has moved to a license-based security model. It is available for download by licensed HPE employees, Certified Partners, and others who are sponsored by an HPE Sales Representative. Partners can download it from the Partner portal.

Note

You can access the HPE Storage Sizing Tool at http://www.hp.com/storage/sizer.

SalesBuilder for Windows

SalesBuilder for Windows (SBW) is the premier HPE configuration and quotation tool for the sales force and channel partners (Figure 8-9). This downloadable tool is designed to support the complete HPE product portfolio of servers, storage, and related services.

Components of the SBW include

- **Configurator**—The configuration tool prepares the configuration and verifies it against the built-in configuration rules.
- **Whiteboard**—The center of the SBW screen shows the technical solutions and is the unit of quoting and storing.

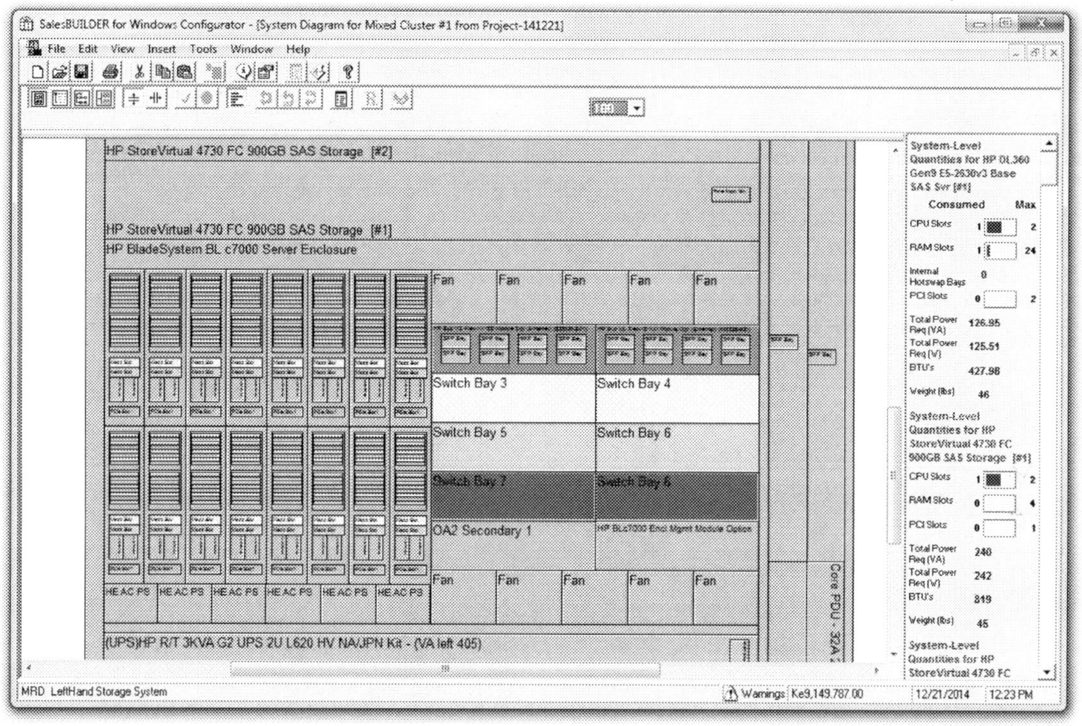

Figure 8-9 SalesBuilder for Windows Configurator

- **Config Worksheet**—As the heart of the configuration tool, the worksheet enables you to configure complete technical solutions.
- **System Diagram**—This is a graphical view of your configuration that you can modify and export to Visio.
- **Quoter**—This component that prepares a budgetary quotation that shows part numbers, descriptions, and prices of all the products in the solution.
  - **Price Book**—Data files containing the latest product descriptions and prices.
  - **Knowledge Base**—A data file that contains the rules and product modeling used by SBW so you can check configurations.

With the Consolidated Model Selection Form (CMSF), you can configure a total solution including HPE Industry-Standard Servers (ISS), Business-Critical Solutions (BCS), and Network Storage Solutions (NSS) hardware and associated software and services. You can also quote a solution with current pricing and ensure that the hardware and software choices are compatible.

SBW features include

- **Fast configuration**—Use the system diagram, configuration worksheet, or part list to quickly configure, upgrade, or add on to systems
- **Menu-driven interfaces**—Find detailed help and built-in configuration rules to leverage existing configurations to create and modify configurations for one or multiple systems
- **System diagram and modifications**—View the system diagram and any changes you made
- **Accurate and timely order delivery**—Place orders by exporting information from SBW

SBW reduces the need for the printed HPE Configuration and Ordering Guides, significantly reducing the time required to select all of the components necessary to build an HPE solution. The configuration engine checks the selections and alerts you to potential conflicts such as an operating system that does not support a desired peripheral.

HPE distributors, channel partners, and selected enterprise customers can use SBW externally.

Note

HPE partners can access SBW at the HPE Partner Portal: http://www.tools.hp.com.

An HPE Passport account is required to access the HPE Partner Portal.

Using the SBW tool

SBW has two main components: the Configurator and the Quoter.

Configurator

The SBW Configurator displays the system diagram and modifications for HPE clusters, servers, and storage running HP-UX, Windows, Linux, or mixed environments (Figure 8-10). You can use the Configurator to configure and customize technical solutions for new systems, upgrades, and add-ons.

The Configurator is used to configure and customize technical solutions. It provides access to the Configuration Worksheet, System Diagram, and Whiteboard.

Quoter

The Quoter is used to prepare a legal quote for a client, showing product numbers, descriptions, and prices of all the products included in the technical solutions (Figure 8-11).

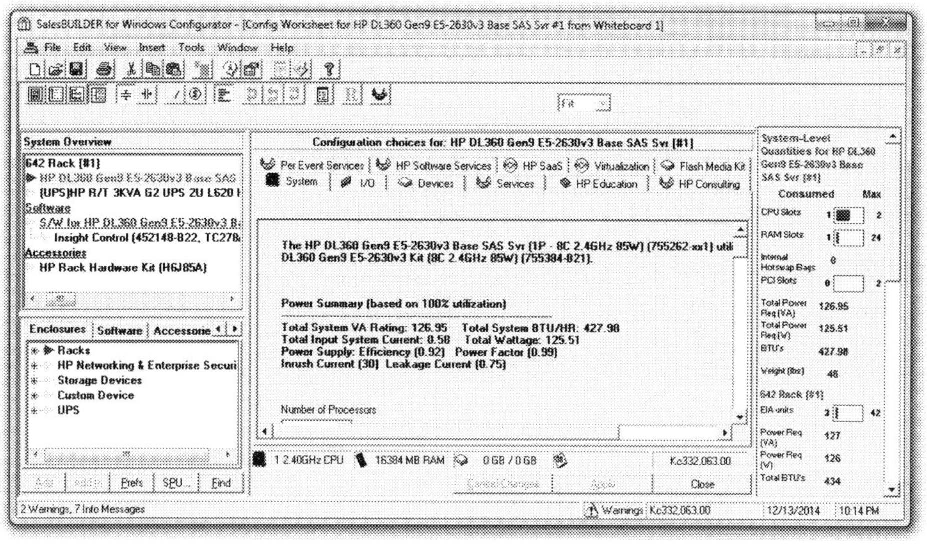

Figure 8-10 SBW Configurator view

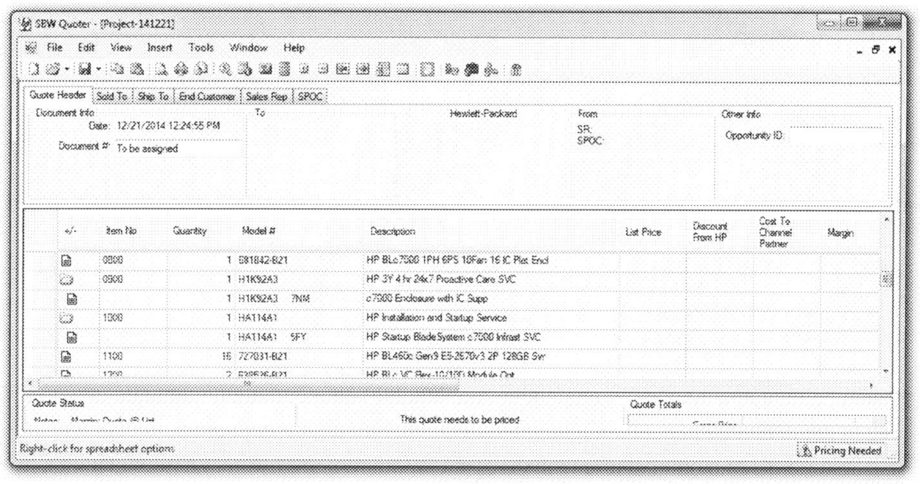

Figure 8-11 SBW Quoter view

Note

Access to the quotation tool might not be available in all versions of SBW.

Example—Using the HPE SBW to configure storage

Following are several examples showing how to use SBW to prepare a configuration and quotation. They are all based on the following customer scenario:

- Headquarters in your country
- Branch in the UK requires a local consolidated storage for two hypervisors
- Electricity: Single phase, 220V
- Shared SAS Storage for two hypervisors
- SAS 12 Gb/s interface
- HPE Rack in place (642 series) with 12U of free space
- Power protection in place (HPE R3000 G4)

Configure an HPE MSA 2040 solution

Assume that sizing steps already performed with the customer produced the following result:

- MSA2040 Dual Controller
- SAS interface
- 2.5 in drives, 15K
- 14 TiB raw capacity, RAID 50
- 24x 600 GB drives
- 12 Gb/s
- Up to 256 snapshots

Following are key points for configuring an MSA solution with SBW.

- Start the HPE SBW Configurator using the icon on the desktop (Figure 8-12).

Figure 8-12 SBW Configurator

- The available price books need to be accepted (Figure 8-13).

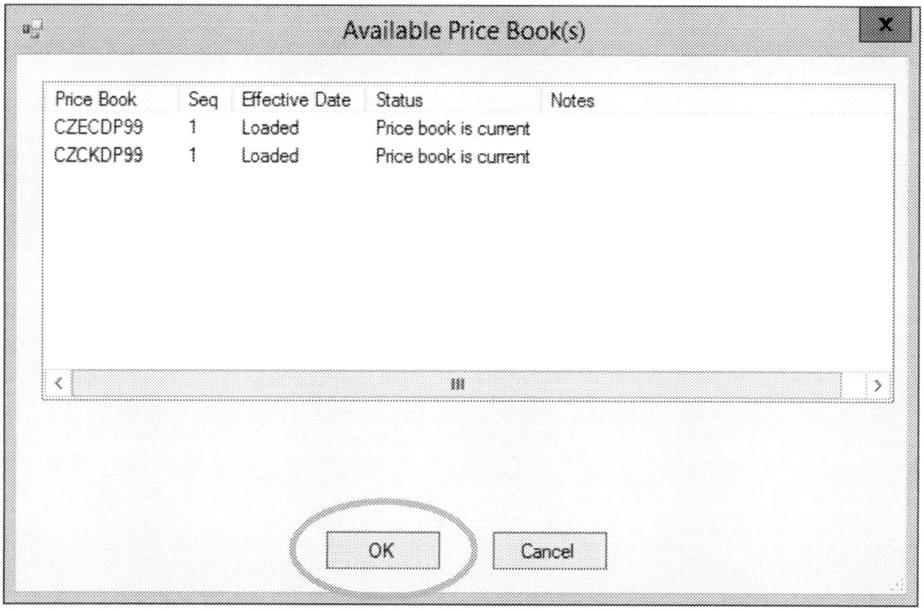

Figure 8-13 Available price books

- To configure storage products, change the product category to **Storage** (Figure 8-14).

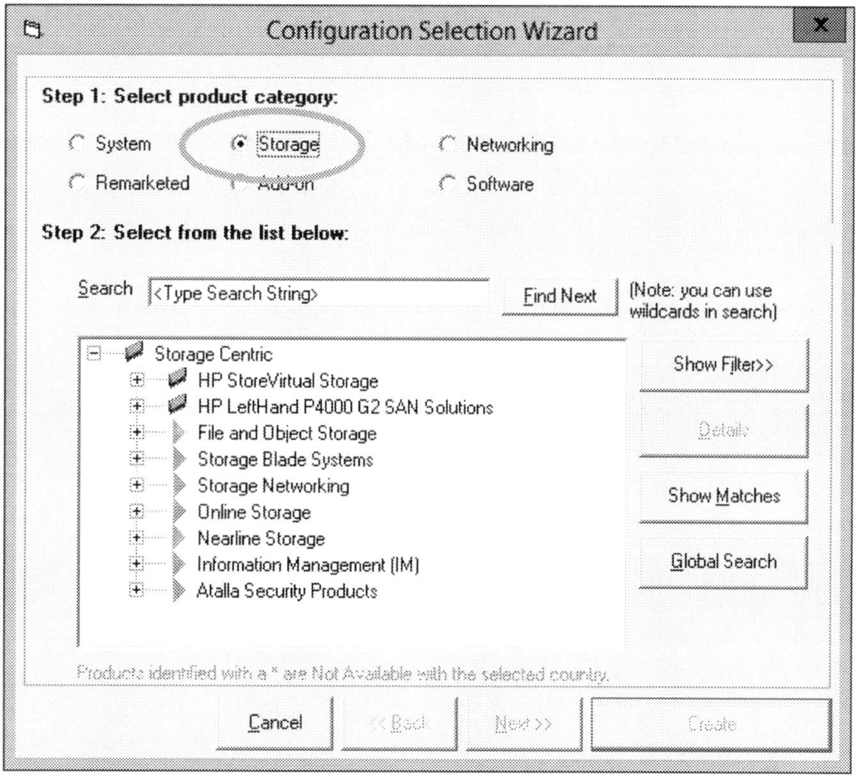

Figure 8-14 Configuration Selection Wizard—Storage

- You can use the Search field to locate a product (Figure 8-15).

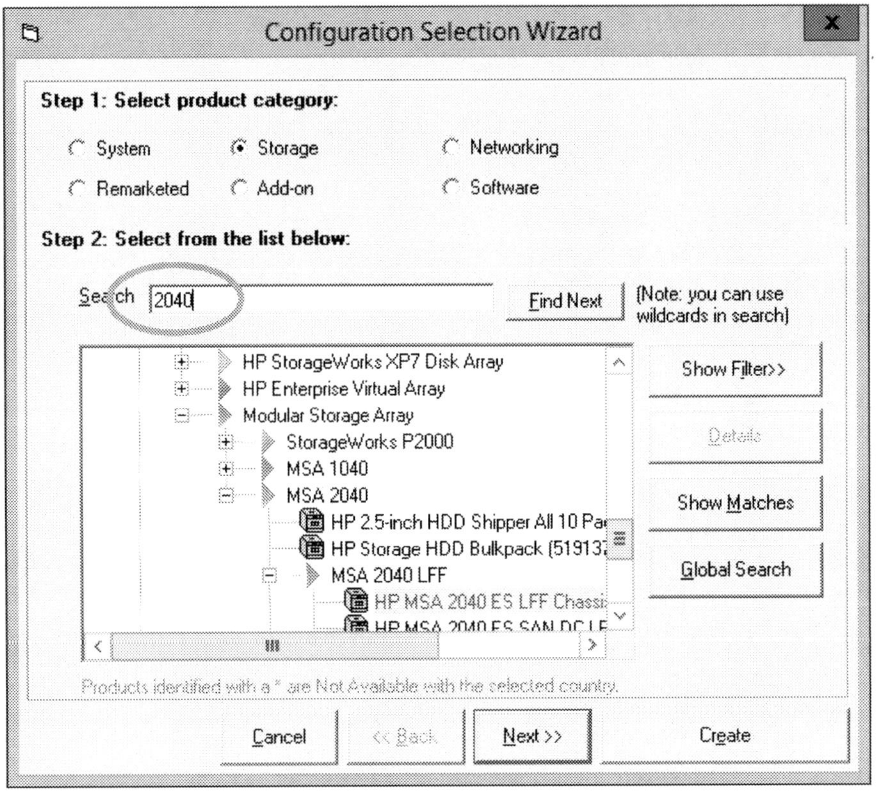

Figure 8-15 Configuration Selection Wizard—Search

Then you can select the particular product from search results (Figure 8-16).

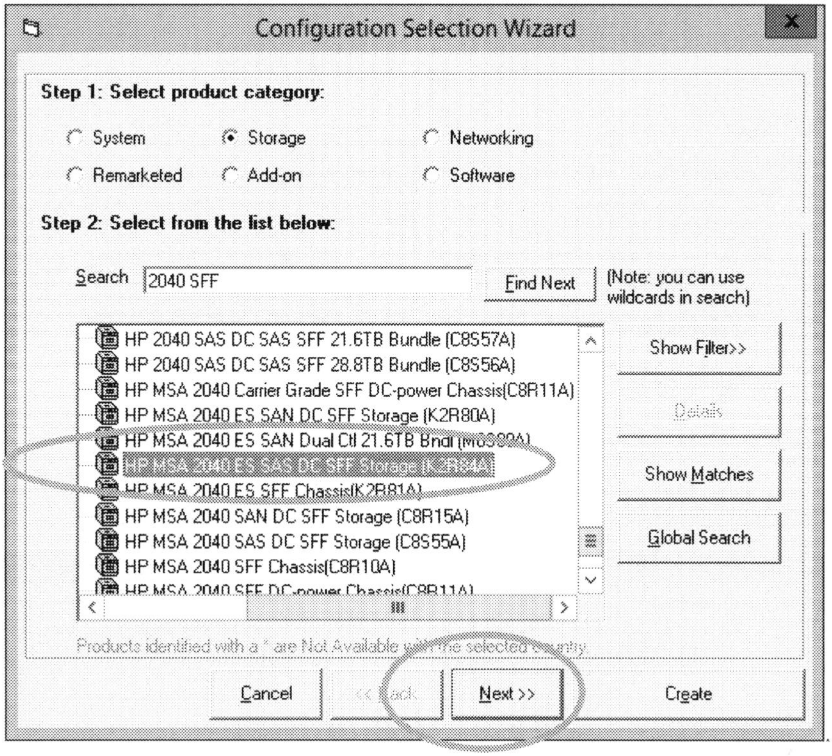

Figure 8-16 Configuration Selection Wizard—Product catagory

- You can select the region, language, rack, and power accessories from options and drop-down lists to create your solution (Figure 8-17).

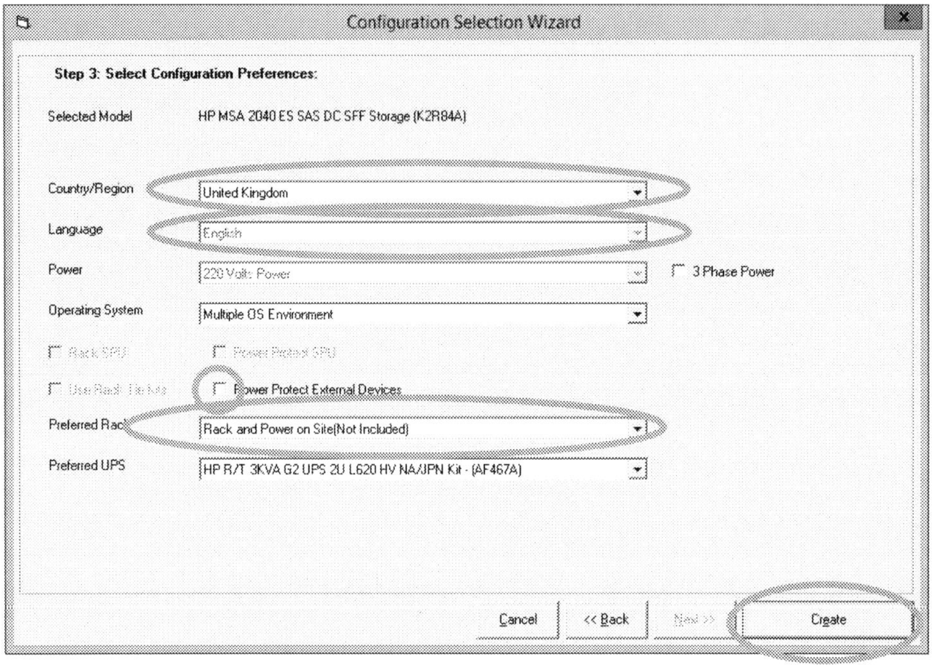

Figure 8-17 Configuration Selection Wizard—select the region, language, and so forth

- SBW will display warning messages if anything in your configuration needs to be checked or fixed (Figure 8-18).

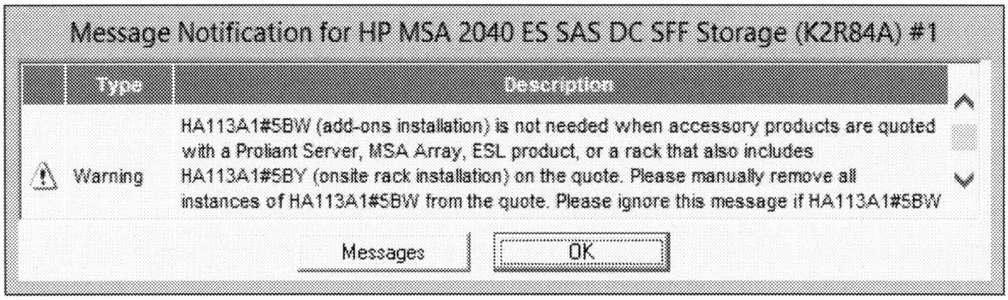

Figure 8-18 Display warning messages

- Additional details such as software that can also be configured in SBW (Figure 8-19).

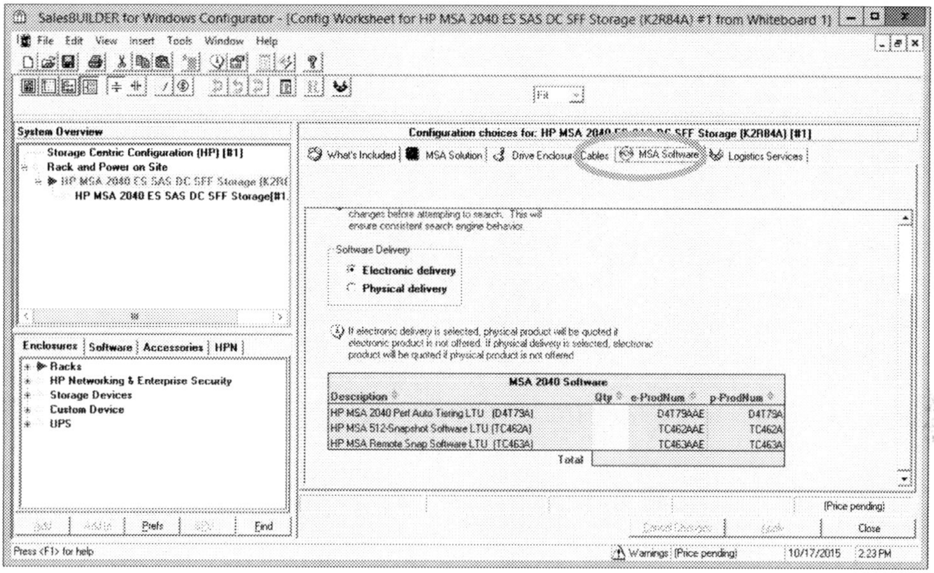

Figure 8-19 Additional details such as software

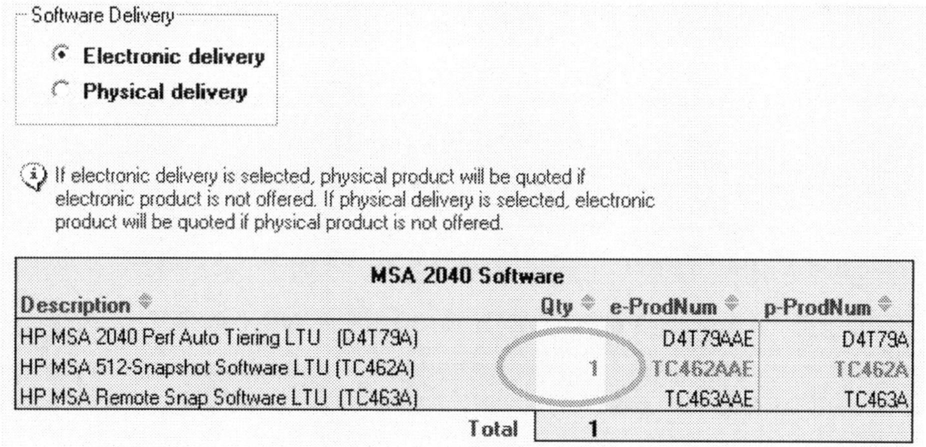

Figure 8-20 Software delivery

- Configuring drives for HPE MSA requires simple editing (Figure 8-21).

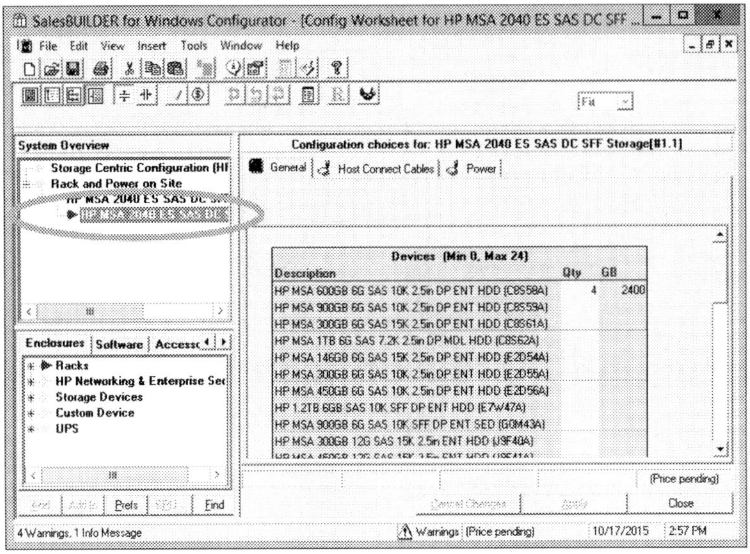

Figure 8-21 Configuring drives for HPE MSA

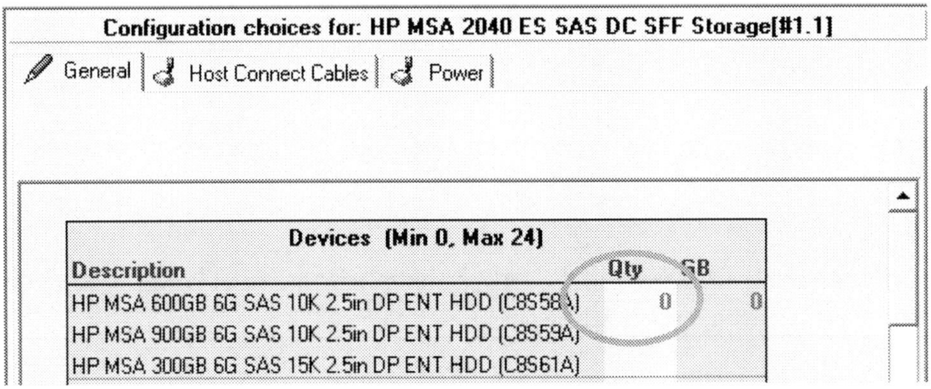

Figure 8-22 Configuring drives for HPE MSA—Devices

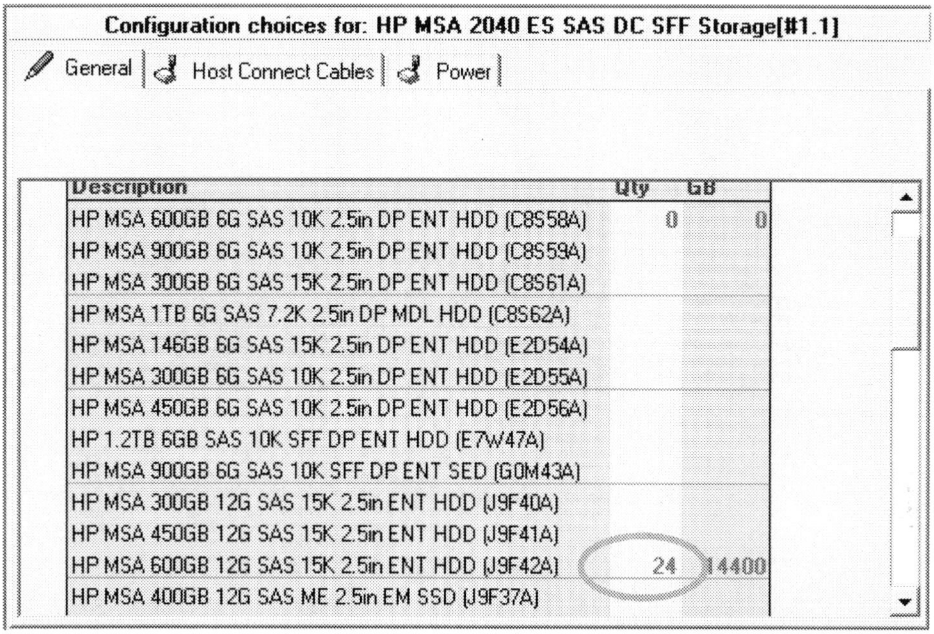

Figure 8-23 Configuring drives for HPE MSA—Choices

- When the configuration is complete, SBW can convert it to quote (Figure 8-25).

Figure 8-24 SalesBUILDER for Windows Configurator—convert Whiteboard to quote

- The SBW Quoter assembles all necessary part numbers and puts them in the quote (Figure 8-25).

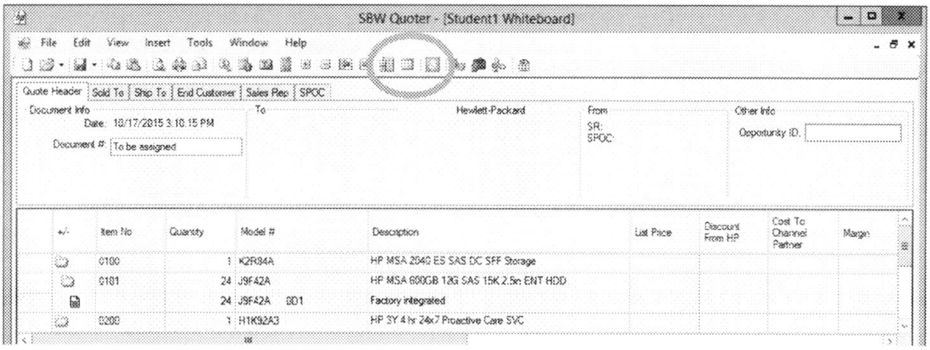

Figure 8-25 SBW Quoter assembles all necessary part numbers

Configure a B-Series fabric switch

SBW can be used to configure and quote the MSA 2040 solution just described with SAN controllers and B series Fibre Channel switches (Figure 8-26).

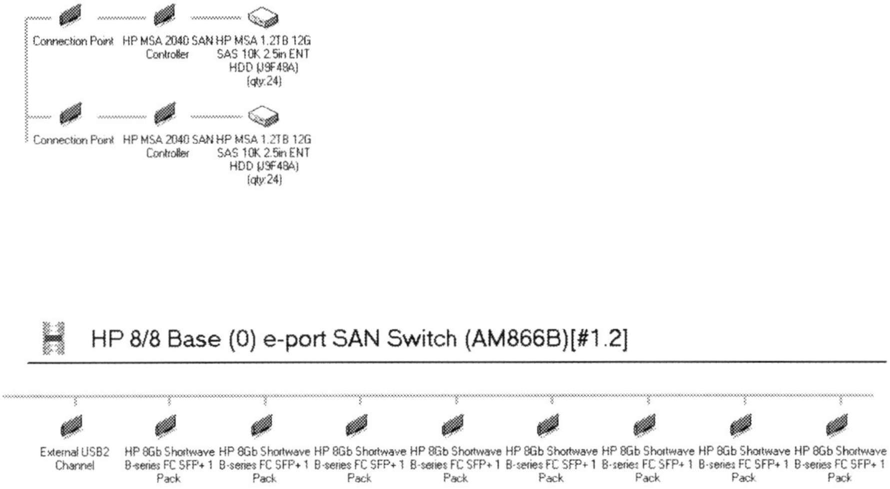

Figure 8-26 Configure a B-Series fabric switch

Configure an HPE StoreEasy 1650 Storage system

You can also use SBW to configure StoreEasy products (Figure 8-27).

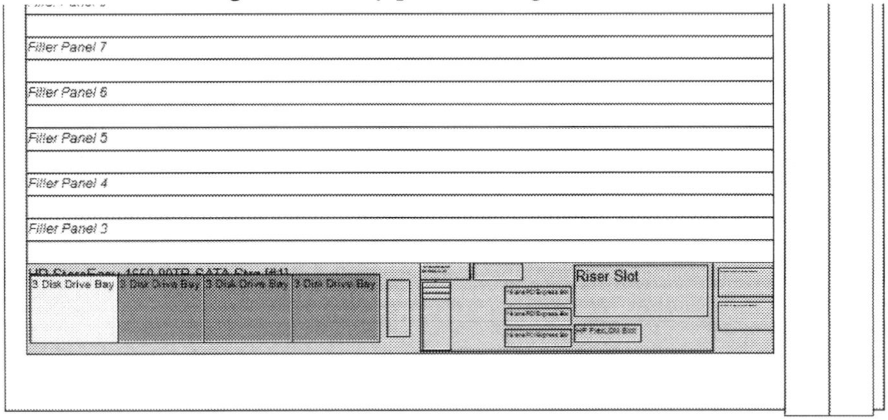

Figure 8-27 Configure an HPE StoreEasy 1650 Storage system

Configure an HPE 3PAR StoreServ 8450 Storage system

SBW can be used to configure HPE 3PAR StoreServ products (Figure 8-28).

+/-	Item No	Quantity	Model #	Description
	0100	1	BW904A	HP 42U 600x1075mm Enterprise Shock Rack
		1	BW904A 001	HP Factory Express Base Racking Service
	0200	1	H6Z25A	HP 3PAR StoreServ 8450 4N Stor Cent Base
		1	H6Z25A 0D1	Factory integrated
	0201	8	K2P88A	HP 3PAR 8000 480GB SAS cMLC SFF SSD
		8	K2P88A 0D1	Factory integrated
	0202	1	L7C17A	HP 3PAR 8450 OS Suite Base LTU
		1	L7C17A 0D1	Factory integrated
	0203	16	L7C18A	HP 3PAR 8450 OS Suite Drive LTU
		16	L7C18A 0D1	Factory integrated
	0204	1	L7C29A	HP 3PAR 8450 Virtual Copy Base LTU
		1	L7C29A 0D1	Factory integrated
	0205	16	L7C30A	HP 3PAR 8450 Virtual Copy Drive LTU
		16	L7C30A 0D1	Factory integrated
	0206	1	L7C19A	HP 3PAR 8450 Data Opt St v2 Base LTU
		1	L7C19A 0D1	Factory integrated
	0207	16	L7C20A	HP 3PAR 8450 Data Opt St v2 Drive LTU
		16	L7C20A 0D1	Factory integrated
	0208	1	L7D49A	HP Smart SAN for HP 3PAR 8xxx LTU
		1	L7D49A 0D1	Factory integrated

Figure 8-28 Configure HPE 3PAR

VisioCafe stencils for SAN design

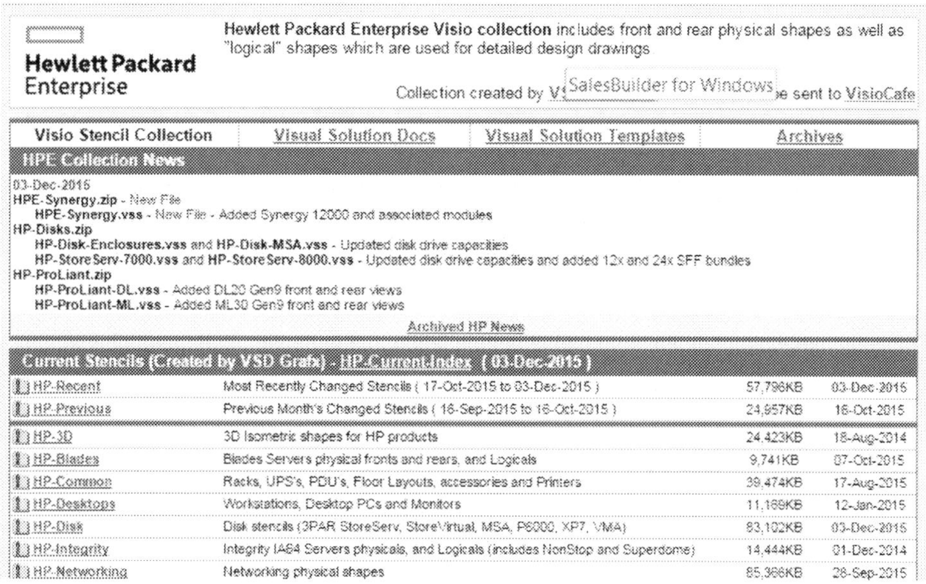

Figure 8-29 VisioCafe interface

Some customer configurations can be complex and difficult to present (Figure 8-29). The VisioCafe website provides graphics that you can use with any version of Microsoft Visio to create diagrams, storage network cable management plans, and schemas for your proposed solutions. HPE not only provides storage subsystem graphics in this tool, but it also provides graphics of servers and networking devices. All graphics are free to download from the VisioCafe website.

Note

To access the HPE VisioCafe site, go to: http://www.visiocafe.com/hp.htm.

VisioCafe can be especially useful when you are presenting the proposal because it enables you to show the customer a graphical representation of a future solution, not just part numbers and quotations. This representation can enhance the selling process and influence the decision-makers.

HPE Simple Configurator solutions

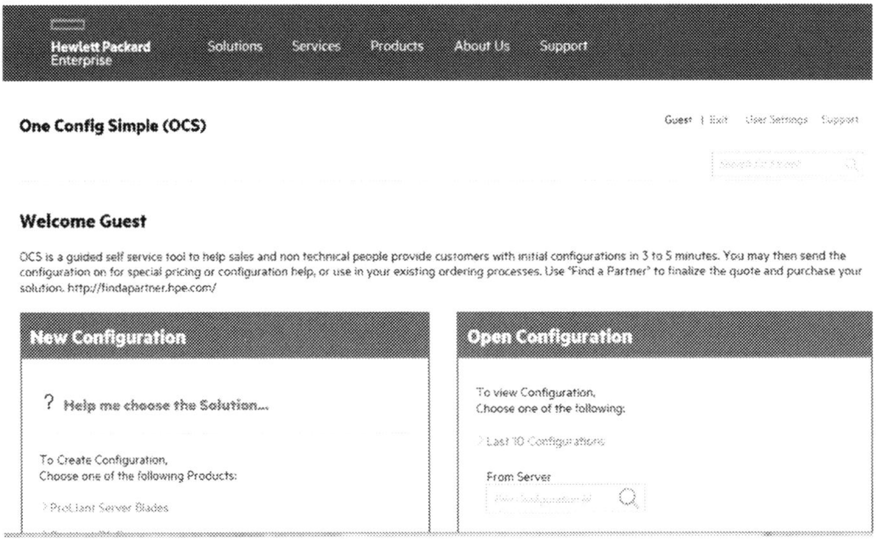

Figure 8-30 One Config Simple (OCS) website

The One Config Simple (OCS) is replacing a number of older configuration tools (Figure 8-30). This is an important step in the ongoing HPE efforts to make configuring solutions easier, quicker, and more responsive to your needs. This new configuration application has a number of advantages over previous HPE tools:

- Easy-to-use, self-service environment
- Fast—3 to 5 minutes to initial solution configurations
- Localized part and SKU numbers
- Available in 10 languages
- Local list pricing
- Guidance to proper solutions
    - Solution wizards based on applications
    - Listings of popular configurations that can be modified
- Ability to export configurations to a Microsoft Excel file format

---

Note

To access the OCS, go to https://h22174.www2.hp.com/SimplifiedConfig/Index.

HPE Product Bulletin

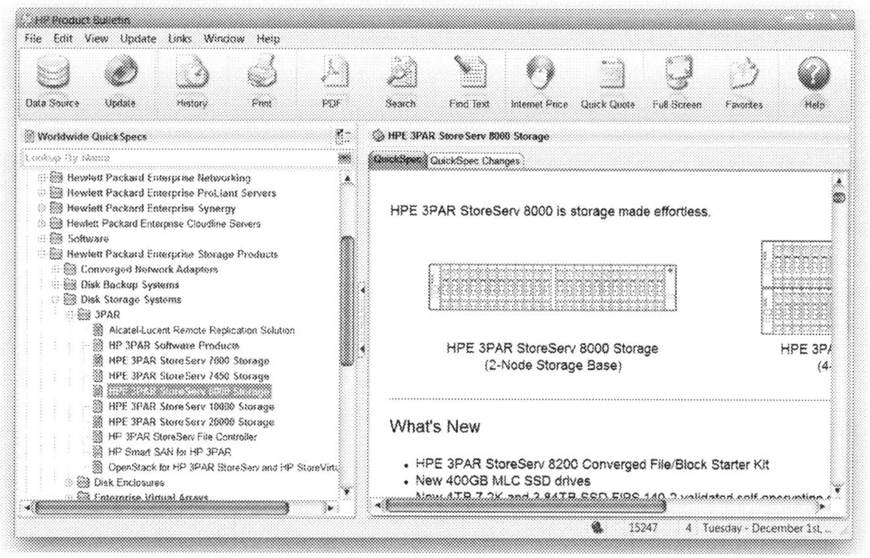

Figure 8-31 HPE Product Bulletin interface

The HPE Product Bulletin website is a convenient central resource that provides technical overviews and specifications for HPE hardware and software (Figure 8-31). The downloadable HPE Product Bulletin application is loaded with features to aid with the purchase, sale, and support of HPE products. The Product Bulletin contains the following useful features:

- Updated QuickSpecs
- Quick Quotes
- Product photos
- "Locate by Name" search feature
- Advanced search capabilities
- Favorites
- Retired products
- Tip of the Day

Note

To access and download the HPE Product Bulletin, visit: https://www.hpe.com/info/qs.

Alinean ROI and TCO analysis tools

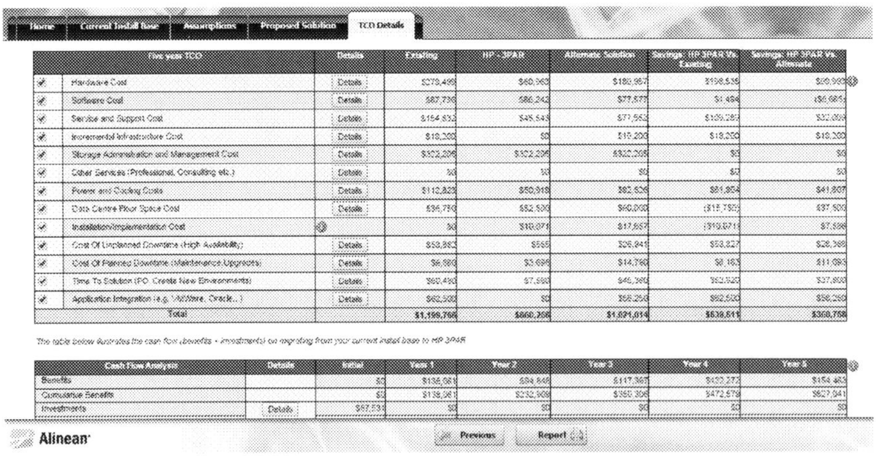

Figure 8-32 Alinean 3PAR TCO Analysis

The Alinean ROI and TCO calculators make it easy to build a business case to position the value of an HPE Converged Infrastructure solution (Figure 8-32). The tools produce values for ROI, payback, net present value (NPV), and internal rate of return (IRR) that are often required by chief financial officers to compare different projects. For example, your project for a new IT infrastructure might be competing for funds that are also being considered for other, quite different projects such as the refurbishment of company buildings (Figure 8-33).

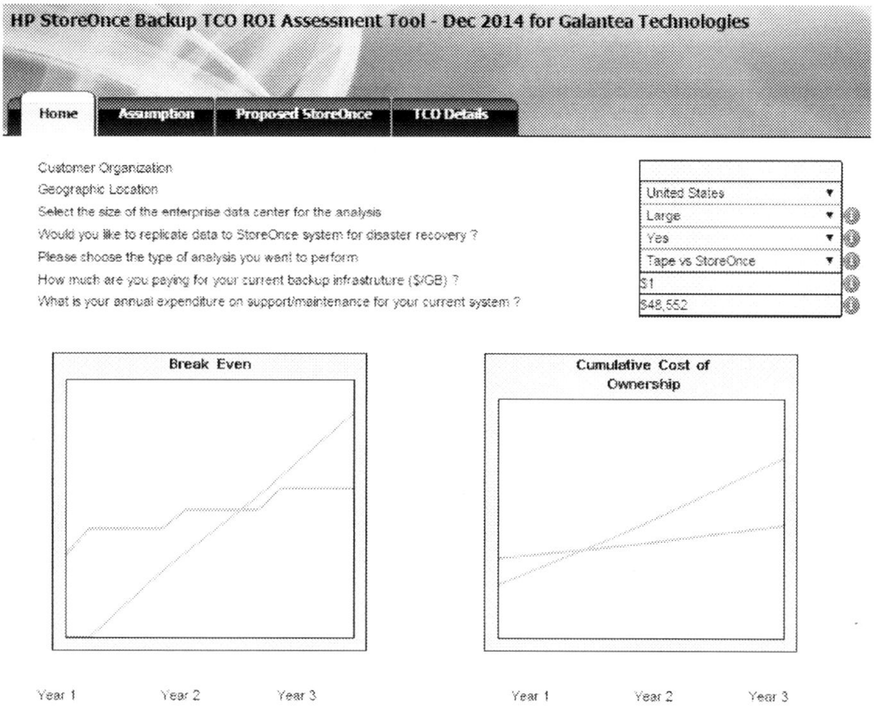

Figure 8-33 HPE StoreOnce Backup TCO ROI Assessment Tool screen

Alinean output

Figure 8-34 Steps for using the calculators with customers

The Alinean tools also enable comparison against many potential competitor configurations because the templates that HPE uses include key competitive information (Figure 8-34). The ROI/TCO tools also include a large amount of default information, factored for geography and location.

Other sizing and planning tools

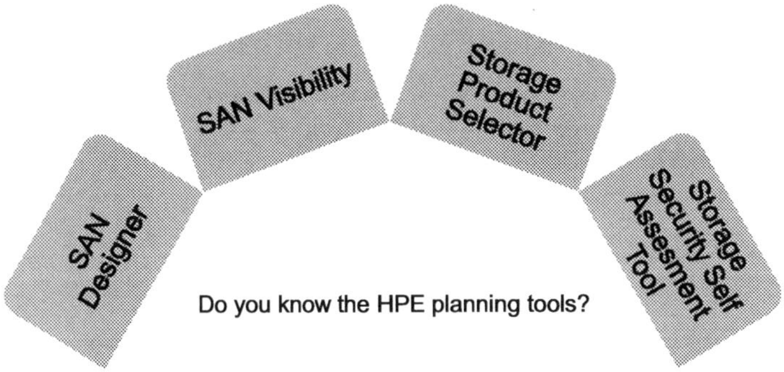

Figure 8-35 Key sizing and planning tools

HPE offers several other sizing and planning tools (Figure 8-35).

- **SAN Designer**—SAN Designer is a powerful application to help customers design a customized SAN with HPE components. They can configure and design a SAN based on performance, cost, and future growth requirements. The tool helps customers to select the fabric characteristics, as well as the SAN elements, for a complete SAN design. The reporting feature of SAN Designer generates topology diagrams, a list of required SAN components, and additional recommendations for the SAN design.

SAN Designer is a complimentary software utility for HPE customers and runs on a Windows desktop or laptop.

- **SAN Visibility**—SAN Visibility is a complimentary software utility for HPE customers that helps with SAN analysis, diagnostics, and optimization. SAN Visibility saves considerable time, money, and effort by automating inventory activities and providing a quick and accurate view of the SAN topology.

SAN Visibility has an automated report generation feature that produces recommendations, topology diagrams, and informative SAN element reports for switches, host bus adapters (HBAs), and storage array connectivity. This utility runs on any Windows desktop or laptop that has network connectivity to a switch within the SAN.

SAN Visibility also documents the server, storage, and storage network infrastructure and offers firmware update recommendations.

- **Storage Product Selector**—The Storage Product Selector enables you to navigate through the HPE storage product portfolio, visually reducing the number of storage options based on your selected criteria.
- **Storage Security Self-Assessment Tool**—The Storage Security Self-Assessment Tool helps you understand how well a business is prepared to manage risks confronting sensitive data in its storage and backup environment.

Note

HPE is constantly optimizing its portfolio of presales tools, introducing new products and retiring older solutions.

Note

For a complete list of available tools, check the ActiveAnswers site at http://www.hp.com/go/activeanswers.

Note

SAN Visibility can be downloaded from: http://h18006.www1.hp.com/products/storageworks/san/index.html.

Note

To access the HPE Storage Product Selector, go to: http://h22193.www2.hp.com/data-storage/index.html.

Solution Demo Portal

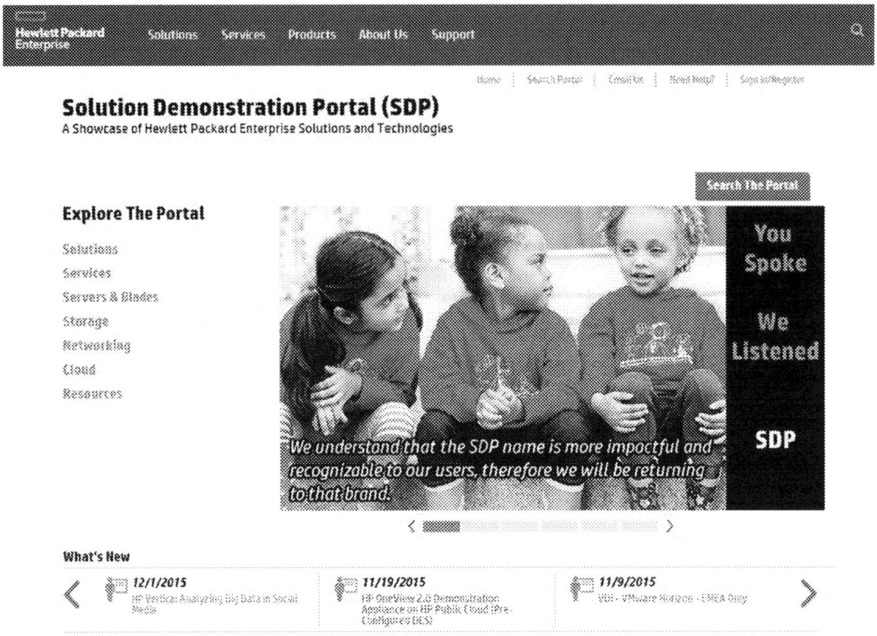

Figure 8-36 HPE Solution Demo Portal

The Solution Demo Portal, or SDP (formerly, the Virtual Resource Portal, or VRP), showcases how HPE technologies lead, innovate, and transform businesses. This portal provides a central location for all demonstrations, webinars, and supporting collateral.

Live and prerecorded demonstrations, or "demos," feature HPE hardware, software, services, and partnerships in exciting multimedia formats to show how HPE can help solve business and IT problems. The demos are high-level overviews of a product or solution, and they are an excellent way to learn more about HPE and partner products or solutions.

The Solution Demo Portal is an interactive tool that allows you to navigate quickly and easily to find the information most relevant to you. Six main areas comprise the first level of the portal:

- Products
- Business and IT services
- Infrastructure solutions
- Business solutions
- Industry solutions
- Partners

Choose the area you are interested in and then select a category within the area to view all available live and prerecorded demonstrations as well as supporting collateral. Supporting collateral is located in the same area with each demo.

Live demonstrations are typically tailored to business and IT concerns and are scheduled through an HPE account representative. A live demo event can be held on-site at an HPE Solution Center demonstration facility or using a virtual web conference with live support from an HPE or partner technical expert. A live demo can provide valuable information for specific business and IT needs.

HPE NinjaSTARS for 3PAR

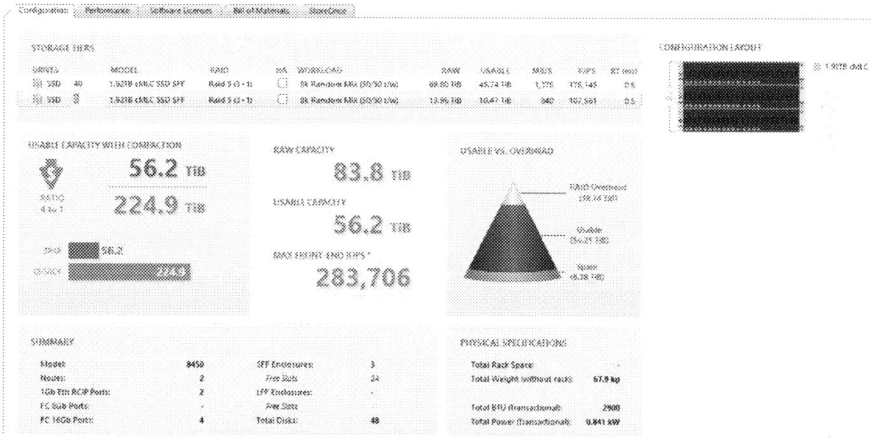

Figure 8-37 HPE NinjaSTARS for 3PAR

NinjaSTARS for 3PAR replaces NinjaThin 7000. It includes measured data from the HPE Storage Optimizer and other new features for increased accuracy and performance modeling. NinjaSTARS was developed to help worldwide channel partners to assess legacy storage environments for customers interested in HPE 3PAR StoreServ 7000 and 8000 arrays.

NinjaSTARS can also be used to determine whether customers qualify for the Get Thin Guarantee Program. Offering your customers 50% or more guaranteed capacity reduction gives you another key advantage over the competition.

NOTE

To download the tool, navigate to the **Get Thin Guarantee** page on your regional partner portal. From the landing page, go to **Business Units → HPE Enterprise Group → HPE Storage, Programs and Promotions → HPE 3PAR Get Thin Guarantee Program & NinjaThin Tool**. This page also provides access to NinjaSTARS for 3PAR technical training. For tool support and feedback, contact ninjastars@hpe.com.

HPE NinjaSTARS for 3PAR has built-in templates and wizards that help you to

- Assess a customer's current storage solution
- Provide StoreServ configuration and sizing recommendations
- Meet customers' performance and growth needs
- Drive better customer conversations based on their unique IT environment

HPE NinjaSTARS for 3PAR includes (Figure 8-38) the following:

- Powerful editing tools and direct import of EVA configuration data
- Highly accurate performance sizing based on measured or lab-tested data
- Written data utilization and TCO projections similar to the NinjaThin 7000 tool
- Storage Optimizer measured data for increased accuracy and performance modeling

---

Note

Partners can access the NinjaSTARS tool and training on the partner portal:

Home → Enterprise Group → Storage → Programs and Promotions → HPE 3PAR Get Thin Guarantee Program & NinjaThin Tool.

Alternatively, search for "NinjaSTARS."

Figure 8-38 Partner portal for HPE NinjaSTARS for 3PAR

NinjaProtected Tool and Get Protected Guarantee

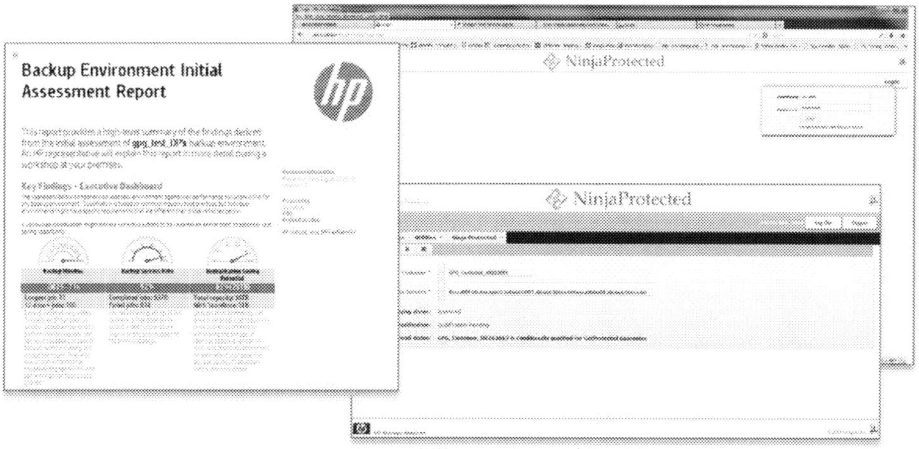

Figure 8-39 NinjaProtected Tool

NinjaProtected is the new name for what used to be called SizerAdviser and Get Protected Guarantee and is a marketing program being offered worldwide (Figure 8-39). The NinjaProtected Tool can be thought of as the engine behind a web interface and reports.

The NinjaProtected Tool is responsible for managing customer information, analyzing and parsing it, and generating useful reports from that data. This engine supports two reporting processes for HPE products:

- **The Assessment and Analysis Report**—This report is synonymous with what the SizerAdviser had been delivering for almost two years.
- **The Get Protected Guarantee Program Report**—This report works in conjunction with Assessment and Analysis Report and supports the Get Protected Guarantee marketing program. It also delivers immediately useful information about the customer's backup environment to assist in driving BURA conversations at any level.

Note

You can access the NinjaProtected Tool at http://www.hp.com/go/ninjaprotected.

What is the NinjaProtected Tool?

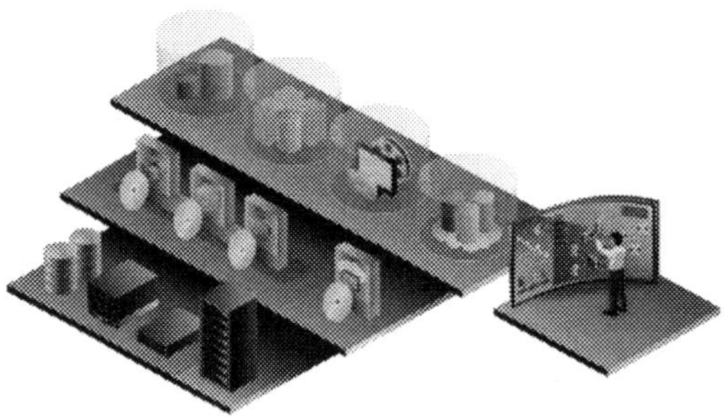

Figure 8-40 NinjaProtected Tool

The NinjaProtected Tool is a process culminating in an assessment and analysis of a customer's existing backup environment (Figure 8-40).

- Collects metadata
- Analyzes and parses data
- Generates reports
- Integrates into sizer tools (but is **not** a standalone sizer)
- Enables conversations about BURA

What does it do? What does it deliver?

What does it do?

NinjaProtected performs noninvasive, agentless metadata collection from existing backup servers and data post-processing analysis of KPIs, and compiles it into a comprehensive report.

What are the deliverables?

A workshop to present results and recommendations on the next steps while a high-level summary report about backup environment status can be presented to customers.

Reports contain information such as daily data backed up, throughput data, total capacity used for backup retention, client information, top 10 backup job information, and data types being backed up.

Why use the NinjaProtected Tool?

With the NinjaProtected Tool, customers can receive valuable, relevant insights into the current backup environment and discover other issues such as backup windows and failed backups. The NinjaProtected Tool can help find opportunities for cost reductions (such as retention times) and identify potential deduplication-driven benefits.

DD Analyzer Tool for HPE Storage Presales

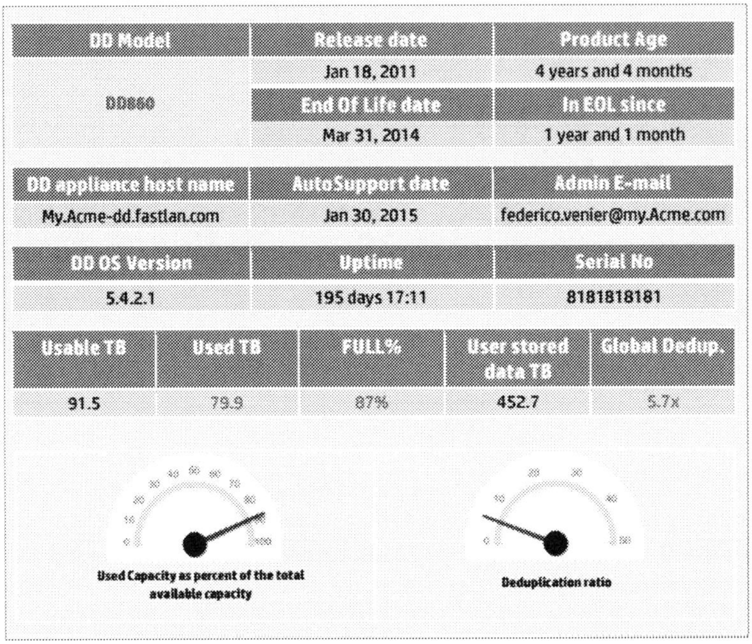

Figure 8-41 EMC Data Domain configuration summary

The DD Analyzer is a tool to analyze the "autosupport" file of an installed base EMC Data Domain (DD; Figure 8-41). It generates a detailed report about the existing EMC DD, with a swap-out proposal and sizing to an HPE StoreOnce solution.

		User Data (Pre Comp.)	Used Capacity (Post Comp.)	Global Comp. Factor (dedup)	Local Comp. Factor (zip)	Total Comp. Factor	
						Factor	%
Active tier Stored data	GiB	286,555	32,072	-	-	8.9x	89%
	TB	307.7	34.4				
Written data last 7 days	GiB	14,929	229	26.4x	2.5x	65.1x	99%
	TB	16.0	0.2				
Written data last 24h	GiB	3,907	36	41.1x	2.6x	106.0x	99%
	TB	4.2	0.0				
Archive Tier Stored data	GiB	135,093	42,301	-	-	3.2x	69%
	TB	145.1	45.4				
Total Stored data	GiB	421,649	74,374	-	-	5.7x	82%
	TB	452.7	79.9				

Figure 8-42 Deduplication ratio and recent workload example

The DD Analyzer tool is not downloadable, but instead is provided as an email autoreply service (Figure 8-42). Simply send the customer's "EMC autosupport file" to the support email at DD_assessment@hpe.com to receive the report. Partners cannot send the report directly, but must go through their HPE representative.

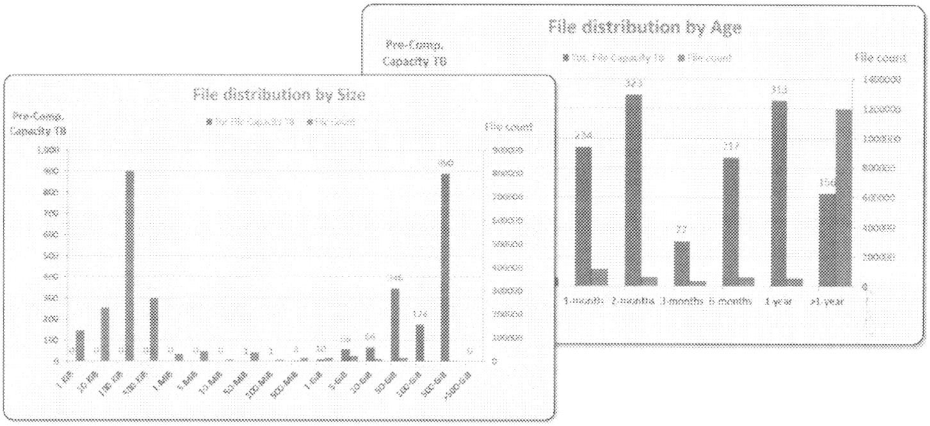

Figure 8-43 File System details example

A sample DD Analyzer tool output report can be downloaded from the HPE partner portal (Figure 8-43). This report was generated by an automated HPE assessment tool using data from the EMC Data Domain weekly report.

Learning check

The following questions will help you to measure your understanding of the material presented in this chapter. Read all the choices carefully because there might be more than one correct answer. Select or write the correct answer for each question.

1. The HPE Storage Sizing Tool requires a license to run.

   ☐ True

   ☐ False

2. Which tool or resource will you use to find design rules concerning SAN topologies and supported configurations?

   a. SAN visibility

   b. *HPE Product Bulletin*

   c. *SAN Design Reference Guide*

   d. Alinean ROI and TCO tools

3. List two main components of SalesBuilder for Windows.

   ....

   ....

   ....

   ....

4. Which product series can you size with NinjaSTARS for 3PAR?

   ....

   ....

   ....

   ....

For answers, see Chapter 8 Answers on page 443.

Practice Test

INTRODUCTION

The HPE ATP – Storage Solutions V2 certification validates you can consult, architect, design and propose solutions using today's HPE Storage portfolio products primarily focused on SMBs. Certification demonstrates you understand foundational storage architectures and technologies. These include the characteristics, performance, advantages and uses of different storage drive and interconnect technologies, storage transport components, SAN topologies, storage virtualization, and backup and recovery systems. It verifies that you can position and support HPE storage products, solutions, and warranty offerings that form HPE's Converged Storage portfolio. Additionally, certification validates you can help your customers achieve their transformation priorities with HPE Storage Solutions.

This certification validates you have the following:

- Foundational knowledge of HPE 3PAR StoreServ, HPE StoreEasy, HPE StoreVirtual, HPE StoreOnce, and other HPE federated storage solutions
- Skills to help design information protection, retention and analytics across many applications and environments
- Foundational server and networking knowledge supporting design and deployment in a converged infrastructure

MINIMUM QUALIFICATIONS

To achieve the HPE ATP – Storage Solutions V2 certification, you must pass the HPE0-J74 exam or HPE0-J75 delta exam. This exams tests your ability to demonstrate the technical essentials of storage and storage area networks along with principles of backup, data protection, the HPE Storage portfolio, and converged storage concepts. It also tests your understanding of the convergence of siloed IT infrastructure domains and the importance of the holistic approach to IT.

Exams are based on an assumed level of industry-standard knowledge that may be gained from the training, hands-on experience, or other prerequisite events.

HPE0-J74 EXAM DETAILS

This exam has 60 questions. Here are types of questions to expect:

- Multiple choice (multiple responses), scenario based
- Multiple choice (single response), scenario based

- Matching
- Multiple choice (multiple responses)
- Multiple choice (single response)

Exam duration is 1 hour 30 minutes and passing score is 70%. No online or hard copy reference material will be allowed at the testing site. This exam may contain beta test items for experimental purposes. During the exam, you can make specific comments about the items (i.e. accuracy, appropriateness to audience, etc.). HPE welcomes these comments as part of our continuous improvement process.

Rather than emphasize simple memorization, HPE exams attempt to assess whether you have the knowledge and skills that a Storage Solutions Professional requires on the job. Therefore, many test items present a scenario, which outlines a particular environment or problem. You may also need to use exhibits to complete the test items successfully.

Take the time to read the entire question and consider all of the options carefully before you answer. If the question indicates that it features an exhibit, study the exhibit and reread the question. Make sure to select the answer that correctly responds to the question that is asked — not simply an answer that includes some correct information. If the question asks for more than one answer, remember to select each correct answer. You will not receive partial credit for a partially correct answer.

HPE0-J74 testing objectives

This exam validates that you can successfully perform the following objectives. Each main objective is given a weighting, indicating the emphasis this item has in the exam.

42% Identify and describe foundational storage architectures and technologies

- Explain the characteristics, performance, advantages and uses of different drive technologies
- Explain interconnect technologies
- Explain transport technology
- Explain storage transport components
- Explain and evaluate RAID levels
- Evaluate the different SAN topologies
- Identify internal storage virtualization technologies.
- Describe backup systems technology.

17% Describe the functions, features, and capabilities of HPE storage products, solutions, and warranty service offerings

- Identify the components that form the Converged Storage Strategy
- Correctly position the HPE storage portfolio
- List, describe, and map HPE proactive resources to customer requirements

3% Describe the storage market and recognize competitive opportunities to apply HPE storage solutions to meet customer needs

- Identify the strength of HPE storage solutions

23% Plan and design HPE storage solutions

- Discover opportunities
- Plan and design
- Size the solution
- Review and validate the proposal
- Present to the customer

7% Performance-tune, optimize, and upgrade HPE storage solutions

- Identify and compare the existing solution design to the documentation

8% Manage, monitor, administer, and operate HPE storage solutions

- Demonstrate base functionality of SMB storage solutions

Test preparation questions and answers

The following questions help you measure your understanding of the material presented in this book. Read all of the choices carefully since there may be more than one correct answer. Choose all correct answers for each question.

Questions

1. What are the three characteristics of software-defined storage? (Select three.)
   a. Hardware and vendor dependent
   b. Hardware agnostic
   c. Hypervisor independent
   d. Federated and autonomic
   e. Used in Public Cloud only
   f. Used in Private Cloud only

2. What matters when transforming to a hybrid infrastructure? (Select three.)
   a. Open-standards-based solutions
   b. IT dedicated to "keeping the lights on"
   c. Rigid IT environment
   d. Easy control of infrastructure and apps
   e. The right capabilities across people, processes, and governance
   f. Technical and organizational silos

3. A customer wants a system with servers, storage, networking, and virtualization preconfigured. This customer is already using VMware vSphere on their rack-based servers and prefers a solution that is deployable in minutes. Which HPE solution should you recommend?
   a. HPE Hyper Converged 250 for Microsoft
   b. HPE Hyper Converged 250 System
   c. HPE BladeSystem c7000 enclosure and HPE Helion OpenStack
   d. HPE MSA 2040

4. Which solution protects against cyber threats?
   a. HPE OneView
   b. HPE Fortify
   c. HPE ArcSight
   d. HPE Data Protector

5. Which of the following are types of Serial Attached SCSI (SAS) devices? (Select three.)
   a. Initiator
   b. Backplane
   c. SAS cable
   d. Low-level switch
   e. Target
   f. Expander

6. What is a benefit of third-generation SAS generation (SAS-3), compared to SAS-2?
   a. Speed of 6 Gb/s
   b. Speed of 12 Gb/s
   c. Speed of 22.5 Gb/s
   d. Speed of 24 Gb/s

7. Why does a drive array increase storage performance?
   a. Multiple drives are accessed serially.
   b. Multiple drives are accessed simultaneously.
   c. Drives in an array benefit from using the backplane instead of cables.
   d. Multiple drives can be connected to a single SAS cable.

8. A customer is selecting a RAID level and requires maximum write performance. A LUN will be used as a temporary storage for intermediate calculation results, and the customer is not concerned about data loss. Which RAID level should you recommend?
   a. RAID 50
   b. RAID 1
   c. RAID 0
   d. RAID 5

9. A customer is selecting a RAID level for a LUN on array containing 10 hard drives. This customer requires redundancy to protect against single-drive failures, and they require the maximum possible capacity. Which RAID level will meet these requirements?

   a. RAID 0
   b. RAID 1+0
   c. RAID 5
   d. RAID 6

10. A customer is selecting a RAID level for a LUN on an array containing 12 hard drives. This customer requires redundancy to protect against the simultaneous failure of any two drives, and they require the maximum possible capacity. Which RAID level will meet these requirements?

    a. RAID 0
    b. RAID 1+0
    c. RAID 5
    d. RAID 6

11. What is the advantage of direct-attached storage (DAS)?

    a. Ease of deployment
    b. Centralized storage management
    c. Data replication
    d. Unlimited scalability

12. A customer is designing a storage infrastructure and is selecting a protocol. Requirements for the infrastructure include bandwidth over 10 Gb/s, cable distance of more than 25 meters, and of the capability to connect multiple vendors to the network. Which protocol should you recommend?

    a. SAS-2
    b. SAS-3
    c. Fibre Channel
    d. SCSI

13. Which of the following are typical components of a DAS solution? (Select three.)
    a. Server
    b. FC switch
    c. Array controller
    d. Network-attached storage
    e. Cables
    f. FC hub

14. Which interface can be used to access data on the HPE D6000 Disk Enclosure?
    a. 6 Gb/s SAS
    b. 8 Gb/s FC
    c. 3 Gb/s SATA
    d. 6 Gb/s SATA

15. What is the difference between the HPE D2600 Disk Enclosure and the HPE D2700 Disk Enclosure?
    a. Bandwidth to the enclosure
    b. Physical size
    c. Drive form factor
    d. Dual domain support

16. A customer is purchasing a disk enclosure. The customer requires 12 Gb/s SAS connectivity, Solid-State Drive (SSD) support, and small form factor (SFF) drives. Which disk enclosure will meet these customer needs?
    a. D2600
    b. D2700
    c. D3600
    d. D3700

17. A customer is considering DAS for an HPE BladeSystem solution. The customer requires 6 Gb/s SAS connectivity and 120 TB capacity. Which disk enclosure will meet these requirements?

    a. D2700

    b. D2220sb

    c. D6000

    d. D3700

18. A customer without a data center requires a network-attached storage (NAS) solution in a tower form factor. Which of the following HPE StoreEasy products will meet this requirement?

    a. HPE StoreEasy 1450 Storage

    b. HPE StoreEasy 1550 Storage

    c. HPE StoreEasy 1850 Storage

    d. HPE StoreEasy 3850 Gateway Blade Storage

19. A customer requires a NAS solution supporting 90 TB of storage. Which of the following HPE StoreEasy products will meet this requirement?

    a. HPE StoreEasy 1450 Storage

    b. HPE StoreEasy 1650 Storage

    c. HPE StoreEasy 1850 Storage

    d. HPE StoreEasy 1550 Storage

20. A customer requires a NAS solution supporting 2000 users and large form factor (LFF) drives. Which of the following HPE StoreEasy products will meet this requirement?

    a. HPE StoreEasy 1450 Storage

    b. HPE StoreEasy 1650 Storage

    c. HPE StoreEasy 1850 Storage

    d. HPE StoreEasy 1550 Storage

21. A customer is designing a storage solution to allow users to access data on a legacy FC SAN array over the Common Internet File System/Server Message Block (CIFS/SMB) protocol. The solution must be highly available using clustering, and it must support up to 3000 users. Which HPE StoreEasy product will meet these requirements?

    a. HPE StoreEasy 1450 Storage

    b. HPE StoreEasy 1650 Storage

    c. HPE StoreEasy 1850 Storage

    d. HPE StoreEasy 3850 Storage

22. Which of the following are the key features of HPE MSA Storage solutions? (Select three.)

    a. Simple

    b. Unlimited scalability

    c. Centralized management

    d. Fast

    e. Network File System (NFS) support

    f. Affordable

23. Match the virtualization features of HPE MSA 1040 Storage to their description.

Virtualization feature	Description
Thin provisioning	Better use of storage resources; moving data that has not been accessed for long periods of time to more cost-effective midline SAS HDDs
Wide striping	Allocation of physical storage resources only after they are consumed by an application
Automated Tiering	Simplified volume expansion to facilitate the full utilization of all resources allocated to a specific volume

24. Which of the following are possible host connectivity options of HPE MSA 2040 Storage? (Select three.)

    a. 56 Gb/s InfiniBand

    b. 16 Gb/s FC

    c. 10 Gb/s iSCSI

    d. 12 Gb/s SATA

e. 12 Gb/s SAS

f. 10 Gb/s CIFS/SMB

25. A customer is impressed by your HPE StoreVirtual presentation and considers using HPE StoreVirtual features on their current x86 platforms. Which HPE StoreVirtual product should you recommend?

    a. HPE Hyper Converged 250 System

    b. HPE StoreVirtual 4335 Hybrid Storage

    c. HPE StoreVirtual VSA

    d. HPE StoreVirtual 4330

26. Which of the following management applications can be used to configure an HPE StoreVirtual solution and to assign volumes to servers?

    a. HPE Systems Insight Manager

    b. Centralized Management Console

    c. HPE Storage Management Utility

    d. HPE StoreServ Management Console

27. A customer is interested in an HPE StoreVirtual solution. The customer requires 1.5 TB storage and the most cost-effective solution. Which of the following HPE StoreVirtual solutions will meet these requirements?

    a. HPE StoreVirtual 4130

    b. HPE StoreVirtual 4330

    c. HPE StoreVirtual 4330FC

    d. HPE StoreVirtual 4730

28. A customer is interested in an HPE StoreVirtual solution. The customer requires 10 TB of storage, 10 Gb/s iSCSI connectivity, and Adaptive Optimization. Which of the following HPE StoreVirtual solutions will meet these requirements?

    a. HPE StoreVirtual 4130

    b. HPE StoreVirtual 4335

    c. HPE StoreVirtual 4330FC

    d. HPE StoreVirtual 4730

29. Which of the following are features of the HPE 3PAR Gen5 ASIC? (Select two.)
    a. Built in 16 Gb/s FC controller
    b. Embedded AES 512 encryption
    c. Built in thin technologies and deduplication
    d. End-to-end data protection against media and transmission errors
    e. Support for HPE 3PAR StoreServ 7000

30. A customer is sizing an HPE 3PAR StoreServ solution. The customer requires 500 TiB raw capacity and the most cost effective solution. Which HPE 3PAR StoreServ model will meet these requirements?
    a. HPE 3PAR StoreServ 8200
    b. HPE 3PAR StoreServ 8400
    c. HPE 3PAR StoreServ 8440
    d. HPE 3PAR StoreServ 8440 AFA

31. A customer is sizing an HPE 3PAR StoreServ solution. The customer requires 200 TiB raw capacity and 1 million IOPS. Which HPE 3PAR StoreServ model will meet these requirements?
    a. HPE 3PAR StoreServ 8200
    b. HPE 3PAR StoreServ 8400
    c. HPE 3PAR StoreServ 8440
    d. HPE 3PAR StoreServ 8440 AFA

32. A customer wants to extend an existing HPE 3PAR StoreServ solution with file access for 5000 concurrent users. Which HPE solution should you recommend?
    a. HPE 3PAR File Persona Software
    b. HPE 3PAR File Controller
    c. HPE 3PAR Application Software Suite for Exchange
    d. HPE 3PAR StoreServ 8440 AFA

33. Which of the following are HPE StoreFabric FC switch product lines? (Select three.)
    a. H-Series
    b. HPE MPX series
    c. B-Series
    d. C-Series
    e. HPE StoreAll Series
    f. HPE StoreOnce Series

34. Which of the following is a valid FC switch differentiator?
    a. Automatic traffic rerouting
    b. Maximum amount of ports
    c. Maximum capacity of shared storage
    d. Maximum amount of concurrent users

35. Match the backup terminology to its description.

Backup terminology	**Description**
Full backup	Copies all files that were changed after the last complete backup
Incremental backup	Performs a complete backup of the entire server or client
Differential backup	Copies all files that were changed after the last backup, regardless of what kind of backup it was

36. What is the benefit of HPE StoreOnce deduplication?
    a. Reduces the disk space required to store backup data sets without impacting backup performance
    b. Reduces the disk space required to store backup data sets while increasing the time of the backup
    c. Replicates backups through the LAN or WAN to other StoreOnce systems
    d. Decreases the range of recovery point objectives available

37. What is the benefit of LTO-7 tape comparing to LTO-6?
    a. Increased capacity to 2.5 TB of raw data
    b. Increased capacity to 2.7 TB of raw data

c. Increased capacity to 6 TB of raw data

d. Increased capacity to 7.6 TB of raw data

38. A customer requires 24 TB of raw capacity and a disk-based backup solution. Which product will meet these requirements?

   a. HPE StoreOnce 3100

   b. HPE StoreOnce 3520

   c. HPE StoreEasy 1450

   d. HPE StoreOnce 2900

39. What is the typical write performance of an HPE StoreOnce backup system? (Consider models without Catalyst technology from the StoreOnce 3100 to StoreOnce 4900 models.)

   a. 1.3 TB/hr to 8.5 TB/hr

   b. 1.5 GB/hr to 12 GB/hr

   c. 15 TB/hr to 32.5 TB/hr

   d. 120 GB/hr to 844 GB/hr

40. Which tool or resource provides design rules for SAN topologies and supported configurations?

   a. SAN visibility

   b. *HPE Product Bulletin*

   c. *SAN Design Reference Guide*

   d. Alinean ROI/TCO calculators

41. Which tool or resource will you use to design a storage array through the wizard interface?

   a. HPE Storage Sizing Tool

   b. *HPE Product Bulletin*

   c. *SAN Design Reference Guide*

   d. Alinean ROI/TCO calculators

42. What are the user interfaces of HPE SalesBuilder for Windows? (Select two.)

   a. Configurator

   b. Product Bulletin

c. Quoter

d. Sizer

43. Which tool or resource provides specifications for HPE hardware?

a. SAN visibility

b. HPE Product Bulletin

c. HPE SalesBuilder for Windows

d. Alinean ROI/TCO calculators

44. Which tool or resource will you use to design an HPE 3PAR StoreServ solution and to import current HPE EVA configuration data?

a. HPE Product Bulletin

b. HPE SalesBuilder for Windows

c. HPE Storage Sizing Tool

d. HPE NinjaSTARS for 3PAR

Answers

1. What are the three characteristics of software-defined storage? (Select three.)

   B, **C** and **D** are correct. Software-defined storage can run on virtually any x86 based hardware. True software-defined storage technologies provide long-term investment protection by not locking you into any one proprietary hypervisor technology. Software-defined storage solutions seamlessly scale out as your needs grow over time, providing federation and autonomic management from day 1.

   A is incorrect, software-defined storage can run on virtually any x86 based hardware. **E** and **F** are incorrect, software-defined storage can be used in both public and private cloud environment.

   For more information, see chapter 1.

2. What matters when transforming to a hybrid infrastructure? (Select three.)

   A, **D** and **E** are correct. To transform to the new style of business, customers must focus on:

   – Agility and flexibility—A converged and virtualized hybrid infrastructure scales easily and delivers continuous value to make IT a service provider.

   – Workload optimization—Modern infrastructure offers better utilization, adjusting performance and availability dynamically.

– Simplicity and intuitiveness—Software-defined controls, along with automation and converged management, free up IT resources.

– Flexible investments—Open-standards-based systems and new IT consumption models enable continuous business innovation.

B, C and **F** are incorrect. Customers are struggling with inefficiencies and lack of collaboration that means IT is dedicated to "keeping the lights on". Legacy hardware scales poorly and slows deployment of apps and workloads.

For more information, see chapter 1.

3. A customer wants a system with servers, storage, networking, and virtualization preconfigured. This customer is already using VMware vSphere on their rack-based servers and prefers a solution that is deployable in minutes. Which HPE solution should you recommend?

   B is correct. The Hyper Converged 250 family eliminates complex solution stacks. It combines a powerful Apollo server platform and mature HPE StoreVirtual data services from Hewlett Packard Enterprise into one building block to simplify the delivery of virtualized servers and desktops. The Hyper Converged 250 family supports VMware virtualization technologies, and continues to provide a turn-key virtualization solution for medium-sized businesses, enterprises, and IaaS providers.

   A is incorrect because customer is already using VMware vSphere rather Microsoft System Center and Microsoft technologies. **C** is incorrect because, it's not based on rack technologies. **D** is incorrect because it only contains storage component.

   For more information, see chapter 1.

4. Which solution protects against cyber threats?

   B is correct, HPE Fortify protects against cyber threats.

   A, **C** and **D** are incorrect. HPE ArcSight and managed security services identify potential and successful security and compliance breaches rather than protect. HPE OneView is a converged infrastructure management solution and HPE Data Protector is a backup and recovery software.

   For more information, see chapter 1 and 7.

5. Which of the following are types of Serial Attached SCSI (SAS) devices? (Select three.)

   A, **E** and **F** are correct. There are three types of SAS devices: initiators, targets, and expanders. Initiator devices include host bus adaptors (HBAs) and controllers. The initiator attaches to one or more targets, forming a SAS domain. The initiator is also responsible for providing commands and data. Target devices include SAS hard disk drives (HDDs) or solid state drives (SSDs), SATA HDDs or SSDs, and SAS tape drives. The target waits for commands, manages the commands provided by the initiator, and performs the operations. Using expanders (low-cost, high-speed switches), you can increase the number of targets attached to an initiator to create a larger SAS domain.

   B, C and **D** are incorrect, Backplane and cables can complement a SAS solution, but are not considered SAS devices. Low-level switch is not used in SAS solutions.

   For more information, see chapter 2.

6. What is a benefit of third-generation SAS generation (SAS-3), compared to SAS-2?

   B is correct, SAS-3 doubled the rate to 12.0 Gb/s in 2013. **A**, **C** and **D** are incorrect and represent speeds of older or newer SAS generations.

   For more information, see chapter 2.

7. Why does a drive array increase storage performance?

   B is correct. Because multiple drives are accessed simultaneously, the process of writing (or reading) a file across multiple drives is much faster than writing to or reading from a single drive.

   A is incorrect, drives are not accessed serially in array. **C** and **D** are incorrect, using backplane, single or multiple cables does not have a performance impact.

   For more information, see chapter 2.

8. A customer is selecting a RAID level and requires maximum write performance. A LUN will be used as a temporary storage for intermediate calculation results, and the customer is not concerned about data loss. Which RAID level should you recommend?

   C is correct. Data striping (RAID 0) is faster than conventional file writing to a single disk; however, there is no fault tolerance if any single drive fails. If one disk should fail, all data on the array would be lost.

   A, B and **D** are incorrect, these RAID levels have performance penalty compared to RAID 0.

   For more information, see chapter 2.

9. A customer is selecting a RAID level for a LUN on array containing 10 hard drives. This customer requires redundancy to protect against single-drive failures, and they require the maximum possible capacity. Which RAID level will meet these requirements?

   C is correct, RAID 5 protect against single-drive failures.

   A, B and **D** are incorrect. RAID 0 doesn't protect against drive failures. RAID 1+0 and RAID 6 utilizes less capacity from 10 hard drives than RAID 5.

   For more information, see chapter 2.

10. A customer is selecting a RAID level for a LUN on an array containing 12 hard drives. This customer requires redundancy to protect against the simultaneous failure of any two drives, and they require the maximum possible capacity. Which RAID level will meet these requirements?

    D is correct, RAID 6 protect against the simultaneous failure of any two drives.

    A, B and **C** are incorrect, RAID 0 and RAID 5 doesn't protect against the simultaneous failure of any two drives. RAID 1+0 utilizes less capacity from 12 hard drives than RAID 6.

    For more information, see chapter 2.

11. What is the advantage of direct-attached storage (DAS)?

    A is correct. Advantages of DAS include ease of deployment as there's no complexity of building a dedicated storage network or configuring a specialized server-storage device.

    B and **C** are incorrect, there's usually no centralized storage management for multiple DAS solutions and there are usually no common replication technologies available for most DAS solutions. **D** is incorrect as amount of internal drive positions and ports on additional storage controllers are limited.

    For more information, see chapter 2.

12. A customer is designing a storage infrastructure and is selecting a protocol. Requirements for the infrastructure include bandwidth over 10 Gb/s, cable distance of more than 25 meters, and of the capability to connect multiple vendors to the network. Which protocol should you recommend?

    C is correct, FC will meet requirements with speed up to 16 Gb/s and cable distance higher than 25 meters.

    A, B and **D** are incorrect, SAS and SCSI cables are usually shorter than 10 meters.

    For more information, see chapter 2.

13. Which of the following are typical components of a direct-attached storage (DAS) solution? (Select three.)

    A, **C** and **E** are correct. DAS solutions typically are connected directly to the attached server using the SAS protocol and can be either an internal part of the server or implemented as external enclosures. Typical components of a DAS solution are server, storage or array controller, connectivity options such as cables and enclosures.

    B and **D** are incorrect, NAS devices are part of NAS solutions and FC hubs/switches are part of SAN solutions.

    For more information, see chapter 3.

14. Which interface can be used to access data on the HPE D6000 Disk Enclosure?

    A is correct. HPE D6000 Disk Enclosure supports 6 Gb/s SAS.

    B, C and **D** are incorrect, there's no possibility to change the host interface to SATA or FC.

    For more information, see chapter 3.

15. What is the difference between the HPE D2600 Disk Enclosure and the HPE D2700 Disk Enclosure?

    C is correct. The D2000 enclosures are 6 Gb/s SAS low-cost, high-capacity, tiered and modular, external storage systems. HPE D2600 Disk Enclosure supports up to 12 LFF drives and HPE D2700 Disk Enclosure UP TO 25 SFF drives, drive form factor is the primary difference.

    A, **B** and **D** are incorrect, both enclosures are 6 Gb/s SAS, 2U high and support dual domain support with supported controllers.

    For more information, see chapter 3.

16. A customer is purchasing a disk enclosure. The customer requires 12 Gb/s SAS connectivity, Solid-State Drive (SSD) support, and small form factor (SFF) drives. Which disk enclosure will meet these customer needs?

    D is correct. D3700 supports 12 Gb/s SAS connectivity, Solid-State Drives (SSDs), and small form factor (SFF) drives.

    A and **B** are incorrect, D2600 and D2700 support 6 Gb/s SAS. **C** is incorrect, D3600 supports LFF drives.

    For more information, see chapter 3.

17. A customer is considering direct-attached storage for an HPE BladeSystem solution. The customer requires 6 Gb/s SAS connectivity and 120 TB capacity. Which disk enclosure will meet these requirements?

    C is correct, D6000 supports 6 Gb/s SAS connectivity and 120 TB capacity.

    A, **B** and **D** are incorrect as these enclosures cannot scale to required capacity.

    For more information, see chapter 3.

18. A customer without a data center requires a network-attached storage (NAS) solution in a tower form factor. Which of the following HPE StoreEasy products will meet this requirement?

    B is correct, HPE StoreEasy 1550 Storage is a network-attached storage (NAS) solution in a tower form.

    A, **C** and **D** are incorrect, these StoreEasy systems are either in a blade or rack form factor.

    For more information, see chapter 4.

19. A customer requires a NAS solution supporting 90 TB of storage. Which of the following HPE StoreEasy products will meet this requirement?

    B is correct. HPE StoreEasy 1650 Storage can easily scale to 90 TB of storage.

    A, C and **D** are incorrect, these StoreEasy systems cannot scale to required capacity.

    For more information, see chapter 4.

20. A customer requires a NAS solution supporting 2000 users and large form factor (LFF) drives. Which of the following HPE StoreEasy products will meet this requirement?

    B is correct. HPE StoreEasy 1650 Storage is using LFF drives and can scale up to 3400 users.

    A and **D** are incorrect, HPE StoreEasy 1450 Storage and HPE StoreEasy 1550 Storage can scale up to 1300 users. **C** is incorrect, HPE StoreEasy 1850 Storage is using SFF data drives.

    For more information, see chapter 4.

21. A customer is designing a storage solution to allow users to access data on a legacy FC SAN array over the Common Internet File System/Server Message Block (CIFS/SMB) protocol.

The solution must be highly available using clustering, and it must support up to 3000 users. Which HPE StoreEasy product will meet these requirements?

D is correct. HPE StoreEasy 3850 Storage supports up to 20000 users and provides gateway services for your array or Fibre Channel SAN.

A, **B** and **C** are incorrect, these StoreEasy systems do not provide gateway services.

For more information, see chapter 4.

22. Which of the following are the key features of HPE MSA Storage solutions? (Select three.)

    A, **D** and **F** are correct. Key features of HPE MSA Storage solutions are simplicity, speed — Proven fourth-generation MSA architecture built for speed with faster processors, affordable price points and future proof design.

    B, C and **E** are incorrect, HPE MSA Storage do not offer unlimited scalability and centralized management. HPE MSA Storage solutions are not offering built-in file based access.

    For more information, see chapter 5.

23. Match the virtualization features of HPE MSA 1040 Storage to their description.

    Thin provisioning allows storage allocation of physical storage resources only after they are consumed by an application. By leveraging virtualized storage "pools," HPE MSA users can leverage **wide striping** technology to simplify volume expansion and to facilitate full utilization of all resources allocated to a specific volume. **Automated tiering** offers better use of storage resources; moving data that has not been accessed for long periods of time to more cost-effective midline SAS HDDs.

    For more information, see chapter 5.

24. Which of the following are possible host connectivity options of HPE MSA 2040 Storage? (Select three.)

    B, **C** and **E** are correct. HPE MSA 2040 Storage features two high-performance controllers with support for the industry's latest Fibre Channel, iSCSI, and SAS host interfaces.

    A and **D** are incorrect, there's no SATA and InfiniBand support in HPE MSA Storage. **F** is incorrect, HPE MSA Storage solutions are not offering built-in file based access.

    For more information, see chapter 5.

25. A customer is impressed by your HPE StoreVirtual presentation and considers using HPE StoreVirtual features on their current x86 platforms. Which HPE StoreVirtual product should you recommend?

    C is correct, HPE StoreVirtual VSA works with all major hypervisors and any x86 server. No hardware required means lower costs and less complexity.

    A, **B** and **D** are incorrect. HPE Hyper Converged 250 System, HPE StoreVirtual 4335 Hybrid Storage and HPE StoreVirtual 4330 contain hardware components.

    For more information, see chapter 5.

26. Which of the following management applications can be used to configure an HPE StoreVirtual solution and to assign volumes to servers?

    B is correct. Centralized Management Console is the intuitive, GUI-based administrative interface for HPE StoreVirtual solutions. It is used to configure and manage storage volumes spanning clustered storage nodes and provides a single graphical layout of the storage environment.

    A, **C** and **D** are incorrect. HPE Systems Insight Manager does not provide management capabilities related to StoreVirtual volumes. HPE Storage Management Utility and HPE StoreServ Management Console are management solutions for HPE MSA and HPE 3PAR StoreServ products.

    For more information, see chapter 5.

27. A customer is interested in an HPE StoreVirtual solution. The customer requires 1.5 TB storage and the most cost-effective solution. Which of the following HPE StoreVirtual solutions will meet these requirements?

    A is correct. HPE StoreVirtual 4130 supports four 600 GB SFF SAS.

    B, C and **D** are incorrect. HPE StoreVirtual 4330, HPE StoreVirtual 4330FC and HPE StoreVirtual 4730 do support capacity requirements, but HPE StoreVirtual 4130 is generally more cost-effective.

    For more information, see chapter 5.

28. A customer is interested in an HPE StoreVirtual solution. The customer requires 10 TB of storage, 10 Gb/s iSCSI connectivity, and Adaptive Optimization. Which of the following HPE StoreVirtual solutions will meet these requirements?

    B is correct. HPE StoreVirtual 4335 support customer requirements of 10 TB of storage, 10 Gb/s iSCSI connectivity, and Adaptive Optimization.

    A and **D** are incorrect HPE StoreVirtual 4130 and HPE StoreVirtual 4730 do not support adaptive optimization. **C** is incorrect, HPE StoreVirtual 4330FC is using primarily FC as host interface.

    For more information, see chapter 5.

29. Which of the following are features of the HPE 3PAR Gen5 ASIC? (Select two.)

    C and **D** are correct. HPE 3PAR Gen5 ASIC has built in thin technologies, deduplication and supports end-to-end data protection against media and transmission errors.

    A, B and **E** are incorrect. There are no built-in 16 Gb/s FC controllers or AES 512 encryption engine. HPE 3PAR StoreServ 7000 is using Gen4 ASIC.

    For more information, see chapter 5.

30. A customer is sizing an HPE 3PAR StoreServ solution. The customer requires 500 TiB raw capacity and the most cost effective solution. Which HPE 3PAR StoreServ model will meet these requirements?

    A is correct. HPE 3PAR StoreServ 8200 supports 500 TiB raw capacity as the most cost-effective solution.

B, C and **D** are incorrect HPE 3PAR StoreServ 8400, HPE 3PAR StoreServ 8440 and HPE 3PAR StoreServ 8440 AFA does not meet custome requirements for cost-effective solution compared to 8200.

For more information, see chapter 5.

31. A customer is sizing an HPE 3PAR StoreServ solution. The customer requires 200 TiB raw capacity and 1 million IOPS. Which HPE 3PAR StoreServ model will meet these requirements?

    D is correct. HPE 3PAR StoreServ 8440 AFA provides 1 million IOPS and required capacity.

    A, B and **C** are incorrect. Only all-flash arrays are providing required IOPS.

    For more information, see chapter 5.

32. A customer wants to extend an existing HPE 3PAR StoreServ solution with file access for 5000 concurrent users. Which HPE solution should you recommend?

    B is correct. HPE 3PAR File Controller supports up to 20000 users per node and provides file access functionality.

    A is incorrect, HPE 3PAR File Persona Software scales up to 3000 users. **C** and **D** are incorrect. HPE 3PAR Application Software Suite for Exchange and HPE 3PAR StoreServ 8440 AFA (without File Persona or File Controller) do not provide files based access.

    For more information, see chapter 5.

33. Which of the following are HPE StoreFabric FC switch product lines? (Select three.)

    A, **C** and **D** are correct. HPE StoreFabric FC switches portfolio consists of three product lines: H Series, B Series and C Series.

    B, E and **F** are incorrect. HPE MPX series, HPE StoreAll products and HPE StoreOnce products are not FC switches.

    For more information, see chapter 6.

34. Which of the following is a valid FC switch differentiator?

    B is correct. Total number of Fibre Channel ports—fabric port count— needed to connect servers and storage to the fabric is one of the most important differentiators.

    A, C and **D** are incorrect. Automatic traffic rerouting, maximum amount of concurrent users and maximum capacity of shared storage are not features of FC switches.

    For more information, see chapter 6.

35. Match the backup terminology to its description.

    Full backup performs a complete backup of the entire server or client. **Incremental backup** copies all files that were changed after the last backup, regardless of what kind of backup it was. **Differential backup** copies all files that were changed after the last complete backup.

    For more information, see chapter 7.

36. What is the benefit of HPE StoreOnce deduplication?

    A is correct. HPE StoreOnce deduplication reduces the disk space required to store backup data sets without impacting backup performance.

    B is incorrect, HPE StoreOnce deduplication does not increase the time of the backup. **C** and **D** are incorrect, deduplication does not perform replication and is not connected to changes to recovery point objectives.

    For more information, see chapter 7.

37. What is the benefit of LTO-7 tape comparing to LTO-6?

    C is correct. LTO-7 is the seventh generation of LTO tapes introduced in 2015 with 6 TB of raw data.

    A, B and **C** are incorrect, LTO-7 is using tapes with 6 TB of raw data.

    For more information, see chapter 7.

38. A customer requires 24 TB of raw capacity and a disk-based backup solution. Which product will meet these requirements?

    B is correct. HPE StoreOnce 3520 is designed for small to medium-size data centers and as a replication target device for up to remote and branch offices. The StoreOnce 3520 delivers a scalable 2U solution from 7.5 to 15.5 TB of usable capacity (12-24 TB RAW) using an upgrade license.

    A, C and **D** are incorrect. HPE StoreOnce 3100 supports 8 TB raw capacity, HPE StoreEasy 1450 is not a disk-based backup solution and HPE StoreOnce 2900 is a previous generation product.

    For more information, see chapter 7.

39. What is the typical write performance of an HPE StoreOnce backup system? (Consider models without Catalyst technology from the StoreOnce 3100 to StoreOnce 4900 models.)

    A is correct, **B, C** and **D** are incorrect. StoreOnce 4900 supports 8.5 TB/hr write performance without catalyst.

    For more information, see chapter 7.

40. Which tool or resource provides design rules for SAN topologies and supported configurations?

    C is correct. The HPE SAN Design Reference Guide provides information on storage area network (SAN) design philosophy and considerations. The guide includes design rules concerning SAN topologies and supported configurations.

    A is incorrect, SAN Visibility is a complementary software utility for HPE customers that helps with SAN analysis, diagnostics, and optimization.

    B is incorrect, the HPE Product Bulletin is a convenient central resource that provides technical overviews and specifications for HPE hardware and software.

D is incorrect, the Alinean return on investment (ROI) and total cost of ownership (TCO) calculator tools make it easy to build a business case to position the value of an HPE Converged Infrastructure solution. The tools produce values for ROI, payback, net present value (NPV), and internal rate of return (IRR) that are often required by chief financial officers (CFOs) to compare different projects.

For more information, see chapter 8.

41. Which tool or resource will you use to design a storage array through the wizard interface?

A is correct. The HPE Storage Sizing Tool is a downloadable sizing tool that enables you to work with your customers to design a storage infrastructure to meet their needs.

B is incorrect, the HPE Product Bulletin is a convenient central resource that provides technical overviews and specifications for HPE hardware and software.

C is incorrect. The HPE SAN Design Reference Guide provides information on storage area network (SAN) design philosophy and considerations. The guide includes design rules concerning SAN topologies and supported configurations.

D is incorrect, the Alinean return on investment (ROI) and total cost of ownership (TCO) calculator tools make it easy to build a business case to position the value of an HPE Converged Infrastructure solution. The tools produce values for ROI, payback, net present value (NPV), and internal rate of return (IRR) that are often required by chief financial officers (CFOs) to compare different projects.

For more information, see chapter 8.

42. What are the user interfaces of HPE SalesBuilder for Windows? (Select two.)

A and **C** are correct. SalesBuilder for Windows has two main components: the Configurator and the Quoter.

B is incorrect, the HPE Product Bulletin is a convenient central resource that provides technical overviews and specifications for HPE hardware and software.

D is incorrect, the Sizer is not a name of HPE Salesbuilder component.

For more information, see chapter 8.

43. Which tool or resource provides specifications for HPE hardware?

B is correct. The HPE Product Bulletin is a convenient central resource that provides technical overviews and specifications for HPE hardware and software.

A is incorrect, SAN Visibility is a complementary software utility for HPE customers that helps with SAN analysis, diagnostics, and optimization.

C is incorrect, SalesBuilder for Windows (SBW) is the premier HPE configuration and quotation tool for the sales force and channel partners. SBW is a downloadable tool designed to support the complete HPE product portfolio of servers, storage, and related services.

D is incorrect, the Alinean return on investment (ROI) and total cost of ownership (TCO) calculator tools make it easy to build a business case to position the value of an HPE

Converged Infrastructure solution. The tools produce values for ROI, payback, net present value (NPV), and internal rate of return (IRR) that are often required by chief financial officers (CFOs) to compare different projects.

For more information, see chapter 8.

44. Which tool or resource will you use to design an HPE 3PAR StoreServ solution and to import current HPE EVA configuration data?

    D is correct. NinjaSTARS for 3PAR includes measured data from HPE Storage Optimizer, as well as other new features, for increased accuracy and performance modeling. NinjaSTARS was developed for worldwide channel partners to assess legacy storage environments for customers interested in HPE 3PAR StoreServ 7000/8000 arrays. HPE NinjaSTARS for 3PAR includes powerful editing tools and direct import of EVA configuration data.

    A is incorrect. The HPE Product Bulletin is a convenient central resource that provides technical overviews and specifications for HPE hardware and software.

    B is incorrect, SalesBuilder for Windows (SBW) is the premier HPE configuration and quotation tool for the sales force and channel partners. SBW is a downloadable tool designed to support the complete HPE product portfolio of servers, storage, and related services.

    C is correct. The HPE Storage Sizing Tool is a downloadable sizing tool that enables you to work with your customers to design a storage infrastructure to meet their needs.

    For more information, see chapter 8.

Answer Key

1. B, C and D
2. A, D and E
3. B
4. B
5. A, E and F
6. B
7. B
8. A
9. C
10. D
11. A
12. C

13. A, C and E
14. A
15. C
16. D
17. C
18. B
19. B
20. B
21. D
22. A, D and F
23.

Virtualization feature	Description
Thin provisioning	Allocation of physical storage resources only after they are consumed by an application
Wide striping	Simplified volume expansion to facilitate the full utilization of all resources allocated to a specific volume
Automated Tiering	Better use of storage resources; moving data that has not been accessed for long periods of time to more cost-effective midline SAS HDDs

24. B, C and E
25. C
26. B
27. A
28. B
29. C and D
30. A
31. D
32. B
33. A, C and D
34. B
35.

Backup terminology	Description
Full backup	Performs a complete backup of the entire server or client
Incremental backup	Copies all files that were changed after the last backup, regardless of what kind of backup it was
Differential backup	Copies all files that were changed after the last complete backup

36. A
37. C
38. B
39. A
40. C
41. A
42. A and C
43. B
44. D

Answers to Learning Checks

Chapter 1 Answers

1. Describe today's idea economy.

   In today's world, the ability to turn an idea into a new product, capability, business, or industry has never been easier or more accessible.

   This presents an opportunity and a challenge for most enterprises. Creating and delivering new business models, solutions, and experiences requires harnessing new types of applications, data, and risks as well as implementing new ways to build, operate, and consume technology.

2. What should organizations focus on to create better business outcomes through IT transformation? (Select three.)

   - Continuously creating and delivering new services
   - Software differentiating products and services
   - Providing real-time insight and understanding

3. Which solution protects against cyber threats?

   HPE Fortify

4. Complete the action plan to empower a data-driven enterprise.

   - Modernize the enterprise data warehouse
   - Deploy a Big Data platform
   - Deliver actionable business value
   - Enable best-in-class data management

5. What matters when transforming to a hybrid infrastructure? (Select three)

   - Open-standards-based solutions
   - Easy control of infrastructure and apps
   - The right capabilities across people, processes, and governance

Chapter 2 Answers

1. Name the three main approaches for storing data.

   DAS, NAS, and SAN

2. Name at least three components of the Fibre Channel (FC) solution.

   Any three of the following:

   - Host bus adapters (HBAs)
   - SFP transceivers
   - Storage and FC drive arrays
   - FC array controllers
   - FC switches
   - Tape libraries
   - FC cables

3. The simplest FC topology is point to point.

   True

4. iSCSI stack uses UDP.

   False

5. What is the name of the embedded Virtual Connect management tool that is accessible by both GUI and CLI?

   Virtual Connect Manager (VCM)

Chapter 3 Answers

1. List the typical components of a direct-attached storage (DAS) solution.

   Server, storage or array controller, connectivity/cables

2. Would you suggest the HPE D2700 Disk Enclosure over the D2600 to a customer requiring maximum capacity?

   No

Why or why not?

The HPE D2700 uses SFF drives that usually have smaller capacity (maximum 30 TB with 1.2 TB drives), while the HPE D2600 supports LFF drives, up to 6 TB each, allowing 72 TB.

3. You can use Solid-State Drives (SSD) in both the HPE D3600 and HPE D3700 disk enclosures.

   True

4. Which interface can be used to access data on the HPE D6000 Disk Enclosure?

   6 Gb/s SAS

5. Where will you find supported drives for the HPE D2700 disk enclosure and its maximum capacity?

   - HPE QuickSpecs
   - *HPE Product Bulletin*
   - HPE Website

Chapter 4 Answers

1. What is a network-attached storage (NAS) solution?

   A NAS solution consists of a specialized appliance that connects directly to the network. A file system is located and managed on the NAS device. Data is transferred to servers and other devices through the LAN and to clients using industry-standard, file-sharing protocols.

2. List two typical StoreEasy use cases.

   Any two the following:

   - Heterogeneous file sharing—File sharing across multiple client platforms
   - Remote office–branch office—Simplifying connected branch office deployment and management
   - Home directory consolidation—Consolidate end-user data for greater security, mobility, and efficiency
   - Applications over SMB—Enabling SAN-like capabilities for Hyper-V and SQL Server at lower cost with simplified management

3. HPE StoreEasy 1000 Storage is based on HPE ProLiant servers.

   True

4. HPE StoreEasy 1850 9.6TB SAS Storage comes with preinstalled Windows Storage Server 2012 R2.

   True

5. Up to four StoreEasy 3850 Gateway Systems can be configured together as a four-node cluster (when using two StoreEasy 3850 Gateway Nodes per chassis).

   False

6. Is the SMB 3 protocol compatible with previous versions, and are all features available? Please explain your answer.

   SMB 3 is compatible with clients using version 1 or 2 of the SMB protocol, but the new features are available only to client operating systems that are also using SMB version 3.

Chapter 5 Answers

1. A storage area network (SAN) requires all components to come from a single vendor.

   False

2. Name at least three components of a SAN solution.

   Any three of the following:

   - Switches
   - Routers
   - Bridges
   - Gateways
   - Storage devices
   - Servers
   - Cabling and connectors
   - SAN management applications

3. What are the differences between HPE MSA 1040 Storage and HPE MSA 2040 Storage?

   Price level, performance, and the possibility of using SSD

4. The maximum capacity of the MSA 1040 device is 16 TB.

   False

5. The HPE MSA 2040 device can be configured to achieve more than 100k IOPS for random reads.

   True

6. Name at least three components of the HPE StoreVirtual solution.

   Any three of the following:

   - Storage system
   - LeftHand OS
   - CMC
   - iSCSI Initiator

7. HPE StoreVirtual Systems are built on the HPE ProLiant platform.

   True

8. HPE 3PAR StoreServ nodes are connected in a ring topology.

   False

9. Recommend two products capable of providing file and object access to HPE 3PAR StoreServ.

   HPE 3PAR File Persona Software Suite and HPE 3PAR StoreServ File Controller

10. HPE NinjaSTARS for 3PAR supports imported performance data from HPE EVA systems.

    True

Chapter 6 Answers

1. Name at least three SAN infrastructure products.

   Any three of the following:

   - SAN switches
   - FC directors
   - HBAs
   - Cables
   - Transceivers
   - Port licenses
   - Software

2. SAS is the common protocol for connecting FC switches and servers.

   False

3. Name at least two FC switch differentiators.

   Any two of the following:

   - Hardware vendor
   - Form factor
   - Protocol support
   - Maximum number of ports
   - Extra features
   - Software

4. The host interface (FC versus Ethernet) is the main differentiator for HPE StoreFabric FC HBAs and CNAs.

   True

5. What is the common HPE FC transceiver speed?

   8 Gb/s or 16 Gb/s

6. Congestion can limit the speed between servers and storage devices.

   True

Chapter 7 Answers

1. Name at least two different types of backup.

   Any two of the following:

   - Full: Normal, Copy
   - Partial: Incremental, Differential

2. Incremental backups back up only the data that changed since the last full backup and reset the archive bit.

   True

3. HPE StoreOnce VSA is HPE StoreOnce storage delivered as a VMware virtual appliance.

   True

4. What is the typical write performance of an HPE StoreOnce backup system? (Consider StoreOnce 3100 to StoreOnce 4900 models without Catalyst technology.)

   1.3 TB/hr to 8.5 TB/hr

5. HPE Data Protector Software can be licensed based on the capacity needed.

   True

6. Name at least three possible backup targets for HPE StoreOnce backup systems.

   Any three of the following:

   - iSCSI VTL
   - FC VTL
   - CIFS share (TCP/IP)
   - NFS share (TCP/IP)
   - Catalyst Store (TCP/IP)

Chapter 8 Answers

1. The HPE Storage Sizing Tool requires a license to run.

   True

2. Which tool or resource will you use to find design rules concerning SAN topologies and supported configurations?

   SAN Design Reference Guide

3. List two main components of SalesBuilder for Windows.

   Configurator and Quoter

4. Which product series can you size with NinjaSTARS for 3PAR?

   3PAR StoreServ 7000/8000

Index

E

O

P

T